An engineering approach to linear algebra

An engineering approach to linear algebra

W. W. Sawyer

*Professor jointly to
the Department of Mathematics
and the College of Education
University of Toronto*

*Cambridge
at the University Press
1972*

CAMBRIDGE UNIVERSITY PRESS
Cambridge, New York, Melbourne, Madrid, Cape Town, Singapore, São Paulo, Delhi

Cambridge University Press
The Edinburgh Building, Cambridge CB2 8RU, UK

Published in the United States of America by Cambridge University Press, New York

www.cambridge.org
Information on this title: www.cambridge.org/9780521084765

First published 1972
This digitally printed version 2008

A catalogue record for this publication is available from the British Library

Library of Congress Catalogue Card Number: 70–184143

ISBN 978-0-521-08476-5 hardback
ISBN 978-0-521-09333-0 paperback

Contents

Preface

This book first took form as duplicated notes handed out to first year engineering students at the University of Toronto. The preparation of this material was undertaken because no published book seemed to meet the needs of first year engineers. Some books were mathematically overweight; in order to prove every statement made, long chains of propositions were included, which served only to exhaust and antagonize the students. Other books, emphasizing applications to engineering, involved mathematics beyond the student's comprehension. This was naturally so, for while linear algebra has a wealth of applications to engineering and science, many of these lie outside the range of the first year student, both in their mathematical and in their scientific content.

Thus problems of treatment arose both on the mathematical and the engineering side.

In some discussions of mathematics teaching, it seems to be assumed that only two courses are open: either the full rigorous proof of every statement from a set of axioms, or a cookbook approach in which the student memorizes some rigmarole. The end result of these two approaches is much the same; in neither case does the average student have any understanding of what he is doing. A third course is in fact possible, in which the student learns to picture or to imagine the things he is dealing with, and then, because his imagination is working, he is able to reason about these things. The objective of this book is to let the student achieve such insight and reasoning ability. I would strenuously oppose any suggestion that this book is less mathematical because it steers in the direction of this third course. Imagination is the source of all mathematics; a mathematician begins with an idea; the details of proofs come later. All the explanations, all the arguments, all the justifications brought forward in this book could, if one desired, be put in a rigorous form that would satisfy a professional mathematician, but it would not help the first year student if this were done.

If a mathematics course for engineers is to succeed, it must give the student the feeling that it will be of practical value in enabling him to master his profession. No mathematical concept has been included in this book without a careful examination of its value for applications, and the right of the student to demand evidence of such applicability has throughout been respected. The difficulty that most actual applications are very complicated has been met by discussing simplified situations, but indicating the existence of more realistic problems.

So far as I know, no other book uses the approach of the opening sections, which start with the real world and show how real situations lead to the basic concepts of linear algebra. The exercises at the end of §3 are intended to show the variety of situations in which these concepts are involved. Students should be warned that they will find these exercises hard. They should be prepared to bypass them to some extent, to go ahead with the rest of the book, and to come back to these exercises from time to time.

Some books are written to provide problems and extra reading for students whose main instruction is derived from a series of lectures. That is not the purpose of this book. It would be interesting to apply the spirit of operational research to the institution of the university lecture. What purpose are lectures supposed to achieve, and to what extent do they in fact achieve it? For several years now thoughtful teachers have raised this question. Stephen Potter in his book *The Muse in Chains* maintains that universities have simply not realized that printing has been invented, and that the lecture has survived from the time when the lecturer had a handwritten copy of Aristotle which he read at dictation speed in order that each student might possess a similar manuscript. Professor Crowther in New Zealand used to maintain that precise communication was possible only through the written or printed word. When we speak, we improvise. When we write, we examine what we have written and amend it many times until it expresses truly what we wish to say. It was with such remarks in mind that this book was composed. The printed material was handed to the students to read. In the class periods and tutorials, students raised difficulties and these were discussed – between students themselves, between students and graduate assistants, and between students and the 'lecturer'. It was in such discussions that the most effective learning was done. A clear explanation is a useful starting point, but it is hardly ever the end. Students are liable to misinterpret or fail to understand the explanation in the most individual and unpredictable ways. It is only through discussion that such misunderstandings can be discovered and cleared up.

September 1971   W. W. S.

1 Mathematics and engineers

Is it justified to have a course in mathematics for engineers? Would it not be better to provide a course in electrical calculations for electrical engineers, chemical calculations for chemical engineers and so on through all the departments? The reason for teaching mathematics is that mathematics is concerned not with particular situations but with patterns that occur again and again. This is most obvious in elementary mathematics. In arithmetic, the number 40 may occur as $40, 40 horsepower, 40 tons, 40 feet, 40 atoms, 40 ohms and so on indefinitely. It would be most wasteful if we decided not to teach a child the general idea of 40, but left this idea to be explained in every activity involving counting or measurement.

Advanced mathematics cannot claim the universal relevance that arithmetic has. There are mathematical topics vital for aerodynamics that leave the production engineer cold. An engineer cannot simply decide to learn mathematics. He must judge wisely what mathematics will serve him best. His aim is not only to find mathematics that will help him frequently now, but also to guess what mathematics is most likely to help in industries and processes still to be invented.

Linear algebra qualifies on both counts; it is already used in most branches of engineering, and has every prospect of continuing to be.

Very many situations in engineering involve the mathematical concept of mapping. We shall restrict ourselves to the simplest class of these, namely *linear mappings*. We shall explain what these are and how to recognize an engineering situation in which they occur. And again in the interests of simplicity, we will not consider linear mappings in general, but restrict ourselves to linear mappings involving only a finite number of dimensions.

2 Mappings

Obviously we can find many applications of the scheme

$$\text{input} \xrightarrow{\text{Process}} \text{output.}$$

The process may involve actual material, as when certain components or ingredients (input) are used to produce some manufactured article (output). Or the process may

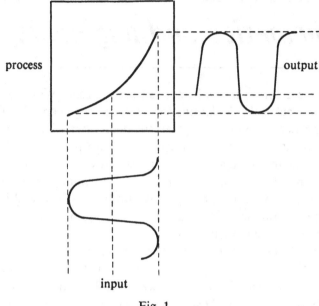

Fig. 1

be one of calculation, the input certain data, the output an answer. Another type of example would be a public address system; the input being the sounds made by a speaker or performer, the process being one of amplification and perhaps distortion, the output the sounds or noise emerging.

Diagrammatically, this last example might be shown as in Fig. 1.

Here distortion is occurring. The input is supposed to be a pure sine wave; the output certainly is not.

This is an example of a *mapping*, but it would not seem sensible to describe it as a linear mapping. We notice, for instance, that the graph representing the process is *not* a straight line.

There are two ways in which we might go about getting a straight-line graph. One way would probably be expensive – to use extremely good apparatus, which would yield a straight-line graph. The other way to avoid distortion is simpler and cheaper – to turn down the volume control. This means that we operate with a very small part of the curve. It usually happens that a small part of a curve is nearly (though of course not exactly) straight.

In fact, most of the examples we think of, that seem to arise in the first way (a graph that is really straight) are probably instances of the second way (using a small piece of a curve). Here we are thinking not only of our particular example – sound reproduction – but of all the cases in science where a straight-line law seems to hold. Someone might suggest, say, Hooke's Law: that for a stretched spring, force is proportional to deformation. But if we allow sufficiently large forces to act, the graph connecting force and deformation looks as is shown in Fig. 2. It seems reasonable to suppose that, if we

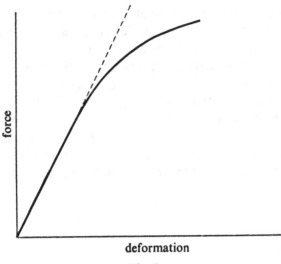

Fig. 2

could make very precise measurements, we would find a slight curvature in every part of the graph. But Hooke's formula is very convenient: it is easy to use and, for forces of moderate size, it gives all the accuracy we need.

In the same way, the formula $V = gt$ for a body falling from rest is linear, and is useful in many small-scale situations, but would be entirely misleading if used by astronauts. Ohm's Law, giving current and voltage proportional, is linear, but heavy currents are liable to produce changes in resistance and ultimately of course to melt the conductor.

Linear algebra then, like calculus, relies on the fact that, very often, a small part of a curve can be efficiently approximated by a straight line, a small part of a surface by a plane, and so on. (Question for discussion: what does 'and so on' mean here?) The qualification 'very often' is necessary; mathematicians have studied curves that are infinitely wriggly; for these no part, however small, can be approximated by a line. The functions corresponding to such graphs could appear only in quite sophisticated engineering problems.

The utility of linear algebra thus depends on a very general consideration – the tendency of functions to have good linear approximations. This argument is not tied to any branch of science, or to any particular department of engineering, or to the present state of the engineer's art; even if the whole of our present technology became obsolete, the argument would, in all probability, retain its force.

Sometimes of course an engineer has to deal with disturbances of such a scale that he cannot regard all the operations involved as effectively linear. He is then confronted with *non-linear problems*, which are much harder. As mentioned earlier, we do not plan to discuss these.

The illustrations so far given are not sufficient to make precise just what we mean by *linearity*. This idea we shall have to discuss for quite a while yet. Before we go into this it may be wise to remove a possible source of confusion. A linear mapping is a generalization of a type of function we meet in elementary algebra, namely $x \rightarrow y$, where $y = mx$ and m is constant. In elementary work, the function defined by $y = mx + b$ is usually called linear. It is important to realize that, in the language used by university mathematicians, $x \rightarrow mx + b$ would *not* be described as a linear function when $b \neq 0$. Our scheme 'input, process, output' may suggest why this is reasonable. We always suppose that if nothing goes in, nothing comes out. Zero input produces zero output. With $y = mx$, taking $x = 0$ makes $y = 0$, which is satisfactory. However, with $y = mx + b$, putting $x = 0$ gives $y = b$, which is not zero when $b \neq 0$. It has been agreed not to attach the label 'linear mapping' in this case.

3 The nature of the generalization

When $x \rightarrow y$ occurs in elementary work, it is understood that both the input x and the output y are real numbers. But the input–output idea is capable of vast generalization. Input and output could be almost anything: all we require is that the input in some way determines the output. For example, in designing an apartment building, we might be concerned about the effects of winds; the input would then be a specification of the wind acting on the building, the output perhaps the extra forces acting on the foundations as the result of such wind pressure. For an electronic computer, the programme fed in could be the input, its response the output. In an automated factory, the materials supplied could constitute the input, the goods produced the output.

It would be easy to produce many more examples; all that is necessary is that x, *what goes in* must determine y, *what comes out*; x and y can stand for extremely complicated objects or collections of data.

Any realistic person will realize that our scheme for a process or mapping, $x \rightarrow y$, is so general that very little information can be derived from it as it stands. Being told only that we have a mapping $x \rightarrow y$ is like being confronted with a machine having many controls and being told only that the machine is consistent – it always reacts in the same way to any particular setting of the controls. It may be comforting to know this, but it is not much to go on.

It is here that the restriction to *linear* mapping is helpful. But what do we mean by *linear* when we are dealing not with 'number $\rightarrow$ number' but with 'anything $\rightarrow$ anything'? Indeed, how has it come about that people have thought of using the term *linear* in such a vague and general situation?

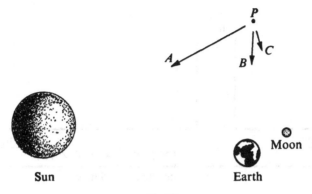

Fig. 3

To answer these questions we have to realize that this terminology arose only after many years' experience in a variety of subjects. In many sciences we meet something called a Principle of Superposition. In the theory of gravitation, for instance, we may want to find the force on an object at the point P, due to the attractions of the Sun, Earth and Moon. Let A be the force the object would experience if the Sun alone acted on it, B the force that would be produced by the Earth alone, and C that due to the Moon alone (Fig. 3). The principle of superposition states that the force due to the Sun, Earth and Moon acting simultaneously is the combined effect of the forces A, B, C. Thus one complex problem is broken into three simpler problems.

Again, in the theory of gravitation, a principle of proportion can be used. If a mass M exerts a force F on an object, then – provided the positions are unchanged – a mass of $3.7M$ will exert a force $3.7F$ on the object (Fig. 4). A mass kM would exert a force kF.

We should not think of the principles of superposition and proportion as applying to everything. For example, they do not apply to traffic.

In the situation shown in Fig. 5, it is possible that 20 cars a minute arriving from the north might pass without delay, if none came from the west. Also 30 cars a minute from the west might pass without delay if none came from the north. It would be rash to assume that, if both streams of traffic were arriving, 50 cars a minute would emerge

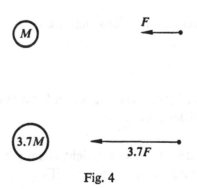

Fig. 4

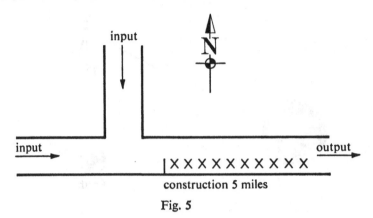

Fig. 5

beyond the construction area. Superposition does not apply. Nor does proportionality. If 10 times as many cars arrive per minute we do not expect to find 10 times as many emerging; rather we expect the rush-hour crawl.

Informally we may define a linear mapping as one arising in a situation to which the principles of superposition and proportionality apply.

Notice that, to make this explanation clear, we have to state carefully what we mean by superposition and proportionality both for input and output. In the gravitation question above, the input consists of a specification of the bodies exerting attraction. To superpose the three situations in which the Sun alone, the Earth alone and the Moon alone act on an object, we consider the Sun, Earth and Moon acting simultaneously. The output is concerned with the force experienced by the object; to apply superposition we must know how to find the combined effect of the three forces A, B and C. Someone unfamiliar with mechanics might be at a loss how to find this combined effect.

A number of situations are given below for consideration. In each case, it is necessary to spell out reasonable interpretations of superposition and proportionality both for the input and for the output. Then one has to consider whether principles of superposition and proportionality would apply (as in the gravitation problem) or would fail (as in the traffic problem).

Students may wish to select the situations that are most familiar to them.

Situations

(1) Purchasing articles at fixed prices (no reduction for quantity).
Input. Specification of articles purchased.
Output. Cost.

(2) Sand, or other heavy material, rests on a light beam (i.e. one of negligible weight). Spring balances measure the reactions at the ends (Fig. 6).

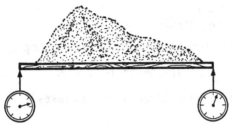

Fig. 6

Input. The distribution of weight along the beam.
Output. The readings of the spring balances.

(3) An electrical system of the type shown in Fig. 7.

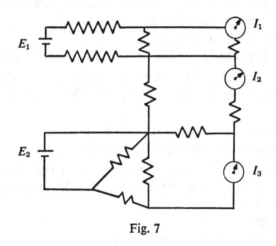

Fig. 7

Input. Voltages E_1, E_2.
Output. Currents I_1, I_2, I_3.

(4) Manufacturing.
Input. Materials required.
Output. Articles manufactured.

(5) Would it make sense to reverse input and output in Situation (4) i.e. to consider?
Input. Articles manufactured.
Output. Materials required.

(6) In order to obtain crude estimates of how quickly the temperature of a furnace is rising, the temperature (T) is measured at unit intervals of time, and the changes (ΔT) are calculated; for example

T	100		400		650		770		840
ΔT		300		250		120		70	

Input. The numbers in the T row.
Output. The numbers in the ΔT row.

(7) A student is given problems in differentiation. For example, given

$$f(x) = x^2 + 5x + 2,$$

he writes $f'(x) = 2x + 5$. (He does not make any mistakes.)
Input. The function $f(x)$.
Output. The function $f'(x)$.

(8) In an experiment, the mileage gone and the speedometer readings of a car are recorded at various times.
Input. Figures for mileage at these times.
Output. Speedometer readings at these times.

(9) In a radio programme listeners are able to phone in their opinions. In order to exclude obscenity, blasphemy, slander, sedition, etc. it is arranged that their remarks are heard on the radio ten seconds after they are spoken.
Input. The sounds made into the telephone.
Output. The sounds heard on the radio, on an occasion when nothing is censored.

(10) A number of pianos are available. A tape recorder deals faithfully with a certain range of notes on the piano. Notes above that it reproduces at half strength, and notes below that not at all.
Input. The sounds made by the pianos.
Output. The sounds as reproduced by the tape recorder.

(11) In pulse code telephony, the voice produces certain vibrations, shown in Fig. 8 as a graph. 8000 times a second a computer measures the ordinates (i.e. the y-values) shown here as upright lines.
Input. The voice graph.
Output. The values measured by the computer.

(12) Some function f, reals to reals is specified. Its values $f(0), f(1), f(2)$ are calculated.
Input. The function f.
Output. The numbers $f(0), f(1), f(2)$.

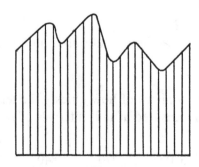

Fig. 8

(13) A function f, reals to reals, is specified. From it we calculate another function, g.
   Input. The function f.
   Output. The function g.

Discuss the following cases.
   (a) $g(x) = f(2x)$.
   (b) $g(x) = f(x - 1)$.
   (c) $g(x) = f(x^2)$.
   (d) $g(x) = [f(x)]^2$.

4 Symbolic conditions for linearity

Superposition and proportionality lend themselves to a very convenient representation by mathematical symbols.

In the gravitation problem, we wanted to find the attraction due to the Sun *and* the Earth *and* the Moon. In beginning arithmetic we associate 'and' with $+$. So it is natural to represent the superposition of Sun, Earth and Moon by $S + E + M$. The forces produced by Sun, Earth and Moon acting separately were indicated by A, B and C; it is natural to represent their combined effect by $A + B + C$, and indeed this notation will already be familiar to most students as vector addition of forces.

Thus from
$$S \to A$$
$$E \to B$$
$$M \to C$$

we conclude
$$S + E + M \to A + B + C.$$

Note that it is a subtle question of language whether we should say that the plus signs on the two sides of this statement have the same or different meanings. Someone could maintain that the meanings were the same because $S + E + M$ stands for 'the combined effect of S and E and M', while $A + B + C$ stands for 'the combined effect of A and B and C'. An opponent of this view would point out that the detailed procedure by which we find the combined effect of three forces is very different from that of considering three massive bodies existing together.

It will be seen from the list of situations given above that superposition applies to a great variety of problems. It would be very confusing to have a different sign for superposition in each of these cases. As we discussed right at the outset, mathematics is concerned with the similarities between different things, so that we learn one idea which can be used in many different circumstances. In order to bring out similarities, we use the same sign, $+$, whenever superposition is involved. But we must bear in mind the different systems that exist: we must never fall into the error of writing an expression that indicates the addition of a number to a force, or a mass to a voltage.

Proportionality can also be represented very simply. S standing for the Sun, $10S$ would indicate a body having the same position as the Sun but 10 times its mass. Again $10A$ would represent a force having the same direction as A but 10 times its magnitude.

The principle of proportionality shows that from the given fact

$$S \to A,$$

we may conclude

$$10S \to 10A.$$

We may now sum up our explanation of linearity.

We suppose we have an input, the elements of which may be denoted by $u, v, w, \ldots$ and an output with elements $u^*, v^*, w^*, \ldots$ In both the input and the output, we have sensible definitions of addition and of multiplication by a number, so we know what is meant by $u + v$ and ku, where k is any real number, and also what is meant by $u^* + v^*$ and ku^*. We have a mapping M which makes $u \to u^*$, $v \to v^*$, $w \to w^*$. This mapping will be called linear if for any two elements, u and v of the input, we find $u + v \to u^* + v^*$, and also that, for any number k and for any u, $ku \to ku^*$.

This is illustrated in Fig. 9.

The statement $u + v \to u^* + v^*$ means that we have two ways of finding where the mapping M sends $u + v$. First there is the obvious way, that works for any mapping. Starting with u and v in the input, we find $u + v$ which is also an element of the input. We then follow the arrow and find which element of the output we land on. The second method, which is not allowable for an arbitrary mapping but works with a linear

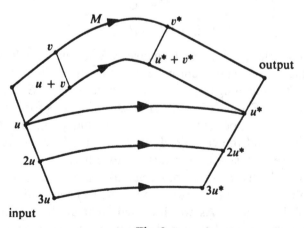

Fig. 9

mapping, is to begin with u and v, follow the arrows from each to u^* and v^*, and *then* add to find $u^* + v^*$. This is an element of the output and is found by the addition rules appropriate to the output.

In the same way, if we want to find the effect of the mapping on an element of the type ku, there are two ways of doing it. Consider for example $k=3$; what does the mapping do to $3u$? We may begin with u, find $3u$, and then follow the arrow. Alternatively, we may begin with u, follow the arrow to u^*, and thence find $3u^*$. That is, it does not matter whether we first multiply by 3 and then map, or first map and then multiply by 3. Similarly for addition, it does not matter whether we first add and then map, or first map and then add.

When we are dealing with a general mapping, knowing $u \to u^*$ does not enable us to say where any other element goes. But with a linear mapping, the information that $u \to u^*$ tells us immediately the fate of infinitely many other points. We know $2u$ must go to $2u^*$, that $3u \to 3u^*$, $7u \to 7u^*$, $0.51u \to 0.51u^*$, $-8u \to -8u^*$ and so on. The fate of any element of the form ku is determined once we know the fate of u.

If we know what a linear mapping does to two points, say $u \to u^*$ and $v \to v^*$, we can make even more extensive deductions. We know, for example, $5u \to 5u^*$ and $3v \to 3v^*$, by proportionality. Superposing these, we deduce $5u + 3v \to 5u^* + 3y^*$. Obviously, there is nothing special here about the numbers 5 and 3. For any numbers a and b we could deduce $au + bv \to au^* + by^*$.

From the strictly mathematical point of view, our explanation is still incomplete. In summing up, above, we said that 'in both the input and the output, we have sensible definitions of addition and of multiplication by a number'. But how do we tell that a definition is sensible? What are the requirements? These questions we must answer later on. We do not deal with them immediately, as we do not wish to distract attention from the important and central idea of a linear mapping, expressed verbally by reference to superposition and proportionality and symbolically by the conditions, if $u \to u^*$ and $v \to v^*$, then $u + v \to u^* + v^*$ and $ku \to ku^*$.

The following two little exercises are intended to fix these conditions in mind, and also to call attention to forms in which they may occur in the literature.

Exercises

1 A mapping is a function. If we denote it by the symbol F, the element to which u goes, instead of being written u^*, will be written $F(u)$. Express the conditions $u + v \to u^* + v^*$, $ku \to ku^*$ in this notation, that is, using F and not making any use of the arrow, $\to$, or the star, $*$.

2 This hardly differs from Exercise 1. If we denote the mapping by M, we may instead of u^* write simply Mu. Express the two conditions for linearity in this notation. When so written, they resemble equations in elementary algebra. With what laws of elementary algebra do you associate the resulting equations?

5   *Graphical representation*

Some of the conclusions of §4 are not at all surprising if we remember the first example of §3, under the heading 'Situations' – 'purchasing articles at fixed prices (no reduction for quantity)'. If in a store you are told that an article costs $10, the salesman will be inclined to think you are feeble-minded if you then proceed to ask 'What would two of them cost? How much would three cost?' If article A costs $10 and article B costs $20, you are expected to understand for yourself what 5 of A and 3 of B would cost.

The purchasing situation is in many ways typical of situations involving linear mappings, and the type of argument just used for finding the cost of $5A + 3B$ can be applied to *any* linear mapping. This is emphasized here, as experience shows that students often fail to appreciate this idea and to use it freely.

Of course the purchasing situation is special in one respect; the output is always simply a sum in dollars; it can be specified by giving a single number. It is easy to devise a situation where this special feature is removed. Suppose we wish not only to purchase certain articles but also to transport them to some rather inaccessible place. We may then be concerned to know not only the price but also the weight. We now need two numbers to specify the output, – price in dollars of the purchase and weight in pounds. But our procedure is not essentially affected. If we know the price and weight of A, and have similar information for B, we can work out the price and the weight for 5 of A and 3 of B. What we are using – though we are hardly conscious of this – are the principles of proportionality and superposition.

It has been said that an engineer thinks by drawing pictures. Pictorial thinking is certainly important for linear algebra, as may be guessed from the very name of the subject; while the second word, 'algebra', suggests calculation, the first word, 'linear' (having to do with straight lines) clearly comes from geometry. Most students will cope most successfully with linear algebra if they use ideas that are suggested by pictures and checked by calculation. There are certain limitations to what we can do graphically. We live in a world of 3 dimensions. We can deal with problems involving 2 numbers by means of graph paper; when 3 numbers occur, we can use or imagine solid models. Beyond that we are stuck.

The only solution to this difficulty seems to be to become thoroughly familiar with those situations which we can draw or represent physically. These help us to remember the essential ideas, and will often suggest an approach to the problems involving more than 3 numbers. Fortunately, many of the results in linear algebra do not depend essentially on the number of dimensions involved. There are strong analogies between the situations in 2 or 3 dimensions, which we can show graphically, and those in higher dimensions which we cannot.

For this reason, it is useful to see how our present topic looks graphically, in the very convenient case, a mapping from 2 dimensions to 2 dimensions. This is the purpose of the questions given below. Each question calls for points on one piece of graph paper to be mapped onto another piece. The investigation is complete when a student has mapped enough points to see the emergence of a pattern. There is no point in undertaking the drudgery of going past this stage. On the other hand, you should not stop until you have reached the stage at which you can predict, *without calculation*, to what positions further points would map. Each question is based on the illustration we used earlier, where we are concerned with the price and weight of purchases involving two articles, *A* and *B*.

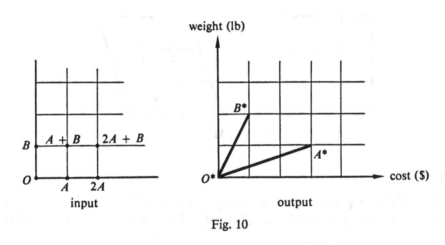

Fig. 10

Questions

1 In Fig. 10 $A \rightarrow A^*$; this is intended to indicate that article *A* costs $3 and weighs 1 pound. Similarly from $B \rightarrow B^*$ we read off that *B* costs $1 and weighs 2 pounds. On the input diagram mark the purchases that other points represent (some are already marked), and on the output diagram mark the points that give the corresponding cost and weight.

2 As in question 1, except that *A* now costs $1 and weighs 2 pounds, while *B* costs $2 and weighs 1 pound.

3 Now *A* costs $1 and has negligible weight, *B* costs $1 and weighs 1 pound.

4 *A* costs $1 and weighs 1 pound, *B* costs $2 and weighs 2 pounds.

5 *A* costs $1 and has negligible weight, and exactly the same is true of *B*.

In Questions 4 and 5, it will be found that several points in the input may go to the same point in the output. It is of interest to mark on the input diagram the set of points that go to some selected point on the output diagram.

In all these questions it will be noticed that we have only used the first quadrants of the graph paper. This is due to the nature of our illustration; the number of articles purchased, their cost and their weight are naturally taken to be non-negative. In most engineering applications we are concerned with quantities that may be positive or negative. If you care to experiment with situations in which the price is negative (e.g. garbage, which you pay people to take) or the weight (balloons filled with hydrogen or helium), you will find this does not bring in any essentially new feature. The network, or pattern of points, that appears in the output diagram still has much the same character as before.

In drawing such diagrams, it is convenient to use integers, since these correspond to the grid of the graph paper. In most applications, fractional values are perfectly possible, and points with fractional co-ordinates are subject to the mapping. However, a diagram such as Fig. 11 does effectively convey to our minds what is happening to *every* point of the plane, not only to the points of the grid that are actually drawn.

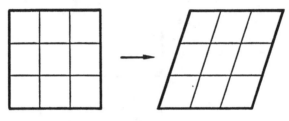

Fig. 11

For discussion

Which of the following are true for the mapping in Questions 1, 2, 3 above? How do your answers have to be modified if we allow also the mappings in Questions 4 and 5?

(1) Every point goes to a point.

(2) Distinct points go to distinct points.

(3) Every point in the output plane arises from some point of the input plane. (*Note.* We suppose throughout that the diagrams have been extended by allowing fractional and negative numbers.)

(4) Points in line go to points in line.

(5) Squares go to squares.

(6) Rectangles go to rectangles.

(7) Parallelograms go to parallelograms.

(8) Right angles go to right angles.

(9) The mapping does not change the sizes of angles.

(10) A circle goes to a circle.

(11) An ellipse goes to an ellipse (a circle being accepted as a special case of an ellipse).

(12) A hyperbola goes to a hyperbola.

(13) A parabola goes to a parabola.

6 Vectors in a plane

Most students have already met vectors and know that it is possible to define operations on them purely geometrically, without reference to any system of co-ordinates. Accordingly, we give these definitions without any introductory explanation.

We suppose a plane given, in which a special point O has been marked. If P and Q are any two points, by $P + Q$ we understand the point that forms a parallelogram with O, P and Q, as in Fig. 12.

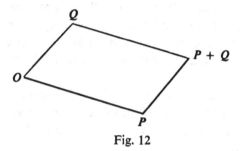

Fig. 12

Again, for any point P, by kP we understand the point that lies in the line OP, and is k times as far from O as P is.

A convention is understood here. If $k > 0$, then kP is on the same side of O as P; if $k < 0$, P and kP are separated by O. If $k = 0$, of course kP is at O. Fig. 13 shows the cases $k = 3$, $k = -2$. (Students sometimes fall into the error of thinking that we get to $3P$ by taking 3 intervals *beyond* P; this is not so, we reach $3P$ by measuring 3 intervals from O, the first interval being from O to P.)

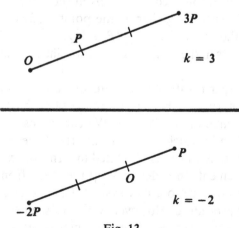

Fig. 13

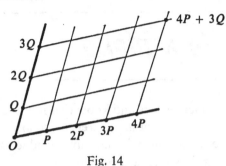

Fig. 14

In mechanics we tend to speak of 'the vector *OP*', and to connect the points *O* and *P* by an arrow symbol. In graphical work, these arrows may complicate and obscure the diagram. As the point *O* is laid down once and for all, it is perfectly satisfactory to do as we did in our first definition and speak of 'the point *P*', 'the point *Q*', 'the point *P* + *Q*'.

There is no mathematical difference involved here. It is simply a question of whether the diagram is clearer when the points *O* and *P* are connected by an arrow, or when they are left as they are. In any situation, we are free to use whichever we think conveys a better picture to the mind.

For instance, it seems much better not to draw the arrows in the situation now to be discussed. Suppose *P* and *Q* are any two points of the plane, such that *O*, *P*, *Q* are not in line. We mark the multiples of *P*, say 2*P*, 3*P*, 4*P*,..., and those of *Q*, such as 2*Q*, 3*Q*,... By drawing parallel lines, we obtain a grid, as in Fig. 14.

The point at the top right of this diagram forms a parallelogram with *O*, 4*P* and 3*Q* and hence must be the point 4*P* + 3*Q*.

The grid in this diagram could be used as graph paper. The point we have just considered would then be specified by its co-ordinates (4, 3), since it is reached from *O* by 4 steps in the direction *P*, followed by 3 steps in the direction *Q*.

This point thus can be specified either as the point (4, 3) in the co-ordinate system with basis [*P*, *Q*] or, in the vector form, as 4*P* + 3*Q*.

Similarly, the other points in the diagram can be specified in these two ways. (Consider some of them.)

Conventional graph paper is rather restricted. Being based on squares, the axes are perpendicular, and the intervals on both axes are the same length. The system based on two vectors, *P* and *Q* is much more flexible. We can choose *P* and *Q* so as to fit the system neatly to a problem in which the most important lines are not perpendicular.

Again it often happens that a problem is stated in terms of axes which are later found not to be the most convenient. In order to keep the work from becoming impossibly difficult, it is necessary to change to a new system. Change of axes is often thought of as being a very difficult procedure. However, we have seen that in the vector form a point can be specified by a very simple expression such as 4*P* + 3*Q*. Manipulating such expressions should not be difficult and in fact, by the vector approach, change of axes can be done very easily, as we shall see in §9.

Note on a superstition

If we compare the co-ordinate specification, $(4, 3)$, with the vector specification, $4P + 3Q$, we see that the same pair of numbers appears in each. This effect is quite general, and we naturally use algebra to state it; if x stands for the first number and y for the second, we may assert that, in the P, Q system, the point with co-ordinates (x, y) is the point $xP + yQ$.

On many occasions, students have been asked in tests to give the vector expression for the point (x, y) and have written the answer $aP + bQ$. Some teacher must have given them the idea that x and y had some special, sacred quality, that these letters always stood for unknowns, and could not appear in an answer. This is a complete misconception. To go back to a first exercise in algebra, the correct answer to the question 'A room contains x boys and y girls, how many children in all?' is $x + y$. This could be regarded as a formula. It is a general result, in so far as x and y may stand for *any* whole numbers. Whether x and y are unknowns or not depends on the circumstances – whether we have seen the room and have had time to count the boys and girls. We cannot use the formula until we have been given the values of x and y.

In the same way, the statement 'the point (x, y) may be specified as $xP + yQ$' can be regarded as a formula or rule. If we are asked to apply it to the point $(8, 5)$, we know x has the value 8 and y the value 5, so $(8, 5)$ may be specified as $8P + 5Q$.

The assertion 'the point (x, y) may be specified as $aP + bQ$' cannot be so used. If we are told that x and y have the values 8 and 5, we still need to ask, 'What are the values of a and b? By what procedure are a and b to be derived from x and y?' Really, such an assertion does not make sense as it stands.

7 Bases

In §6 we considered the grid built on two vectors P and Q. Only part of this grid was shown in our last diagram, but it is clear that, in that diagram, the grid could be extended to cover the whole plane. Now any point covered by the grid can be expressed in the form $xP + yQ$, by reading off its co-ordinates x, y. (These of course may be fractional or negative.) Accordingly, in the situation illustrated, every point of the plane can be specified in the form $xP + yQ$. Also, this can be done in only one way.

For example we can get the point $4P + 3Q$ only by taking $x = 4$, $y = 3$. Any other choice of x and y will land us somewhere else.

An expression such as $4P + 3Q$ is called a *linear combination* of P and Q. More informally, we may refer to it as a *mixture* of P and Q. We can have linear combinations, or mixtures, of any number of elements. If in some system we have elements A, B, C, D, and the expression $5A - 2B + 9.1C + 0.3D$ is meaningful, then that expression is a mixture of A, B, C and D, as would be any expression of the form $xA + yB + zC + tD$, with real numbers x, y, z, t.

The points P and Q are said to form a *basis* of the plane, since they have the following properties: (1) every point of the plane is a mixture of P and Q, (2) each point can be specified as such a mixture in only one way.

It will be seen that a basis for the plane can be chosen in many ways. Almost any two points will do for P and Q. What does 'almost' mean here? First of all, neither P nor Q must be a zero vector, i.e. neither point may be at O. Secondly, P and Q must not be in line with O. If they were, every point $xP + yQ$ would lie on that line; the grid would not fill the whole plane.

Thus there are infinitely many ways of choosing a basis with 2 points (or vectors) in it. Would it be possible to have a basis with 1 or 3 points in it?

If we had only one point, say R, we could only form expressions of the form xR. As we saw at the beginning of §6, xR is defined as a point on the line OR. Thus we would not be able to find an expression to specify the points of the plane not on OR.

What about a basis with 3 vectors? Consider for example the three points, A, B, C in Fig. 15. They certainly have the first property. The point (x, y) of the graph paper is $xA + yB$, so we can get it without using C. We may write it as $xA + yB + 0 \cdot C$ if we want to bring out that it has the required form for a mixture of A, B and C. But this expression is one of many. For example, the point J may be written as $2A + 3B$, or $A + 2B + C$, or $B + 2C$, or $-A + 3C$, or in many other ways.

The vector C is redundant, for $C = A + B$. C is already a mixture of A and B. By mixing A, B and C we get nothing beyond what we could get by mixing A and B. If we

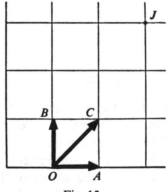

Fig. 15

interpret A as a gallon of red paint and B as a gallon of white paint, C would be a mixture of 1 gallon of red and 1 gallon of white paint. It does not extend in any way the range of tints we can produce.

A collection of vectors is said to be *linearly dependent* if one vector is a mixture of the remaining vectors. Thus A, B and C are linearly dependent. One can see that a linearly dependent collection of vectors is no good as a basis. Property (2) will always fail for it. A basis must consist of linearly independent vectors.

In the plane, as we have seen, it seems that every basis contains the same number of vectors, namely 2. For this reason a plane is said to be 'of 2 dimensions'. Further any 3 vectors in a plane are bound to be linearly dependent. In a space of 2 dimensions it is impossible to find 3 linearly independent vectors.

In a strictly mathematical development, these statements would be proved, and it would be shown that corresponding statements can be made in spaces of $3, 4, 5, \ldots,$ dimensions. The proofs are somewhat tedious, and the time spent on them is somewhat unrewarding, since the results are what we would in any case guess or expect to be true. Later on, we shall return to some of the points involved, in a rather more formal way. For the present we will restrict ourselves to giving a definition of vector space and of basis, and a statement, without proof, of the theorems.

In §2 and §3 we considered mappings input $\rightarrow$ output. In all the situations listed at the end of §3, it was possible to define addition and multiplication by k 'in a sensible way'. Thus each input and each output was an example of a vector space. But in many of the situations it was not possible to specify what was coming in or what was coming out by a finite collection of numbers. Such inputs or outputs were examples of vector spaces of infinite dimension, which we do not intend to discuss further. But there were cases in which the input, or the output, or both, could be specified by a finite list of numbers. (For which inputs was this true? For which outputs?) In several cases we can adapt the situation described so as to make it finitely specifiable. For instance, in Situation 2, instead of putting sand on the light beam, we may suppose 3 weights hung from it, as in Fig. 16.

The input is now specified by giving the 3 weights W_1, W_2, W_3; the inputs form a space of 3 dimensions. The output is specified by giving the 2 reactions; the possible outputs form a space of 2 dimensions. (Strictly speaking, we ought to allow for these

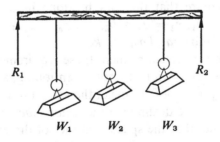

Fig. 16   Situation 2a.

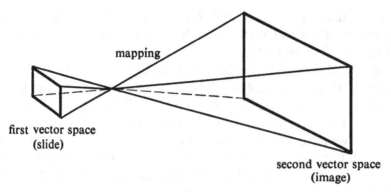

mapping

first vector space
(slide)

second vector space
(image)

Fig. 17

quantities becoming negative: a weight could be replaced by an upward pull, and it might be necessary to hold an end down rather than to support it. If we are only interested in actual weights and in positive reactions we are mapping *part* of a 3-dimensional space to *part* of a 2-dimensional space.)

In reading the formal definitions below, we should have in mind not only the plane and our everyday space of 3-D, but also such systems as occur in Situation 2*a* just discussed, or in other suitable examples resembling the situations given at the end of §3. It should be remembered that a linear mapping connects the input and the output. When we are looking for an example of a vector space we look at *the input alone* or the *output alone*. For instance, when a slide is projected onto a screen, the points of the slide *or* the points of the screen may help us to visualize a space of 2 dimensions. The projection, by which the little picture on the slide produces the big picture on the screen, is then a mapping from one vector space to the other (Fig. 17). It might well be a linear mapping.

We now come to our formal definitions. We suppose we have some system consisting of various objects, situations, ideas or other elements. In this system addition is defined; given any two elements, a and b, there is a definite element c, called the sum of a and b, and written $c = a + b$. Also if k is any real number, we know what is meant by ka. Further, we suppose it is possible to choose a basis within the system, that is, n elements $v_1, v_2, \ldots, v_n$ such that each element can be expressed as a linear combination (or mixture) of them, for example $a = a_1v_1 + a_2v_2 + \cdots + a_nv_n$ where $a_1, a_2, \ldots, a_n$ are real numbers; we also require that this can be done in one way only. Further, if $c = a + b$, then $c_1 = a_1 + b_1, c_2 = a_2 + b_2, \ldots, c_n = a_n + b_n$. Also we require that the numbers occurring in ka be $ka_1, ka_2, \ldots, ka_n$.

(In strict logic, it is necessary to supplement these requirements by statements which many students will feel are too obvious to need mentioning, such as, for example that v_1 and $1 \cdot v_1 + 0 \cdot v_2 + 0 \cdot v_3 + \cdots + 0 \cdot v_n$ mean the same thing.)

In these circumstances, we are dealing with a *vector space of finite dimension*.

We naturally expect to say that the space is in fact of dimension n. But a point has to be cleared up. In a vector space we expect to find many ways of choosing a basis.

For instance in the plane any pair of vectors (with different directions) will do, and all these choices are on an equal footing. Suppose we were to find one basis with the n vectors $v_1, \ldots, v_n$, and another basis, with a different number, m, of vectors, $u_1, \ldots, u_m$. The first basis would lead us to assign n dimensions to the space, the second would indicate m dimensions. We would be very surprised if such a thing did happen – for example, if it turned out that a plane could be the same thing as a space of 3 dimensions. And in fact there is a theorem – all the bases in a vector space contain the same number of vectors. That number we define to be the dimension of the space. This theorem is proved with the help of another theorem – in space of n dimensions it is impossible to have $(n + 1)$ linearly independent vectors.

In some books you will find a vector defined as something that has length and direction. These were in fact the first objects to be considered as vectors – such things as force, velocity, acceleration. But now, as we have seen, 'vector spaces' comprise a great variety of systems. In present usage, a vector means simply an element of such a system.

Questions

1 Usually 3 vectors in 3 dimensions form a basis of the space. In what cases do they not do so? (This question is most easily considered in the everyday, geometrical space of 3 dimensions.)

2 A drawer contains a certain collection of nuts, bolts, washers and nails. Can such a collection be regarded as an element in a vector space? If so, discuss how $a + b$ and ka are defined; the dimension of the space; and a simple example of a basis.

3 For a collection of vectors to form a basis, the collection must satisfy two conditions, (i) every point of space is representable as a mixture of vectors in the collection, (ii) this representation can be done in only one way (i.e. the vectors in the collection must be linearly independent).

For each collection below, state whether it is a basis or not. If it is not, state which condition or conditions it fails to satisfy.

Space of 2 dimensions
(a) $(1, 0)$ (b) $(3, 1), (3, -1)$ (c) $(2, 0), (2, 1), (2, 2)$ (d) $(2, 3), (4, 6)$

Space of 3 dimensions
(e) $(1, 0, 0), (0, 1, 0)$ (f) $(1, 0, 0), (0, 1, 0), (0, 0, 1), (1, 1, 1)$
(g) $(1, 1, 0), (1, -1, 0), (3, 1, 5)$ (h) $(1, 1, 1), (1, 2, 3), (2, 3, 4)$

4 Find a simple way of choosing a basis for any input or output, at the end of §3, that is a space of finite dimension; also for any space of finite dimension that can be obtained by modifying the description of an infinite dimensional space occurring in that list of situations.

8 Calculations in a vector space

In a formal mathematical treatment, the definition of vector space would be followed by a number of theorems about the properties of such spaces: it would be shown that addition was commutative and associative, that subtraction could be defined, that a distributive law held, and so forth. Subsequent calculations would be justified by citing these theorems. It is not a difficult exercise to check that these various properties do hold.

There is a common sense consideration that enables us to see how calculations are made in vector spaces and to feel confident that these calculations will have properties closely resembling those of elementary arithmetic.

Arithmetic, as taught to young children, considers situations involving one kind of article. Children are asked to add 2 apples to 3 apples or $2\frac{1}{2}$ gallons to $3\frac{1}{4}$ gallons. It is only a very small extension of this to consider several articles – add 2 nuts and 3 bolts to 4 nuts and 5 bolts. But now we are essentially dealing with a vector space of 2 dimensions. We have seen this on a number of occasions – the input to Situation 1 at the end of §3, various remarks in §5, and Question 2 at the end of §7. Now if we add a_1 nuts and b_1 bolts to a_2 nuts and b_2 bolts we get $(a_1 + a_2)$ nuts and $(b_1 + b_2)$ bolts. We have only to replace 'nuts' by v_1 and 'bolts' by v_2 to arrive at the formal definition of addition in a vector space of 2 dimensions. Multiplication by k can be arrived at similarly. If instead of 2 articles we have n articles, our statements take longer to write out, but no new idea is involved. This means that work in n dimensions proceeds along much the same lines as work in 2 dimensions. Thus in vector theory we do not often have to refer to the number of dimensions of the space; there are many theorems that hold for a space of any finite number of dimensions. In vector calculations we are using a very simple extension of arithmetic; an expression such as $a_1v_1 + a_2v_2$ can arise in many contexts, but if, when computing with it, we imagine it to mean 'a_1 articles v_1 together with a_2 articles v_2' we shall obtain the correct answer.

There is one important distinction to bear in mind however. If, on ordinary graph paper, we use the point $(2, 3)$ to signify '2 nuts and 3 bolts', then the horizontal and vertical axes acquire a special significance. Any point on the horizontal axis represents a collection of nuts alone; any point on the vertical axis represents bolts alone. All other points represent collections in which both articles are present. Now in a vector space, no direction has any special significance. We may picture a vector space of 2 dimensions as a plane on which the point O has been marked. But nothing else is marked; in particular, no line through O has special significance. One line through O is as good as another. Any two distinct lines through O will serve as axes, and vectors in them will constitute a basis. Any basis is as good as any other basis.

In §6 we met a grid composed of parallelograms and saw that such a grid could be used as graph paper. Most students have experience only of conventional graph paper, based on squares. It is therefore desirable to do some work with this more general kind of graph paper, to acquire familiarity with it and to discover which results remain true for it.

In the worked example below, three approaches are used for dealing with certain straight lines. In the first approach, we simply guess the equations of the lines. In the second approach, vector notation gives a way of specifying the lines. In the third approach, the vector results are used to prove the correctness of our earlier guesses.

Worked example

Investigate the lines $OFMV$ and $PGNW$ shown in Fig. 18, the co-ordinates x and y being measured parallel to the lines OE and OS, with OP and OQ as units in these directions.

First approach The points O, F, M, V have co-ordinates $(0, 0)$, $(1, 1)$, $(2, 2)$, $(3, 3)$. In each of these, the x and y co-ordinates are equal. This *suggests* that $OFMV$ has the equation $y = x$.

The points P, G, N, W have co-ordinates $(1, 0)$, $(2, 1)$, $(3, 2)$, $(4, 3)$. The equation $y = x - 1$ is suggested for the line $PGNW$.

Second approach $OFMV$ is the line joining O to F. By definition, the vector tF lies on this line, and is t times as far from O as F is. If t runs through all real numbers, positive, zero or negative, the point tF will take every possible position on the line $OFMV$. Thus this line may be specified as consisting of all points of the form tF.

The line $PGNW$ results if every point of the line $OFMV$ experiences a displacement equal to the vector OP. For example, the point W arises when V is given such a displacement. We observe that $OVWP$ is a parallelogram, so $W = V + P$. Similar considerations apply to every point of the line $PGNW$; each point can be obtained by adding P to the vector for a point on the line $OFMV$. As the general point of $OFMV$ is tF, the general point of $PGNW$ is of the form $P + tF$, where t is any real number.

An alternative way of seeing this result is the following; if a mass starts at position P and moves with velocity OF for t seconds, it will reach a point on the line $PGNW$.

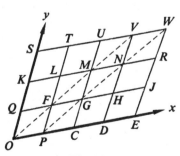

Fig. 18

The velocity OF may be written simply as F, since single letters are understood to denote vectors starting at the origin. Thus after t seconds the mass will be at the position $P + tF$. As time passes, the moving point describes the entire line. (Negative times must be considered to get the points below P.)

Third approach We have $F = P + Q$, so $tF = tP + tQ$. Accordingly tF has the co-ordinates (t, t) by the argument of §6 in which we identified (x, y) with $xP + yQ$. Thus for any point of the line $OFMV$ we have $x = t, y = t$ for some t; hence $y = x$.

The line $PGNW$ consists of all points of the form $P + tF$. As $F = P + Q$, we have $P + tF = P + t(P + Q) = (t + 1)P + tQ$. Thus for any point of $PGNW$ we have $x = t + 1, y = t$ for some real number t. Hence $y = x - 1$ is the equation of the line, as we guessed earlier.

In the work of the second approach, it should be noted that $P + tF$ is not the only correct answer. The moving mass does not have to be at P when $t = 0$; it could equally well be at any other point of the line $PGNW$. Again, we are free to choose the speed with which it describes the line; the velocity could be, for instance $\frac{1}{2}F$ or $-3F$. Thus $G + \frac{1}{2}tF$ and $W - 3F$ are among the many possible correct ways of specifying the general point of this line. We may speak of any such expression as a *vector parametric specification* of the line, with parameter t.

Exercises

(All questions relate to Fig. 18.)

1 For the lines listed below, carry out the procedure just described, that is (i) guess the equation, (ii) find a vector parametric specification, (iii) from your answer to (ii) deduce the equation of the line.

   (*a*) *CHR*   (*b*) *QLU*   (*c*) *SLGD*

(*Hint.* This line is parallel to *QP*. The vector that goes from Q to P is $P - Q$.)

   (*d*) *TMHE*   (*e*) *OQKS*   (*f*) *PFLT*   (*g*) *KLMNR*

2 The position of a moving point at time t is given by $Q + tP$. State (*a*) where the point is at $t = 0$, (*b*) its velocity, (*c*) the line in which it moves, (*d*) its co-ordinates at time t, (*e*) the equation of the line it moves along.

3 Identify on Fig. 18 the lines with the following equations:

   (*a*) $x + y = 3$ (*b*) $y = x + 1$ (*c*) $y = \frac{1}{2}x + 1$ (*d*) $y = 3 - \frac{1}{2}x$ (*e*) $y = 2 - \frac{1}{2}x$.

9 Change of axes

It very often happens that a problem can be immensely simplified by changing to a new set of axes. It is therefore important to know how such a change is made. Fortunately with vector notation changing axes is very simple.

Fig. 19 shows a plane on which two grids have been marked, one based on the points A, B, the other on the points P, Q. The point H has the co-ordinates (2, 1) in the P, Q system and (10, 5) in the A, B system. How could we, by calculation, pass from one set of co-ordinates to the other? More generally, if a point has co-ordinates (X, Y) in the P, Q system, how shall we calculate its co-ordinates, (x, y), in the A, B system?

Vector notation gives us a simple way for making such calculations. It also gives us a much shorter and more convenient way of indicating which system we are using, and also avoids the dangers of error due to mistaking co-ordinates in one system for co-ordinates in the other. Our initial information, that H is (2, 1) in the P, Q system can be written quite shortly as $H = 2P + Q$. Now, looking at P and Q, we see that $P = 4A + B$ and $Q = 2A + 3B$. Substituting these in our first equation, we find $H = 2(4A + B) + (2A + 3B)$, that is $H = 10A + 5B$, from which we can, if we wish, read off the co-ordinates (10, 5) in the A, B system.

We can go through the same steps with any point. Quite generally, suppose the

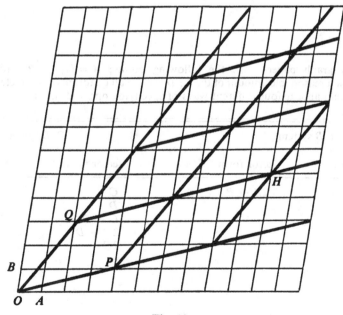

Fig. 19

point S has the co-ordinates (X, Y) in the P, Q system. Then $S = XP + YQ$. Substituting, as before, for P and Q, we find $S = X(4A + B) + Y(2A + 3B)$ so

$$S = (4X + 2Y)A + (X + 3Y)B.$$

Now if S is (x, y) in the A, B system, this means $S = xA + yB$. Comparing the two equations, we obtain

$$x = 4X + 2Y$$
$$y = X + 3Y.$$

These equations tell us how the co-ordinates in the two systems are connected. We can solve these equations, to find X and Y in terms of x and y. We find

$$X = \quad 0.3x - 0.2y$$
$$Y = -0.1x + 0.4y.$$

Thus, given the co-ordinates of any point in one system we can calculate its co-ordinates in the other system.

In the diagram both grids contain what are sometimes called oblique axes – that is, the axes are not at right angles. This did not cause any difficulty in the calculations; in fact, right angles have nothing to do with our present work. In pure vector theory they are not mentioned and are not even defined.

Note that what we called 'change of axes' could equally well have been called 'change of basis'. In the A, B system every point is shown in the form $xA + yB$, that is, we use A, B as a basis. In the P, Q system, every point is shown as $XP + YQ$, a mixture of P and Q; here we are using P, Q as a basis. (See the definition of *basis* in §7.)

Questions

1 On ordinary graph paper mark the following points; $A = (1, 0)$, $B = (0, 1)$, $C = (2, 1)$, $D = (1, 2)$. Draw a reasonable amount of the C, D grid, in all quadrants. Complete the table below: on each line should appear four different ways of specifying one and the same point.

A, B system		C, D system	
Co-ordinates	Vector form	Vector form	Co-ordinates
(2, 1)	$2A + B$	C	(1, 0)
		D	
		$C + D$	
		$2C + D$	
		$2C$	(2, 0)
			(2, −1)
(2, −2)			
(0, −3)			
(−2, −4)			

If $XC + YD = xA + yB$, find equations giving x, y in terms of X, Y and also equations giving X, Y in terms of x, y. Check for accuracy by using the data read off from your diagram and recorded in the table above.

2 In the same way, consider the A, B system and the E, F system where E is $(1, 1)$ and F is $(-1, 1)$ on the original graph paper.

3 We sometimes wish to introduce only one new axis. Do the same work using the A, B system and the A, E system.

4 Co-ordinates x, y are understood to be in the A, B system.
Find
(a) The equation in the C, D system that corresponds to $5x^2 - 8xy + 5y^2 = 9$.
(b) The equation in the E, F system corresponding to $x^2 - xy + y^2 = 1$.
(c) The equation in the A, E system corresponding to $x^2 - 2xy + 3y^2 = 1$.

5 Let $P = 4A + B$, $Q = 2A + 3B$. Find the equations in the P, Q system of the lines whose equations in the A, B system are $x + y = 10$ and $3y = 2x$. Check your result by drawing part of the P, Q grid, with the help of ordinary graph paper.
A diagram in §9 shows A, B, P, Q but with the axes OA and OB not at right angles. Could this diagram be used to check these results?

6 'Quadratic expressions of the type $ax^2 + bx + c$ form a vector space.' Justify this statement and give any definitions needed to make its meaning clear. Of how many dimensions is this space? What is the most obvious basis for it? Do the three quadratics $(x - 1)^2$, x^2, $(x + 1)^2$ form a basis? Give an example of changing from one basis to another in this space, and obtain the relevant equations.
In questions 7, 8 and 9 a 3-dimensional co-ordinate system (x, y, z) is based on vectors A, B, C and a system (X, Y, Z) on P, Q, R.

7 If $P = 3B + 5C$, $Q = A + 6C$, $R = 2A + 4B$, find expressions giving the co-ordinates (x, y, z) in the A, B, C system of the point whose co-ordinates in the P, Q, R system are (X, Y, Z).
The points for which $6x - 3y - z = 0$ form a plane. Find the equation of this plane in the X, Y, Z system.

8 Find the expressions for the (x, y, z) co-ordinates of a point in terms of its (X, Y, Z) co-ordinates if $P = A + 10B + 5C$, $Q = A + 20B - 5C$, $R = A - 10B + 15C$.
Find the equations involving X, Y, Z that correspond to
(a) $15x - y - z = 0$,   (b) $20x - y - 2z = 0$,   (c) $5x - y - z = 0$.

9 Express (x, y, z) in terms of X, Y, Z when, in the A, B, C system of co-ordinates, P is $(1, 1, 1)$, Q is $(1, 2, 3)$ and R is $(1, 4, 9)$.
Find the equations involving X, Y, Z that correspond to
(a) $8y = 5x + 3z$,   (b) $3x - 3y + z = 0$.
What equation in x, y, z corresponds to $Z = 0$?

Worked example

A complicated metal structure is specified by giving the co-ordinates of all the corners of the plates from which it is built. The points $O = (0, 0, 0)$, $C = (-1, 2, 2)$, $D = (2, -1, 2)$, $E = (3, 0, 6)$, $F = (0, 3, 6)$ are known to be the corners of a flat plate bounded by the polygon $ODEFC$. It is also known that the sides OC and OD are perpendicular and of equal length. A scale drawing, from which this plate could be manufactured, is required.

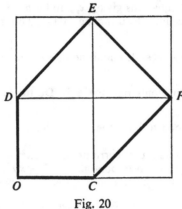

Fig. 20

Solution   We are told that the points E and F lie in the plane OCD. This means that each of them must be expressible in the form $XC + YD$.

If $E = XC + YD$, this means $(3, 0, 6) = X(-1, 2, 2) + Y(2, -1, 2)$, that is $3 = -X + 2Y, 0 = 2X - Y, 6 = 2X + 2Y$. From the first two equations we find $X = 1, Y = 2$. The third equation is satisfied by these values. (If it were not, it would mean that E did not lie in the plane OCD, contrary to the information provided.) Thus $E = C + 2D$. Similarly we may find $F = 2C + D$.

Accordingly, if we take C and D as the basis for conventional graph paper, E is $(1, 2)$ and F is $(2, 1)$. The shape of the plate is thus as shown in Fig. 20.

The points O, C, D having the same meanings as in the worked example above, make drawings of the following shape (all are polygons):

10 $OCMD$, where $M = (1, 1, 4)$.

11 $OCGD$, where $G = (2, 2, 8)$.

12 $OCHD$, where $H = (-1, -1, -4)$.

13 $OCJHK$, where $J = (-3, 3, 0)$, $H = (-1, -1, -4)$, $K = (3, -3, 0)$.

10   Specification of a linear mapping

In §3 and §4 we discussed what was meant by a mapping being linear. In §3 we characterized a mapping as being linear if it enjoyed the principles of proportionality and superposition. In §4 we gave this a more algebraic form; if a linear mapping sends $u \rightarrow u^*, v \rightarrow v^*$ then $c_1u + c_2v \rightarrow c_1u^* + c_2v^*$; that is, if we know what a linear mapping M does to any two vectors u and v, we know what it does to any mixture of u and v.

Suppose now the input is a vector space of n dimensions. This means that every vector is a mixture of the vectors $v_1, \ldots, v_n$ in a basis of the input space. Thus for any vector a of the input we may write $a = a_1v_1 + \cdots + a_nv_n$. Suppose now we know the fate of the basis vectors, that is, we know $v_1 \to v_1^*, v_2 \to v_2^*, \ldots, v_n \to v_n^*$. By proportionality, we know $a_1v_1 \to a_1v_1^*, a_2v_2 \to a_2v_2^*, \ldots, a_nv_n \to a_nv_n^*$. By superposition, we can now deduce

$$a_1v_1 + a_2v_2 + \cdots + a_nv_n \to a_1v_1^* + a_2v_2^* + \cdots + a_nv_n^*.$$

Now all the quantities on the right-hand side are known; we assumed $v_1^*, v_2^*, \ldots, v_n^*$, the point to which the basis vectors go, to be given, and the other quantities, $a_1, a_2, \ldots, a_n$, are the co-ordinates of the point a that is under consideration. Thus, with a linear mapping, if we know what happens to the basis vectors, we know what happens to any vector in the space.

The fate of any basis determines the fate of the entire input space, when the mapping is linear.

The investigations in §5 were intended to prepare for this idea. The questions there were concerned with a plane having A, B and a basis. The diagram for Question 1 shows A^* and B^* in the output space. Thus we knew, in this question, what the mapping did to the basis vectors A and B. What happened to other points, such as $A + B$, $2A + B$, could be determined in the first place by calculation. The graphical work, however, suggested that, in general, the input grid mapped to a very similar grid in the output plane. Thus a very simple geometrical construction enabled us to see the effect of the mapping, once the points A^*, B^* were given. It may help to fix this idea in mind if, at this stage, one or two questions, of the type given in §5 are answered by *purely geometrical means*.

Worked example

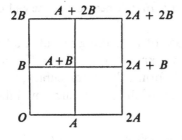 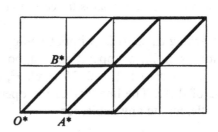

Fig. 21

On one piece of squared paper $A = (1, 0)$ and $B = (0, 1)$. This piece of paper is then mapped onto another piece in such a way that $A^* = (1, 0)$ and $B^* = (1, 1)$ as in Fig. 21.

Without any calculations, by eye alone, draw the network to which the squares of the first piece map, and read off the co-ordinates of the points to which the points $2A$, $A + B$, $2A + B$, $2B$, $A + 2B$ and $2A + 2B$ map.

Solution The parallelogram with sides $O*A*$ and $O*B*$ is completed. The required network of cells congruent to this parallelogram is easily drawn by eye. We now read off $A + B \rightarrow (2, 1)$, $2A \rightarrow (2, 0)$, $2A + B \rightarrow (3, 1)$, $2B \rightarrow (2, 2)$, $A + 2B \rightarrow (3, 2)$, $2A + 2B \rightarrow (4, 2)$.

Exercises

1 Carry out the procedure of the worked example for the following cases.
  (a) $A* = (1, 1)$, $B* = (0, 1)$ (b) $A* = (1, 1)$, $B* = (-1, 1)$
  (c) $A* = (1, 0)$, $B* = (0, -1)$ (d) $A* = (2, 0)$, $B* = (1, -1)$
  (e) $A* = (2, 1)$, $B* = (-1, 1)$

2 Check in the worked example above and in your answers to Question 1 that the geometrical method there used leads to the same results as the algebraic argument, e.g. that the point to which $A + 2B$ maps is in fact the same point as $A* + 2B*$.

Note Our theorem states that the fate of *any* basis settles the fate of the whole plane. It is not necessary for the basis to be the obvious one consisting of A and B. For example, if we are told $P*$ and $Q*$, where $P = A + B$ and $Q = 2A + 3B$, we can deduce $A*$ and $B*$. For $A = 3P - Q$, so $A* = 3P* - Q*$; similarly $B = Q - 2P$, so $B* = Q* - 2P*$.

3 Find $A*$ and $B*$ from the information given in each of the following cases.
  (a) $A \rightarrow A$, $A + B \rightarrow 2A + B$ (b) $A + B \rightarrow A + B$, $A - B \rightarrow B - A$
  (c) $A + B \rightarrow B - A$, $2A + B \rightarrow 2B - A$ (d) $A \rightarrow A + B$, $A + B \rightarrow B$

4 An examination question asks, 'If a linear mapping sends $(1, 0)$ to $(3, 4)$ and $(0, 1)$ to $(5, 6)$ where does it send $(1, 1)$?' A student, observing the sequence of numbers 3, 4, 5, 6 in the question, writes '$(7, 8)$' for his answer. This is simply a guess; it makes no use of the information that the mapping is linear, and in fact the answer is incorrect. What should the answer be, and what argument, using the properties of linear mappings, allows us to arrive logically at the correct answer?

5 A student is asked, 'If a linear mapping sends $(1, 0)$ to $(2, 1)$ and $(0, 1)$ to $(1, 2)$, where does it send $(3, 5)$?' He observes that the mapping increases each co-ordinate of the points mentioned by 1; in effect he assumes the mapping is $(x, y) \rightarrow (x + 1, y + 1)$ and answers '$(3, 5) \rightarrow (4, 6)$'. Why is his argument certainly incorrect? What is the correct answer? (*Hint*, if needed; see the end of §2.)

A student learning calculus is, without being aware of it, making use of the fact that differentiation is a linear operation. Suppose he knows that, for differentiation, $x^3 \rightarrow 3x^2$, $x^2 \rightarrow 2x$, $x \rightarrow 1$, $1 \rightarrow 0$. Now $x^3, x^2, x, 1$ form a basis for cubics; any cubic is a mixture of these. The linearity of the operation of differentiating now allows the student to differentiate any cubic, for

$$a_3x^3 + a_2x^2 + a_1x + a_0 \rightarrow a_3(3x^2) + a_2(2x) + a_1(1) + a_0(0).$$

Some applications of the idea explained in this section will be found in the questions below. Before proceeding to these questions, we will consider an application to mechanics. This application is to such a simple situation that it has no claim to practical value; the problem could be solved equally well without any mention of linearity; it serves simply as an illustration of the basis-mapping principle in a concrete situation.

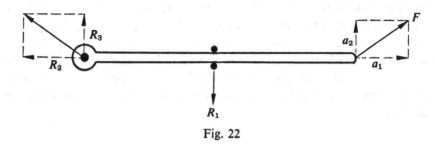

Fig. 22

We suppose a light rod has one end fastened by a smooth pin joint. It is prevented from rotating by smooth nails on either side of its midpoint (see Fig. 22). A force F acts on its free end. We are interested in the reactions produced by this force. The input here is the force F, with horizontal and vertical components a_1, a_2. The output is specified by R_1, R_2, R_3; here R_1 is the reaction at the midpoint, R_2 and R_3 are the horizontal and vertical components of the reaction at the pin joint.

We consider two situations. In the first situation unit, horizontal force acts at the free end (Fig. 23). It is obvious that this will be balanced by a unit horizontal force at

Fig. 23

the pin joint, and that there will be no reaction from the nails. Thus the output is $R_1 = 0, R_2 = 1, R_3 = 0$.

In the second situation, the free end experiences unit vertical force. It is easily seen that the forces on the rod will now be as shown in Fig. 24. That is, the output is $R_1 = 2, R_2 = 0, R_3 = 1$.

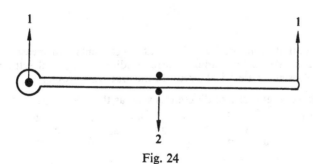

Fig. 24

Now this problem belongs to a large class of statical questions which are known to involve a linear mapping. The unit horizontal and vertical forces form a basis for the input force; the general F is obtained by adding a_1 times the horizontal unit to a_2 times the vertical unit. Taking these multiples of the reactions in the two special cases, we find that the force (a_1, a_2) produces the reactions $R_1 = 2a_2$, $R_2 = a_1$, $R_3 = a_2$.

As already mentioned, this solution could be obtained very easily by the usual methods of statics. It has been included only as a very elementary illustration of the idea that the solution of a complex problem can be found by combining the solutions of several simple cases, provided the problem is one involving a linear mapping from data to solution.

Questions

1   In Situation 2a of §7, the input is specified by the values of W_1, W_2, W_3. What would be a simple basis for the input? To what special situations would the vectors of the basis correspond?

Suppose that in Situation 2a it is impossible to measure any of the lengths involved, but that we can arrange for any loading W_1, W_2, W_3 that we wish and measure the reactions R_1, R_2 produced. Explain what series of experiments you would conduct in order to be able to calculate the reactions caused by any loading W_1, W_2, W_3. (The weights are always hung from the same 3 points.) Give explicit instructions that would enable a person ignorant of mechanics to calculate the reactions after performing the appropriate experiments. (The person may be assumed to understand the use of a formula.)

2   A piece of apparatus has two dials that may be set to any desired values. These settings constitute the input. Three meters indicate the output (Fig. 25). Nothing is known about the apparatus

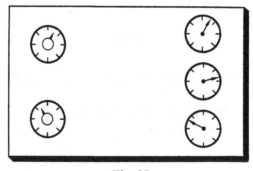

Fig. 25

except that intput → output is linear. Describe what experiments and measurements you would make in order to predict the meter readings corresponding to any dial settings. Embody your method in a formula. Design an electrical circuit to illustrate this problem.

3   In numerical work we meet tables of differences such as the following:

0		1		6		18		40		75		126
	1		5		12		22		35		51	
		4		7		10		13		16		
			3		3		3		3			

The numbers in each row represent the differences of adjacent terms in the row above.

Let p, q, r, s represent the first numbers in the first, second, third and fourth row respectively, and let every number in the fourth row be s. (In our example, $p = 0, q = 1, r = 4, s = 3$.)

It is known that when $p = 1, q = r = s = 0$, every number in the top row is 1. When $p = 0$, $q = 1, r = 0, s = 0$, the numbers in the top row are 0, 1, 2, 3, ... that is, they are given by the expression x.

When $p = 0, q = 0, r = 1, s = 0$ the numbers in the top row are given by the expression $\frac{1}{2}x(x - 1)$, in which the values 0, 1, 2, 3, ... are substituted in turn for x.

When $p = 0, q = 0, r = 0, s = 1$ the numbers in the top row are given by the expression $\frac{1}{6}x(x - 1)(x - 2)$.

It may be assumed that the mapping from the data p, q, r, s to the expression for the top row is linear.

Find the expression that corresponds to arbitrary values of p, q, r, s. Use this result to find the expression corresponding to 0, 1, 6, 18, 40, 75, 126, the numbers in the top row of our example above.

4 In a scientific experiment it is known that there is a law of the form $y = mx + b$. In order to determine m and b, the values of y are measured for two fixed values of x; $x = s$ gives $y = p$ and $x = t$ gives $y = q$. Is the mapping $(p, q) \rightarrow (m, b)$ linear?

When $p = 1, q = 0$ we find $y = (x - t)/(s - t)$.

When $p = 0, q = 1$ we find $y = (x - s)/(t - s)$.

What formula for y corresponds to arbitrary values of p and q?

5 It is known that x and y are related by an equation of the form $y = ax^2 + bx + c$. The values of y corresponding to $x = s$, $x = t$ and $x = u$ are p, q, r.

Find the equation for y corresponding to $p = 1, q = 0, r = 0$. (*Hint.* The Factor Theorem is relevant.)

Write the equations corresponding to $p = 0, q = 1, r = 0$ and to $p = 0, q = 0, r = 1$.

Deduce the equation corresponding to arbitrary values p, q, r.

11 Transformations

In most of the mappings we have considered so far, the output has been different in nature from the input. In the mapping merchandise $\rightarrow$ cost, the input is a certain collection of articles, the output a certain amount of money. In one of the statics problems, the input was a specification of 3 weights, the output was 2 reactions. In an electrical problem, the input was 2 voltages, the output was 3 currents in amperes.

It can happen, however, that the input and output are of the same character. For instance, in the process of differentiation such as $x^2 + 5x + 2 \rightarrow 2x + 5$, the input is a polynomial and so is the output. In a single-transistor amplifier, the input may be specified by means of a voltage and a current, the output also.

Another example occurs in the numerical solution of equations. Suppose we wish to solve an equation $f(x) = 0$ where the graph $y = f(x)$ has the appearance shown in Fig. 26 in a certain interval.

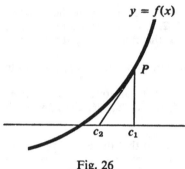

Fig. 26

Let c_1 be any value of x in the interval. Newton's method proceeds as follows. Let P be the point on the graph $y = f(x)$ corresponding to $x = c_1$. Draw the tangent at P, and let this tangent cut the horizontal axis at the point where $x = c_2$. Then c_2 will be a better approximation to the solution of $f(x) = 0$ than c_1.

Here the input is c_1, an approximation to the solution. The output is c_2, also an approximation to the solution (and in fact an improved approximation). Thus in the mapping $c_1 \rightarrow c_2$, the input and the output are of the same nature.

The advantage of such a situation is that we can repeat the operation. If we now take c_2 as our input, we can obtain a still better approximation c_3 by the same process. This type of calculation is particularly suitable for automatic computing. We programme the computer for the operation $c_1 \rightarrow c_2$, and tell it to keep on taking the output and feeding it back in as an input. In this way the computer produces a sequence of better and better approximations $c_1, c_2, c_3, c_4, \ldots$ It is told to stop when c_n and c_{n+1} differ by a sufficiently small amount. The process may be illustrated as in Fig. 27. *Note:* the

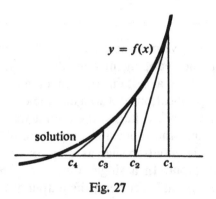

Fig. 27

sequence $c_1, c_2, c_3, \ldots$ does not always converge if the graph $y = f(x)$ is of a different type from that illustrated here.

The method of solution described above is an example of *iteration*. In this example, at each stage the input to the computer consists of a single number. But in general an iteration process may involve several numbers. For example, the initial input might consist of 10 numbers. From these the computer would obtain an output of 10 numbers, which would then form the input for the next stage of the process.

It is obvious that, for iteration to be possible, the output and the input must involve the same number of numbers. If the input is specified by (a_1, a_2, a_3) and the output by (b_1, b_2), we cannot feed the output back into the computer. A computer will not accept 2 numbers if its programme calls for an input of 3 numbers. But for iteration to be meaningful, it is not only necessary for the number of co-ordinates in the input and the output to be equal; they must also have the same significance. For instance, there are machines into which you can put 10 cents and obtain a fluid which is alleged to be coffee. So we have a mapping $10x$ cents produce x cups of coffee: $10x \to x$. But we cannot iterate this mapping; if we invest 100 cents and get 10 cups of coffee, we cannot pour the 10 cups of coffee back into the machine and get anything at all. We have a mapping from the vector space of cents to the vector space of cups of coffee. The amount of cents and the amount of coffee are both specified by a single number; both spaces have the same dimension, namely 1, but this agreement is not enough to permit iteration.

Arrows go from the cents space to the coffee space; arrows do not go from the coffee space anywhere (Fig. 28).

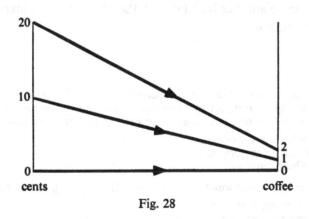

Fig. 28

For iteration to be possible, mapping must be from a vector space *to itself*.

A mapping from a space to itself is called a *transformation*.

A familiar example of a transformation is a rotation. Let T represent the operation of rotating the points of a plane through 30° about the origin. If P is any point of the plane, the transformation T sends P to a point P^*, which is also a point of the same plane (Fig. 29).

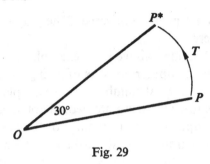

Fig. 29

We can iterate the transformation T if we wish. If we start with any point P_0, applying T again and again we would obtain points $P_1, P_2, P_3 \ldots$ spaced at 30° around a circle. P_{12} would coincide with P_0. We may write $P_1 = TP_0$, $P_2 = TP_1$, $P_3 = TP_2$ and so on. These equations also suggest that we might write $P_2 = TTP_0$, $P_3 = TTTP_0$, so that repeated application of the transformation is naturally written in a way that reminds us of repeated multiplication in elementary algebra. We use the same abbreviations as in elementary algebra, T^2 for TT, T^3 for TTT, and so on. As a rotation of 30°, done 12 times, brings us back to our original position, we have $T^{12} = I$, where I represents the identity transformation.

Of course, the fact that we can iterate T does not mean that we are compelled to. There may be occasions when we are interested in a rotation through 30°, done once and for all. But the fact that powers $T^2, T^3, \ldots$ can be defined is helpful, and can often be exploited to our advantage. *Any* transformation can be raised to a power in this way. Transformations thus differ from more general types of mapping; if M is a mapping from one vector space to another then (as with the cents to coffee mapping) the symbol M^2 is entirely meaningless.

Questions

1 Each of the transformations listed below is periodic, that is, if repeated a certain number of times it brings the plane back to its original position. Thus for some natural number n the transformation T satisfies $T^n = I$. Find the smallest n for each of the following transformations:

(*a*) a rotation of 180° about the origin,
(*b*) a rotation of 60° about the origin,
(*c*) reflection in the horizontal axis.

2 Find T^2 and T^3 for each transformation, T, of the real numbers listed here:

(*a*) doubling every number, that is, $x \rightarrow 2x$,
(*b*) adding 10 to every number, $x \rightarrow x + 10$,
(*c*) squaring every number, $x \rightarrow x^2$.

3 Which of the following transformations, T, of the real numbers are periodic?

(*a*) $x \rightarrow -x$ (*b*) $x \rightarrow x + 1$ (*c*) $x \rightarrow 10 - x$
(*d*) $x \rightarrow 1/x$ (*e*) $x \rightarrow \frac{1}{2}(x + 1)$ (*f*) $x \rightarrow (x - 1)/x$

4 A transformation T acts on every point of the plane except the origin, and sends the point (x, y) to the point with co-ordinates $x/(x^2 + y^2)$, $y/(x^2 + y^2)$. Find the effect of T^2.

5 Would a convergent sequence be obtained if we applied iteration procedure

(a) to the transformation $x \to x/2$,

(b) to the transformation $x \to \frac{1}{2}(1 + x)$,

(c) to the transformation $x \to -2x$?

6 A well known procedure for finding $\sqrt{a}$ is to iterate the transformation $x \to \frac{1}{2}x + \frac{1}{2}(a/x)$. What would happen if we tried to find $\sqrt{(-1)}$ by applying this procedure with $a = -1$?

Worked example

The mapping considered in the worked example on page 29 can be regarded as a transformation if we suppose the two pieces of squared paper to coincide. If this transformation is called T, find T^2 and T^3. What is the effect of T^n?

Solution Transformations are sufficiently specified by their effect on the basis vectors, A and B. As $A^* = A$, the transformation T leaves A where it was, and however many times T may be repeated this will remain so. T sends B to $A + B$. When T acts again, it sends $A + B$ to the point $(2, 1)$, as we found in the earlier worked example; $(2, 1)$ is the point $2A + B$. Thus $T^2A = 2A + B$. When T acts for the third time, $2A + B$ is sent to $(3, 1)$ as we saw earlier, that is, to $3A + B$, so $T^3B = 3A + B$.

Thus T^2 is specified by $A \to A$, $B \to 2A + B$ and T^3 by $A \to A$, $B \to 3A + B$.

T represents a shearing action, which can be demonstrated by pressing suitably on the side of a pack of cards. Points on the horizontal axis, $y = 0$, do not move. Each time T acts, the points on the line $y = 1$ are pushed one further unit to the right. It will be seen that, if T acts n times, points on $y = 1$ move n units to the right, so T^n is specified by $A \to A$, $B \to nA + B$.

7 Find T^2 and T^3 for the transformations corresponding to Question 1, (a), (b) and (c) on page 36. Consider the effect of T^n in these cases.

8 A transformation T acts on polynomials. If the input is $P(x)$ the output is $1 + xP(x)$. Investigate the sequence of polynomials obtained by repeatedly applying T, the initial input being the constant polynomial 1.

9 A transformation T maps the polynomial $P(x)$ to the polynomial $1 + \int_0^x P(t) \, dt$. Discuss the sequence of polynomials obtained by applying T repeatedly with initial input 1. To what well known function is it related? *Note.* The transformation in this question, as also in several earlier questions, is not linear.

12 Choice of basis

We noted in §10 that any linear mapping could be specified by stating what it did to the vectors in any basis. Transformations being a particular kind of mapping, it follows that this remark applies to linear transformations.

While a linear transformation *can* be specified by means of its effect on *any* basis, it is not a matter of indifference which basis is used. One basis may give us a very simple specification, that is easy to calculate with and that allows us to see exactly what the transformation does, while another basis may give a specification that is complicated and difficult to imagine.

If we are given a transformation, there is a procedure for finding a basis in which it will have the simplest form, and later we shall discuss such procedures. For the moment we are merely concerned to give examples which will show the effect that choice of basis can have on the specification of a transformation. How these examples have been selected or constructed need not be considered at this stage.

We begin with an example of a transformation specified in a convenient manner. It is a transformation of a plane, and we use as basis the points A, B on ordinary graph paper, where $A = (1, 0)$ and $B = (0, 1)$. The transformation is completely specified by the information $A \rightarrow A^*$, $B \rightarrow B^*$ where $A^* = (2, 0)$ and $B^* = (0, -1)$.

In order to avoid a confused diagram, Fig. 30 shows separately a number of points

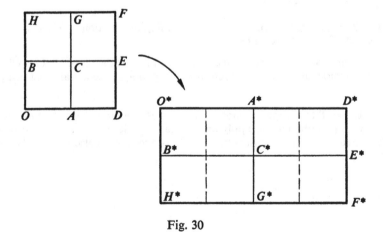

Fig. 30

$O, A, B, C, D, E, F, G, H$ and the points $O^*, A^*, B^*, C^*, D^*, E^*, F^*, G^*, H^*$ to which these go. However, it should be remembered that the transformation maps the plane to itself: we should imagine the right-hand grid drawn on a transparent sheet, which is then placed over the left-hand grid in such a way that O^* covers O and A^* covers D.

Here it is understood that $O*A*$ is twice as long as OA, and that $O*B*$ is as long as OB, but in the opposite direction.

The convenience of this specification arises from the fact that each axis is transformed into itself. If P is any point on the horizontal axis, then $P*$ is also on the horizontal axis. In fact, in vector notation, $P* = 2P$; the scale of the horizontal axis is doubled. Similarly, if Q is on the vertical axis, $Q*$ will also be on the vertical axis. We have $Q* = -Q$. The vertical axis undergoes a reversal of direction with unaltered scale.

Thus the problem of seeing what the transformation does is broken down. We can begin by forgetting all the points except those on the horizontal axis; these are rearranged among themselves. We then consider those on the vertical axis, and these also are rearranged among themselves. Finally, knowing what happens to the points on the axes, we can deduce what happens to any point. Let R be any point. Drop perpendiculars from it to the axes, to give the rectangle $OPRQ$ (Fig. 31).

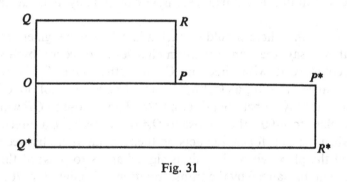

Fig. 31

We know $P \rightarrow P*$ and $Q \rightarrow Q*$ as shown here. Since $R = P + Q$, and this transformation is linear, $R* = P* + Q*$ and $R*$ is in the position shown.

We will now consider a second transformation. This transformation is quite as simple as the transformation just considered, to which indeed it bears a very strong resemblance. This simplicity, however, is obscured by our choice of basis; we shall again use the points A, B of ordinary graph paper. This is an unsuitable choice, the transformation does not fit neatly into it. This transformation sends $A \rightarrow A*$, $B \rightarrow B*$ where $A* = (1, 1)$ and $B* = (2, 0)$.

Fig. 32 shows 8 points of the plane, O, A, B, C, D, E, J, K and the points $O*, A*, B*,$ $C*, D*, E*, J*, K*$ to which the transformation sends them. As before, we have to imagine the two parts of the diagram superposed; $O*$ should cover O, while $B*$ falls on D and $A*$ on C.

Now of course by extending the two grids in this diagram it is possible to determine where any point of the plane goes, but it will be agreed that this diagram does not give us any simple and immediate way of feeling what the transformation does to the plane, or seeing quickly where any particular point goes. However, a way of doing this can be disentangled from the information at our disposal.

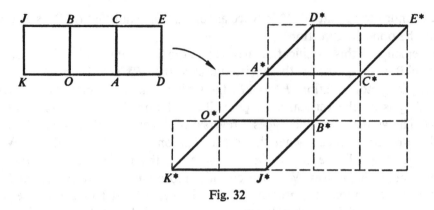

Fig. 32

It may be remembered that in our first example each axis transformed into itself. This suggests that we look at our diagram and see if we can find any line that transforms into itself.

It seems reasonable that there should be such a line. The line segment OA transforms into $O*A*$. That is to say, the transformation changes its direction by $+45°$; it swings this line in the counterclockwise direction. On the other hand OB is transformed to $O*B*$, an angle change of $-90°$, a clockwise swing. Imagine a whole lot of lines sprouting from O in directions lying between OA and OB. Those close to OB will experience a direction change close to $-90°$; those close to OA will experience a direction change of close to $+45°$ when the transformation acts on them. Somewhere between OB and OA we expect to find the place where the minus sign changes to plus; at this point there should be a direction unchanged by the transformation. We might try OC, but this does not quite work; OC is transformed to $O*C*$. This is still a clockwise swing, but it is much smaller than $90°$ in magnitude. We are getting closer to the line we want. If our next guess is OE, we have arrived at the desired place, for $O*E*$ is in the same direction as OE and indeed is exactly twice as long. In vector rotation, $E* = 2E$.

Further search on the diagram will uncover a second line that transforms into itself. It is the line through O and J, for $J, J*$ and the origin are in line. In fact, $J* = -J$.

Compare the equations just found with those we had in our first example. Here we have $E* = 2E, J* = -J$. In our first example we had $P* = 2P, Q* = -Q$. In each case one vector is doubled and one vector is reversed.

In the first example, any horizontal vector was doubled, any vertical vector was reversed. In the second example, any vector lying in the line OE is doubled, any vector lying in the line OJ is reversed (Fig. 33).

This allows us to tell easily where the transformation sends any point S. We draw lines SU and SV parallel to OJ and OE, and take U on the line OE and V on the line OJ. Then $U \to U*$ and $V \to V*$, as shown in the diagram and as $S = U + V$, we can find $S* = U* + V*$ by completing a parallelogram.

It should be noted that this construction is modelled, step for step, on the construction we used for the first transformation. The first transformation fitted neatly to the

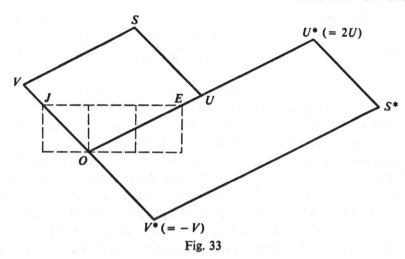

Fig. 33

axes OA and OB, which happened to be perpendicular. The second transformation fitted axes OU and OV, which happened not to be perpendicular. This did not in any way affect our ability to carry through the construction. In our first example, it was purely accidental that $OPRQ$ was a rectangle; all that mattered was that it was a parallelogram. The only use we made of this shape was to write $R = P + Q$, and for such an equation we need only to know that we are dealing with a parallelogram.

Technical terms

The examples we have just considered involve ideas that are fundamental for work with transformations. Naturally, a certain number of words and phrases have been coined, so that these ideas can be referred to without long explanations.

A line that transforms into itself is called an *invariant line*. 'Invariant' means 'unchanging'. *We shall apply this term only to lines through O.*

In our first example, we had a point P for which $P^* = 2P$, and a point Q for which $Q^* = (-1)Q$. Thus the vector P gets doubled when the transformation acts on it, and the vector Q gets multiplied by -1. A vector that gets multiplied by a number is called an *eigenvector* of the transformation. Thus, if V is an eigenvector, $V^* = \lambda V$ for some number λ; the number λ is called the *eigenvalue*. Thus, in our first example, P is an eigenvector with eigenvalue 2, since $P^* = 2P$. Similarly Q is an eigenvector with eigenvalue -1, since $Q^* = (-1)Q$. The symbol λ is nearly always used when an eigenvalue is discussed.

Any point on an invariant line, except the origin, must give an eigenvector. Suppose V is such a point on an invariant line. Then V^* also lies on the line, for the line transforms into itself. But every point of the line is of the form kV. So, for some k, $V^* = kV$; this means that V is an eigenvector with eigenvalue k.

We exclude the origin, O, when we are speaking of eigenvectors. Under any linear transformation, $O \rightarrow O$, so knowing $O \rightarrow O$ does not tell us anything about the transformation. Further, an eigenvector gives us the direction of an invariant line; if V is an eigenvector, the line OV is invariant. Here again, it is no use taking O for the vector; 'the line OO' is meaningless. So we agree that the term 'eigenvector' is not to cover the origin, O.

Note, however, that while an eigen*vector* V is, by definition, not 0, it is perfectly acceptable for the eigen*value* λ to be zero. In that case, $V* = \lambda V = (0)V = 0$. But the *line OV* still has a perfectly good meaning; we can use it as one of our axes, and it will help to give a simple specification of the transformation. In this case the term 'invariant line' may seem strange, for every point of the line then goes to the origin; it seems that the transformation does change the line – the line shrinks to the single point O. However, the definition of *invariant line* is so framed that this case is not excluded; the definition requires that every point of the line maps to some point of the line. We may put it negatively – no point of the line maps to a point outside the line. So to speak, the transformation, so far as the points of the line are concerned, is a domestic matter; we can see where they go without knowing anything about points outside the line. If they all happen to land on the same point, O, which is in the line, this does not conflict with the requirement. The definition, of course, is framed in this way because this is found most convenient in the light of the calculations we have to make. An invariant line gives us an axis that is useful if we are trying to find a simple specification for the transformation.

Students sometimes become confused because they think of $V*$ as the eigenvector. As a rule this does not matter; $V*$ lies on the same invariant line as V, and every point of this line is an eigenvector, as we have seen. But there is trouble in the case where $\lambda = 0$, for then $V* = 0$, and $V*$ is thus not acceptable as an eigenvector. Yet V, which is not 0, is still perfectly good; OV is an invariant line and a useful axis; V is accepted as an eigenvector.

These remarks should be borne in mind when answering questions.

If V is an eigenvector, every multiple kV of V (where $k \neq 0$) is also an eigenvector. For if $V \rightarrow \lambda V$, by proportionality $kV \rightarrow k\lambda V$, so the effect of the transformation on kV is to multiply it by λ. If we are asked to find the eigenvectors of a linear transformation, we often take this principle as understood. Thus, for the first transformation of this section we may say that the eigenvectors are A (with $\lambda = 2$) and B (with $\lambda = -1$). We do not bother to say 'A or any non-zero multiple of A'; we may even speak of the transformation as having 2 eigenvectors. Actually it has an infinity of eigenvectors; what we really mean is that it has 2 linearly independent eigenvectors, or 2 invariant lines.

Questions

In all these questions, the points O, A, B, C, J have the meanings attached to them in the previous section; that is, on conventional graph paper, O is $(0, 0)$, $A = (1, 0)$, $B = (0, 1)$, $C = (1, 1)$, $J = (-1, 1)$. Rotations are given with the angle measured in the usual sense; a clockwise rotation of $30°$ thus receives angle $-30°$. The term 'rotation' should be used only for movements that can be carried out within the plane of the paper. A transformation that sends any point (x, y) to $(x, -y)$

will be called 'a reflection in the horizontal' or (as some students prefer) 'a flip about the horizontal'. At a later stage, it will sometimes be necessary to classify movements as rotations or reflections; it will be confusing if the word rotation is applied to both (though of course in fact a reflection can be produced by rotating through 180° about an axis lying in the plane of the paper).

All rotations are understood to be about the origin, and all reflections in lines through the origin.

1   In the following lists, some linear transformations of the plane are described geometrically, and some are specified in vector notation. In some cases, the same transformation may occur in two or more forms. Group the descriptions and specifications in such a way as to make clear when this happens.

(a) rotation through 90°   (b) rotation through 180°   (c) rotation through 45°
(d) reflection in OA (e) reflection in OB (f) reflection in OC
(g) reflection in OJ (h) $A \to A, B \to -B$ (i) $A \to B, B \to -A$
(j) $A \to -A, B \to -B$ (k) $C \to -C, J \to -J$ (l) $A \to -A, C \to -C$
(m) $C \to J, J \to -C$ (n) $A \to B, C \to J$ (o) $B \to -B, C \to -J$
(p) $A \to -A, C \to J$ (q) $C \to C, J \to -J$ (r) $A \to B, B \to A$
(s) $C \to J, J \to C$ (t) $A \to -B, B \to -A$ (u) $A \to C, C \to B$
(v) $A \to C, B \to J$

2   Investigate the invariant lines of the transformations given in Question 1. Do any of these have no invariant line at all? Do any have an infinity of invariant lines? Discuss the eigenvalues and eigenvectors of these transformations.

3   In a dilation, or dilatation, the origin stays fixed, but the scale of the diagram is changed in the ratio $1:k$, in the manner suggested by Fig. 34. What invariant lines have such a transformation? Consider the cases $k = 2$, $k = 0.1$, $k = 0$, $k = -1$.

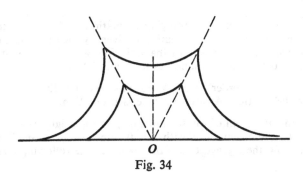

Fig. 34

4   In projection onto the horizontal axis, any point P goes to P^*, the point where the ordinate through P meets OA (Fig. 35). Discuss the invariant lines, eigenvectors and eigenvalues.

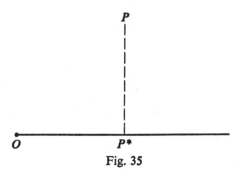

Fig. 35

5 Fig. 36 illustrates a particular case of oblique projection onto the horizontal axis. Discuss invariant lines, eigenvectors and eigenvalues for this projection.

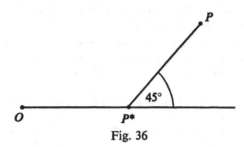

Fig. 36

6 A linear transformation S sends $A \rightarrow A$, $B \rightarrow C$. Make a sketch to show the effect of this transformation. What are the co-ordinates of the points to which S sends the points that were originally at $(1, 1)$, $(2, 1)$, $(3, 1)$? To what point does S send $(n, 1)$? Suppose that lines of slopes 1, $\frac{1}{2}$ and $\frac{1}{3}$ through the origin are drawn before the transformation S is applied. What will be the slopes of the lines to which S sends these? Are there any invariant lines in the plane for this transformation? If so, how many and which lines?

7 A linear transformation sends $(1, 0)$ to $(2, 1)$ and $(0, 1)$ to $(1, 2)$. By trial and error, find the invariant lines, eigenvectors and eigenvalues.

8 A linear transformation T sends $A \rightarrow C$, $C \rightarrow B$. Where does T send B? Investigate the transformations T^3 and T^6.

9 A transformation is obtained by first applying a rotation of $90°$ about the origin, and then reflecting in the horizontal axis; the transformation is given by the final effect of this. Is there any single transformation mentioned in Question 1 that coincides with this transformation? (Suggested approach: consider what the transformation does to A and B.)

10 In question 9, suppose the order of the operations reversed, that is, suppose the reflection applied first, and then the rotation. Answer this amended question.

11 Any line through the origin, all of whose points map to the origin, O, is an invariant line, and its points (other than O) are eigenvectors with $\lambda = 0$. Find the invariant lines that are mapped entirely into the origin, and the eigenvectors with $\lambda = 0$, for the following transformations of the plane.

(a) $A \rightarrow O$, $B \rightarrow B$ (b) $A \rightarrow \frac{1}{2}(A + B)$, $B \rightarrow \frac{1}{2}(A + B)$
(c) $(x, y) \rightarrow (x, x)$ (d) $(x, y) \rightarrow (y, y)$
(e) $(x, y) \rightarrow (x + y, 0)$ (f) $(x, y) \rightarrow (x - y, 0)$

Make sketches, showing the effect of these transformations, and describe the transformations in geometrical terms.

13 Complex numbers

We now come to consider complex numbers. This subject is important to us in two ways. First, complex numbers are used in many engineering calculations and give a great increase in mathematical power. They play a decisive role in all questions concerned with vibrations, whether mechanical, structural or electrical. They are used in the theory of alternating electrical currents. They play an essential part in quantum theory, which is the foundation of all work on atoms and nuclei, and hence of chemical and nuclear engineering. Thus complex numbers are directly used on many occasions, and this is our first reason for being concerned with them. Our second reason is that they throw considerable light on linear algebra. We are going to develop an algebraic theory of linear transformations. Now this seems a very peculiar thing to do. It seems complete nonsense to talk of a reflection being subtracted from a rotation. One naturally asks – how did anyone ever arrive at this strange idea? And once the idea has arisen that we can define addition and multiplication for transformations, the question still remains – how shall we choose suitable definitions? Now complex numbers were developed a couple of centuries before linear algebra, and provided a stimulus, a model and the essential guiding ideas for this later development. When this background is known, it is much easier to see linear algebra as a reasonable subject and to remember how it works. Accordingly we have two aims – to consider how calculations are made with complex numbers, and to see how complex numbers suggested the ideas of linear algebra.

The story begins in Italy, around 1575. A little earlier, a formula had been found for the solution of cubic equations. It was a rather complicated formula, involving two square roots and two cube roots. For example, applied to the equation $x^3 = 15x + 4$ it gave

$$x = \sqrt[3]{2 + \sqrt{-121}} + \sqrt[3]{2 - \sqrt{-121}}.$$

This is particularly unsatisfactory, since this equation has in fact a very simple solution. If we try $x = 4$, we find $x^3 = 64$ and $15x + 4 = 64$, so 4 is a solution. The solution given by the formula is not merely complicated; it contains in two places the 'impossible' symbol, $\sqrt{-121}$. A mathematician called Bombelli decided to try to work with this impossible number. He accepted $\sqrt{-1}$ as a possible number, and applied the usual procedures of algebra to it. We will use modern notation to explain what he did. Today most mathematicians write i for $\sqrt{-1}$; most engineers write j. So, if this number is accepted, we can write $\sqrt{-121} = 11\sqrt{-1} = 11j$. Bombelli then found, no doubt after considerable trial and error, that it was possible to extract the two cube roots. For consider the cube of $2 + j$. We have $(2 + j)^2 = 4 + 4j + j^2 = 3 + 4j$ since $j^2 = -1$. Multiplying again by $2 + j$ we find $(3 + 4j)(2 + j) = 6 + 11j + 4j^2 = 2 + 11j$. This is exactly what stands under the first cube root sign. Accordingly, we can extract this first cube root, and obtain $2 + j$. Similarly, in the second cube root we can check that

$\sqrt[3]{2 - 11j} = 2 - j$. Accordingly the whole formula for the solution in this case boils down to $(2 + j) + (2 - j) = 4$. We know this is what it should be. The final answer does not contain j, so we do not have to know what j means to use it.

For more than two centuries things continued more or less like this. In problem after problem it was found that if you accepted the symbol j and applied the usual rules of algebra to it, you got correct results. Yet no one was able to say why this happened.

Around the year 1800 a new device was brought in which aided calculation with complex numbers and which eventually helped to explain them. As often happens with new discoveries, the same idea occurred at about the same time to men in widely separated places – to Gauss in Germany, to Wessel in Scandinavia and to Argand in France. Wessel's work remained unknown for many years; Gauss was an eminent mathematician whose name was associated with many discoveries; the new idea became known as the Argand diagram. Argand was, in fact, the first to publish this idea. It is essentially a graphical procedure for dealing with complex numbers. If we have two complex numbers, c and z, where $c = a + jb$ and $z = x + jy$, the usual procedures of algebra tell us what we are to understand by $c + z$ and cz, namely

$$c + z = (a + x) + j(b + y),$$
$$cz = (ax - by) + j(bx + ay).$$

To these definitions Argand brought the idea that we could represent complex numbers by means of dots on ordinary graph paper. A dot on the point $(3, 4)$ would indicate that we had in mind $3 + 4j$; a dot on (a, b) would indicate $a + jb$; a dot on (x, y) would indicate $x + jy$.

So far, of course, we have not made any mathematical advance; we have simply devised a kind of signalling code for indicating what complex number we have in mind. The question then arises – will the algebraic definitions of addition and multiplication have simple geometrical constructions corresponding to them when the Argand diagram is used? Addition certainly does. As Fig. 37 shows, the points $c, z, c + z$ and the origin form a parallelogram. Thus complex numbers are added by the usual process of vector addition. It must be remembered that in 1800 the idea of vector was still unknown. The

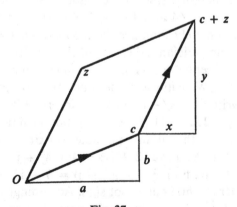

Fig. 37

concept of vector addition must have been formed partly from experience with adding complex numbers, and partly from adding forces in mechanics.

For the geometrical interpretation of multiplication it is best to proceed by stages and consider particular cases.

First, suppose $x + jy$ is to be multiplied by a real number a. Algebraically,

$$a(x + jy) = ax + jay.$$

Thus multiplication by a sends the point (x, y) on the Argand diagram to the point (ax, ay). This point is on the line from the origin to (x, y), but is a times as far away from the origin as (x, y) (Fig. 38(a)). (This clearly suggests the later definition for multiplying a vector by a number.)

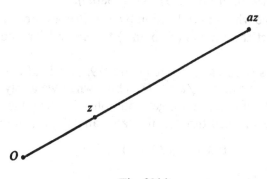

Fig. 38(a)

Next consider j multiplying $x + jy$. By algebra $j(x + jy) = jx + j^2y = -y + jx$. Thus multiplication by j sends (x, y) to $(-y, x)$. By observing the triangles in Fig. 38(b), we see that the second could be obtained from the first by rotating through 90°. Thus multiplication by j does not affect the length of a vector, but rotates the vector through a right angle (Fig. 38(c)).

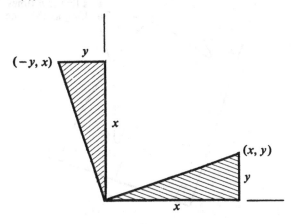

Fig. 38(b)

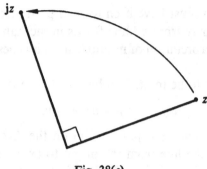

Fig. 38(c)

We can now see the effect of multiplication by jb. For jbz can be obtained from z by multiplying by b and then multiplying by j (Fig. 38(d)).

Now finally we can see what multiplication by $a + jb$ does. For $(a + jb)z = az + jbz$. So given z, we construct the vectors (or points) for az and jbz and then apply vector addition to these.

This construction has been carried out in Fig. 39. If z is at a distance L from the origin, $(a + jb)z$ is at a distance $L\sqrt{a^2 + b^2}$. Thus, when we apply the transformation $z \rightarrow (a + jb)z$, the length of every vector gets multiplied by the same number, $\sqrt{a^2 + b^2}$. Further every vector gets turned through the same angle θ, for in the figure

$$\tan \theta = (bL)/(aL) = b/a,$$

which does not depend on z.

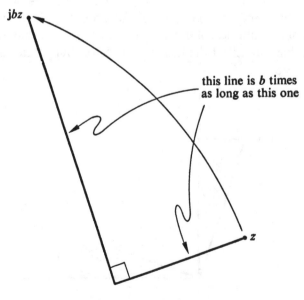

this line is b times
as long as this one

Fig. 38(d)

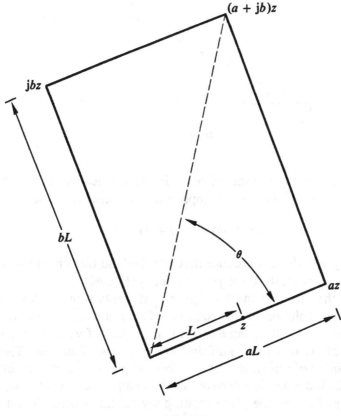

Fig. 39

If we consider the polar co-ordinates of the point $a + jb$, we see that $r = \sqrt{a^2 + b^2}$ and $\tan \theta = b/a$, Fig. 40. Thus, by observing the position of $a + jb$ on the Argand diagram we can immediately read off the angle through which multiplication by $a + jb$ rotates every vector and the change of scale it produces.

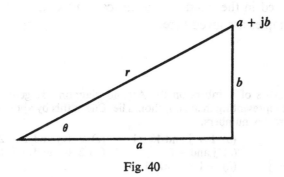

Fig. 40

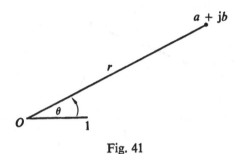

Fig. 41

It is not surprising that r and θ should occur in this connection. For on the Argand diagram the number 1, which is $1 + j \cdot 0$, appears at the point $(1, 0)$. Now

$$(a + jb) \cdot 1 = a + jb.$$

So multiplication by $a + jb$ must produce that rotation and that change of scale that is needed to bring the point $(1, 0)$ to the position (a, b) (Fig. 41).

Earlier, we used the fact that $\tan \theta = b/a$, and students sometimes point out that there is more than one solution of this equation. The argument just given shows that we need not be bothered by this; $\tan \theta = b/a$ is only part of what we know about θ. From the diagrams above we can read off $\sin \theta$ and $\cos \theta$ also if we wish. There remains of course the possibility of replacing θ by $\theta + 2\pi n$, where n is an integer, but this does not affect our work. We are only interested in the mapping $z \rightarrow (a + jb)z$; we do not mind whether people imagine the plane rotating several times before it comes to rest; we are only concerned with the final result.

Our line of thought so far has been the following. The experience of several centuries suggests that the usual rules of algebra, applied to complex numbers, lead to correct results. If we agree to let the point (x, y) be used as a graphical representation of the complex number, $x + jy$, we find that both addition and multiplication of complex numbers can be carried out by simple geometrical operations. Addition is done by drawing parallelograms; multiplication by means of rotation and change of scale, as just explained.

It is desirable to practise these geometrical operations in the Argand diagram until they have become fixed in the mind. A student can make up routine exercises of this kind for himself; samples are given here.

Exercises

1 Plot the following pairs of numbers on the Argand diagram. By geometrical considerations, predict where the point representing their sum should lie. Check this by applying the usual processes of arithmetic to the complex numbers.

 (a) 1 and j   (b) $1 + j$ and $1 - j$ (c) $2 + j$ and $1 + 2j$
 (d) 1 and -1 (e) j and $-j$ (f) $2 + j$ and $-2 + j$
 (g) $2 + 2j$ and $-1 - j$ (h) $-1 + j$ and 1

2   In the same way, for the pairs below, predict geometrically where the product should be and check arithmetically. For each complex number occurring it will of course be necessary to consider its distance, r, from the origin and the angle, θ, at which it occurs.

(*a*) $1 + j$ and j (*b*) $2j$ and j (*c*) $1 + j$ and $1 + j$

(*d*) $1 + 2j$ and $2 + j$ (*e*) $3 + 4j$ and $4 + 3j$ (*f*) $2 + j$ and $2 - j$

(*g*) $2 + j$ and $-2 + j$

3   On the Argand diagram the powers of a number, $1, z, z^2, z^3, \ldots$, display a characteristic pattern. Calculate and plot the powers of z for the values given below. Check that their arrangement agrees with that expected on geometrical grounds.

(*a*) j (*b*) $-j$ (*c*) $1 + j$

(*d*) $(1/\sqrt{2}) + j(1/\sqrt{2})$ (*e*) $(\tfrac{1}{2}) + j(\tfrac{1}{2})$ (*f*) $(\sqrt{3}) + j$

Where would you expect the negative powers $z^{-1}, z^{-2}, z^{-3}, \ldots$ to lie?

4   We know that, if c lies at angle θ and distance r, multiplication by c has a simple geometrical effect. What about division by c?

Division by c is multiplication by c^{-1}. Where would you expect to find c^{-1} on the Argand diagram?

What is the product of $\sqrt{3} + j$ and $(\tfrac{1}{4}\sqrt{3}) - j(\tfrac{1}{4})$? Relate this result to the earlier parts of this question

5   Express in the form $a + jb$ the point of the Argand diagram that lies at angle θ and distance r from the origin.

6   To obtain a square root of -1 we have to introduce a new symbol j. To get the square root of j do we need to introduce another new symbol?

14   Calculations with complex numbers

In work with complex numbers we have two approaches at our disposal – by algebra and by geometry. At each stage of the work we should consider which approach will be simpler and more effective. If, for instance, we wish to calculate a product such as $(2 + 3j)(4 + 5j)$ the direct arithmetical treatment is much shorter and easier than the geometrical; by geometry we would have to determine the angles corresponding to $2 + 3j$ and $4 + 5j$ and then deal with their sum, which would clearly involve much more work than simply multiplying out the original expression.

It is quite different if we wish to extract a root of a complex number. A certain type of linear differential equation, with important engineering applications, leads to algebraic equations with complex roots. Thus we may find ourselves confronted with an equation such as $m^4 = -1$ or $m^3 = j$. By algebra we would probably begin by writing

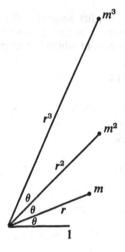

Fig. 42

$m = x + jy$; this would lead to an awkward pair of simultaneous equations. By geometry it is possible to solve these equations mentally after a little practice. The reason for this is the existence of a simple geometrical pattern formed by the positions of the powers $m, m^2, m^3, \ldots$ To solve, say, $m^3 = j$, we consider where we would have to put m to ensure that m^3 landed on j.

If m is at angle θ and distance r from O, multiplication by m rotates the plane through θ and enlarges the scale r times. If we multiply by m twice, the effect is to rotate through 2θ and to enlarge r^2 times; similarly, multiplication three times by m gives a rotation 3θ and an enlargement r^3 times. Suppose we apply these operations to the point 1; we obtain the points shown in Fig. 42.

Thus m^3 appears as the point at angle 3θ and distance r^3. Now we wish to solve $m^3 = j$, that is, to choose r and θ in such a way that angle 3θ and distance r^3 will land us on j. Now j is at distance 1 from O, so evidently we must choose $r = 1$; for $r > 1$ makes m^3 lie outside the unit circle and $r < 1$ makes m^3 lie inside the unit circle. (By the unit circle we mean the circle centre O, radius 1.) Now j lies on the vertical axis, thus at an angle $\frac{1}{2}\pi$. Thus our first thought is to take $\theta = \frac{1}{6}\pi$ and produce the situation shown in Fig. 43.* We can see that the co-ordinates of the point m are $\frac{1}{2}\sqrt{3}$, $\frac{1}{2}$, thus this point represents the complex number $(\frac{1}{2}\sqrt{3}) + j(\frac{1}{2})$, which we may write $(\sqrt{3} + j)/2$. It can be verified arithmetically that the cube of this number is in fact j.

In solving an equation we need to find *all* the solutions. Now $(\sqrt{3} + j)/2$ is a solution, but it is not the only solution, as may be easily seen. For consider $(-j)^3$. We have $(-j)^3 = -j^3 = -j^2 \cdot j = j$ since $j^2 = -1$. Thus $-j$ satisfies $m^3 = j$. In geometrical terms, how does it manage to do it? For $-j$ we have $r = 1$, $\theta = -\pi/2$. The cube involves r^3 and 3θ; thus $(-j)^3$ will be at distance 1, angle $-3\pi/2$, and this is an alternative

* For theoretical results such as $\sin \theta = \theta - (\theta^3/6) + \cdots$ it is essential for θ to be in radians, not in degrees. For description of a geometrical figure, such as those used here, it is possible to specify by degrees; j at 90°, m at 30°.

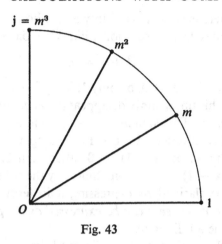

Fig. 43

way of describing where j lies. Thus to find the cube roots of a number, we have not only to consider the obvious specification of its position, but to take into account every angle that is associated with its position. Thus for j we have to consider every angle of the form $(\pi/2) + 2n\pi$, where n is an integer. If $3\theta = (\pi/2) + 2n\pi$ we have

$$\theta = (\pi/6) + (2n\pi/3).$$

If various integral values of n are tried, it will be found that we obtain only 3 distinct positions for m, as shown in Fig. 44.

It is not surprising that we cannot get more than 3 solutions, for it can be proved that in any field a cubic equation has at most 3 solutions. However there is perhaps some novelty in a number having 3 cube roots. When we are working with real numbers, we think of 2 as being *the* cube root of 8. However, if we write the equation $m^3 = 8$ this leads us to $m^3 - 8 = 0$, which may be written $(m - 2)(m^2 + 2m + 4) = 0$. The quadratic factor leads to complex solutions, $-1 + j\sqrt{3}$ and $-1 - j\sqrt{3}$, so when complex numbers are accepted we must regard 8 as having 3 cube roots.

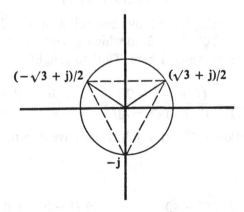

Fig. 44

With real numbers there is some uncertainty about how many solutions an equation will have. A quadratic may have 2 solutions, like $x^2 - 5x + 6 = 0$, or none, like

$$x^2 + 1 = 0.$$

A cubic may have 3, like $x^3 - x = 0$, or only 1, like $x^3 - x + 10 = 0$. When we work with complex numbers, this uncertainty disappears; *an equation of the nth degree always has its full set of n solutions*. This statement has to be interpreted in a certain way, for there is still the possibility of repeated roots. For example, even with complex numbers we cannot find any solution of $(x - 1)^3 = 0$ other than 1. As the equation may be written $(x - 1)(x - 1)(x - 1) = 0$ we sometimes say the solution 1 occurs 3 times. Our earlier statement may be clarified by expressing it in terms of factors; *with complex numbers, any polynomial of degree n can be expressed as the product of n linear factors.* These factors need not be all different.

If the cube roots of 8 are plotted on the Argand diagram, it will be seen that they form an equilateral triangle. We saw the same thing with the cube roots of j. (Does a similar remark apply to the 4th roots of any complex number? To the *n*th roots?)

We have shown in some detail how to find the cube root of a number by geometrical means. The same idea is easily adapted to finding a square root, a 4th root, or any other root.

Exercises

Solve the following equations, and plot the solutions on the Argand diagram.

(a) $m^2 = j$ (b) $m^2 = -1$ (c) $m^2 = -j$ (d) $m^3 = -1$
(e) $m^3 = 1$ (f) $m^4 = 1$ (g) $m^4 = -1$ (h) $m^6 = 1$
(i) $m^2 = 2 + j \cdot 2\sqrt{3}$ (j) $m^3 = -2 + 2j$ (k) $m^3 = -8$ (l) $m^4 = -324$

Fractions

The simplification of fractions is as a rule more convenient by the arithmetical or algebraic approach than by geometry. If we had to deal with a fraction such as

$$(1 + \sqrt{3})/(5 + 2\sqrt{3})$$

we would use the trick of multiplying above and below by $5 - 2\sqrt{3}$. The new denominator would then be $5^2 - (2\sqrt{3})^2$ or 13, in which $\sqrt{3}$ does not appear. As j stands for $\sqrt{-1}$, the same trick can be applied here. Thus to simplify $(1 + j)/(5 + 2j)$ we might proceed

$$\frac{1 + j}{5 + 2j} = \frac{(1 + j)(5 - 2j)}{(5 + 2j)(5 - 2j)} = \frac{7 + 3j}{25 - 4j^2} = \frac{7 + 3j}{29}.$$

If we write this last fraction as $(7/29) + j(3/29)$ we have it in the standard form $a + jb$.

Exercises

Find in standard form

(a) $(1 + j)/(2 + 3j)$ (b) $5/(3 + 4j)$ (c) $(1 - j)/(1 + j)$
(d) $1/(1 + j)$ (e) $1/(\cos \theta - j \sin \theta)$ (f) $1/(1 + j \tan \theta)$

15 Complex numbers and trigonometry

In this section we shall show how several results in trigonometry can be obtained very simply by using complex numbers. Much of this section may be already familiar. It is put here in order to illustrate the power that the use of complex numbers gives us. We shall obtain an even better impression of that power if we remember that the present section does not represent the most important application of complex numbers, and in fact constitutes only a very small part of what can be done with them.

All our results flow directly from the following three obvious remarks concerning rotations about the origin.

(1) The inverse of a rotation through θ is a rotation through $-\theta$.

(2) A rotation through θ combined with a rotation through ϕ gives a rotation though $\theta + \phi$.

(3) A rotation through θ done n times gives a rotation through $n\theta$.

To express these statements in terms of complex numbers, we have only to recall that the mapping $z \rightarrow cz$ corresponds to a rotation through θ and a change of scale in the ratio $r:1$. If $r = 1$, this mapping will be a pure rotation. Under the mapping $z \rightarrow cz$, the point 1 goes to c. Thus, we can find the value of c for a given rotation by seeing where that rotation sends the point 1.

A rotation through θ sends 1 to the point P whose co-ordinates are $(\cos \theta, \sin \theta)$ (see Fig. 45). Thus P represents the complex number $\cos \theta + j \sin \theta$. We note that $r = 1$ for P. Thus multiplication by $\cos \theta + j \sin \theta$ gives a rotation through θ.

A rotation through $-\theta$ would bring 1 to Q with co-ordinates $(\cos \theta, -\sin \theta)$. Thus Q corresponds to $\cos \theta - j \sin \theta$, and multiplication by this complex number gives rotation through $-\theta$.

We are now in a position to translate our three statements about rotations into statements about complex numbers.

The first statement gives a very familiar result. Rotation through θ combined with

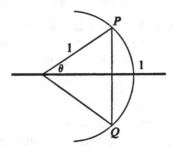

Fig. 45

rotation through $-\theta$ leaves us where we started. Leaving us where we started corresponds to multiplication by 1. So we have

$$\left.\begin{array}{r} (\cos\theta + j\sin\theta)(\cos\theta - j\sin\theta) = 1 \\ \cos^2\theta - j^2\sin^2\theta = 1 \\ \cos^2\theta + \sin^2\theta = 1. \end{array}\right\} \qquad (1)$$

This of course we could prove quite easily by other methods. We have not yet demonstrated any increase of power. But statements (2) and (3) are more rewarding.

Statement (2) considers a rotation through θ combined with a rotation through ϕ to produce a rotation through $\theta + \phi$. This gives

$$(\cos\theta + j\sin\theta)(\cos\phi + j\sin\phi) = \cos(\theta + \phi) + j\sin(\theta + \phi). \qquad (2)$$

Multiplying out we find

$$\cos(\theta + \phi) + j\sin(\theta + \phi) = (\cos\theta\cos\phi - \sin\theta\sin\phi) + j(\cos\theta\sin\phi + \sin\theta\cos\phi).$$

The equality sign indicates that both expressions give the same point on the Argand diagram. Now points coincide only when both co-ordinates agree. If (a, b) coincides with (x, y) we must have $x = a$ and $y = b$. Thus from $a + jb = x + jy$ we can conclude $x = a$ and $y = b$. This is often referred to as 'equating real and imaginary parts', an expression that goes back to the time when j was considered an 'imaginary' or 'impossible' number.

Accordingly we have

$$\cos(\theta + \phi) = \cos\theta\cos\phi - \sin\theta\sin\phi$$
$$\sin(\theta + \phi) = \cos\theta\sin\phi + \sin\theta\cos\phi.$$

Thus we have obtained the addition formulas for cosine and sine.

Students may have met this development in the reverse order – the addition formulas being proved (usually rather awkwardly), and then equation (2) above, deduced from these. But considerations of complex numbers show that equation (2) is obvious, and can be used to give the addition formulas quickly and easily.

Statement (3) leads to a result usually associated with the name of De Moivre, although in fact De Moivre never gave it in the form we know. Using the same ideas as we did earlier, we find it gives directly

$$(\cos\theta + j\sin\theta)^n = \cos n\theta + j\sin n\theta.$$

This can be useful for finding the trigonometric functions of multiple angles. For example, if we wish to recall $\cos 3\theta$ and $\sin 3\theta$ we may write $n = 3$. It is convenient to use the abbreviation c for $\cos\theta$ and s for $\sin\theta$. Then

$$\cos 3\theta + j\sin 3\theta = (c + js)^3 = c^3 + 3jc^2s - 3cs^2 - js^3 = (c^3 - 3cs^2) + j(3c^2s - s^3).$$

From this we read off $\cos 3\theta = c^3 - 3cs^2$ and $\sin 3\theta = 3c^2s - s^3$. If one wishes, these may be put in other forms by using $c^2 + s^2 = 1$.

This treatment of trigonometry can be taken a stage further. There is a remarkable result, which makes trigonometry simply a branch of algebra. To derive this result in a strictly logical manner is a considerable undertaking, and would demand more time and knowledge than can reasonably be expected from an engineering student at this stage. The result itself is valuable. Accordingly in the next section an account of it will be given, and the considerations that lead to it will be sketched, but no claim will be made that this constitutes rigorous deduction.

16   Trigonometry and exponentials

In elementary mathematics we meet exponentials, such as 2^x, 10^x and later e^x, and trigonometric functions such as sine and cosine. The two types of function differ both in origin and behaviour. Exponentials arise from arithmetic and can be handled with the help of a few simple formulas, such as $a^m \cdot a^n = a^{m+n}$ and $(a^m)^n = a^{mn}$. Trigonometric functions, on the other hand, are introduced with the help of geometrical ideas (lengths, right angles, circles) and involve a large number of rather complicated formulas. It was therefore a matter for considerable surprise when Euler, around the year 1750, showed that sines and cosines were simply exponentials in disguise, and that the complicated results of trigonometry could be deduced from the simple properties of exponentials.

Yet engineers and scientists might well have suspected some link between the two. For the engineer can visualize the meaning of sine and exponential by associating these with certain physical situations. If a mass is attached to a spring, it vibrates to and fro. The graph of its motion would be a curve that repeats again and again a rise and fall. One would naturally guess that the function belonging to this graph would be the sine, and so indeed it is. Sines we associate with simple harmonic motion – a mass on a spring, or a pendulum.

Exponentials we associate with unstable equilibrium. We may with difficulty persuade a pole to balance in a vertical position above its support. But the slightest disturbance will cause it to lean, and the forces acting on it will cause it to move more and more rapidly away from its position of equilibrium. A good approximation to the early stages of this motion is provided by the exponential function. Another illustration would be a particle resting on a rotating disc. If it is at the exact centre, it can remain there indefinitely. But if it moves ever so little it will experience a centrifuge effect and be moved outwards with ever increasing violence, its distance from the centre and its velocity increasing exponentially.

Now stable and unstable equilibria are not two totally separated things. It is possible to begin with a stable system and then alter it gradually until it becomes unstable, like a foolish captain who takes on so much cargo that his ship capsizes.

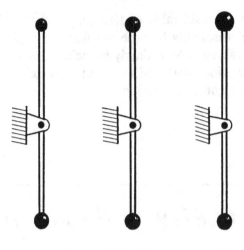

Fig. 46

For instance, we may imagine a bar with a large weight at its lower end and a small weight at the top (Fig. 46). With a pin at its midpoint, it could oscillate like a pendulum. If we gradually increase the weight at the top, the oscillation will become slower and slower and finally a point will be reached where it ceases to be a small oscillation at all; the overloaded top plunges towards the depths.

It is characteristic of the unstable situation that the acceleration carries the system away from equilibrium, and the further the system goes, the greater the acceleration is. Since d^2s/dt^2 or $\ddot{s}$, the second derivative, represents acceleration, it is not surprising that we frequently meet the equation $\ddot{s} = k^2s$ in connection with unstable systems. For such a system, $s = e^{kt}$ is a possible motion.

In a stable system, we usually find acceleration proportional to displacement but in this case the acceleration is pulling the object back towards equilibrium, and the equation is $\ddot{s} = -c^2s$; a possible motion is $s = \sin ct$.

Both the equations above are of the form $\ddot{s} = ms$ where m is a constant. We pass from unstable to stable as m passes from positive to negative. Now $\ddot{s} = -c^2s$ would become identical with $\ddot{s} = k^2s$ if we were allowed to write $k^2 = -c^2$ or $k = jc$. Thus $s = e^{kt}$ would be replaced by $s = e^{jct}$, and we would be led to conjecture, on engineering grounds, that if our calculations led to an exponential with an 'imaginary' exponent, this indicated that we were dealing with some kind of oscillation. This idea plays a large part in investigations of stability, and is among the most important applications of complex numbers to practical problems.

The connection between oscillations and imaginary exponents was in fact discovered by the mathematician Euler around 1750. At that time mathematicians were very much interested in physical applications, and the argument just given may well have occurred to him at some stage of his work. The connection, however, can be demonstrated by purely mathematical formalisms, which we now proceed to indicate.

One way of arriving at this connection is by means of the series for e^x, $\sin x$ and $\cos x$. By quite elementary calculus, using only integration, we can obtain the results

$$e^x = 1 + x + (x^2/2!) + (x^3/3!) + (x^4/4!) + (x^5/5!) + \cdots$$
$$\cos x = 1 \qquad - (x^2/2!) \qquad\quad + (x^4/4!) \qquad\qquad - \cdots$$
$$\sin x = \qquad x \qquad\quad - (x^3/3!) \qquad\qquad + (x^5/5!) - \cdots$$

These results are obtained of course for real values of x. In looking at these results one can hardly fail to be struck by the fact that e^x seems to be built from the same ingredients as $\cos x$ and $\sin x$; e^x contains all the terms of the form $x^n/n!$, $\cos x$ contains the even ones of these and $\sin x$ the odd ones; in e^x all the signs are positive, while in $\cos x$ and $\sin x$ the signs are plus and minus alternately. We could simply observe this as an interesting coincidence and leave it at that. But for a scientist or a mathematician an unexplained coincidence is an indication that some new law or theorem is waiting to be discovered.

If the signs in $\cos x$ and $\sin x$ were all positive we would simply add these series, for $\cos x + \sin x$ agrees with e^x apart from the presence of minus signs. If we are to get genuine agreement, we must find some way of bringing minus signs into the exponential series. If we compare the signs in $\cos x$ with the even terms of the series for e^x, we find that $\cos x$ has a minus sign in the terms containing x^2, x^6, x^{10}, ..., that is, the odd powers of x^2. In fact the signs in the cosine series are as in the sequence $-x^2$, $(-x^2)^2$, $(-x^2)^3$, $(-x^2)^4$, ... This might suggest putting $y^2 = -x^2$, or $x = jy$. And in fact if we put $x = jy$ in the series for e^x, we immediately achieve our goal; the calculation leads directly to the equation $e^{jy} = \cos y + j \sin y$.

Earlier we saw that $\cos \theta + j \sin \theta$ was associated with rotation through θ; now we see that it may be written in the compact from $e^{j\theta}$, and instead of using statements (1), (2), (3) about rotations we can use the algebraic properties of exponents.

Another way of looking at e^x leads to the same conclusion. The original idea of e^x was related to compound interest, and from this point of view it was found natural to define e as the limit of $[1 + (1/n)]^n$ as $n \to \infty$, and then to prove $e^x = \lim [1 + (x/n)]^n$ as $n \to \infty$. If x is a complex number, this last definition is still workable. For any whole number n we can find $[1 + (x/n)]^n$; this merely involves multiplication of complex numbers. We can then see whether the result approaches some fixed point when n is made very large.

So, in order to see what $e^{j\theta}$ should mean, we first consider $[1 + (j\theta/n)]^n$. Multiplication by $1 + (j\theta/n)$ brings 1 to the point with co-ordinates $(1, \theta/n)$. If we keep multiplying by $1 + (j\theta/n)$, we shall obtain a succession of points lying on a spiral, as in Fig. 47. In this figure $OA = 1$ and $AP_1 = \theta/n$. Thus, by Pythagoras,

$$r = \sqrt{1 + \frac{\theta^2}{n^2}} \quad \text{and} \quad \tan \alpha = \theta/n.$$

In the Argand diagram, A represents 1, P_1 represents $1 + (j\theta/n)$, and the number we are interested in is the nth power of $1 + (j\theta/n)$. Thus it occurs at distance r^n and angle $n\alpha$.

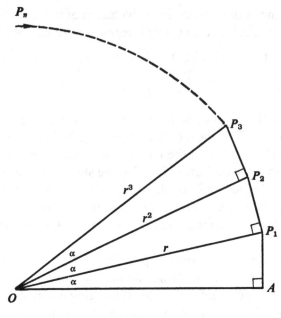

Fig. 47

Now r is slightly larger than 1, and so r^n also must be larger than 1. However, in the expression for r we see $1 + (\theta^2/n^2)$, and the n^2 in the denominator has a more powerful effect than the exponent n in r^n. It can be proved that, as n grows, r^n approaches 1. Thus the *limit* point is at a distance 1 from O, and lies on the unit circle.

For a small angle ϕ, $\tan \phi$ and ϕ are approximately equal. In fact ϕ is slightly smaller than $\tan \phi$. Thus from the equation $\tan \alpha = \theta/n$ we know that α is very nearly equal to θ/n, but is just a little less. Accordingly $n\alpha$ is nearly θ, but is slightly less. Here again it can be proved that $n\alpha$, which is $n \tan^{-1}(\theta/n)$, does tend to θ when $n \to \infty$.

Thus the *limit* of $[1 + (j\theta/n)]^n$ as $n \to \infty$ is at distance 1 and angle θ. Its co-ordinates are accordingly $(\cos \theta, \sin \theta)$ and the complex number it represents is $\cos \theta + j \sin \theta$. But this limit is, by definition, what we mean by $e^{j\theta}$. Thus once more we are led to the conclusion

$$e^{j\theta} = \cos \theta + j \sin \theta.$$

Both arguments above may in places go beyond a student's knowledge. In the second argument, the procedure for calculating the limits may not be known. In the first argument, the actual calculations are not difficult but questions of interpretation arise: what do we mean by an infinite series of complex numbers? When does it converge? How do we define e^z, $\sin z$ and $\cos z$ if $z = x + jy$, a complex number? A whole course might be devoted to answering these questions thoroughly. For the present we are not trying to go beyond the position of the eighteenth-century mathematician – that it seems very reasonable to identify $e^{j\theta}$ with $\cos \theta + j \sin \theta$, and that correct and extremely useful results follow when we do so.

We have already mentioned one important application – the study of oscillations and the solution of the differential equations arising in connection with oscillations. This topic belongs to calculus and will not be further developed here.

It was mentioned at the beginning of this section that all the formulas of trigonometry could be deduced from the simple properties of exponentials. This remark will now be explained.

In our first argument we obtained the equation $e^{j\theta} = \cos \theta + j \sin \theta$ by putting $x = j\theta$ in the series for e^x. If we put $x = -j\theta$ we are led to the equation

$$e^{-j\theta} = \cos \theta - j \sin \theta.$$

We thus have two equations involving $\cos \theta$ and $\sin \theta$, and we may regard these as simultaneous equations with $\cos \theta$ and $\sin \theta$ as unknowns. Solving these we find $\cos \theta$ and $\sin \theta$ in terms of $e^{j\theta}$ and $e^{-j\theta}$. We have

$$\cos \theta + j \sin \theta = e^{j\theta}, \tag{1}$$

$$\cos \theta - j \sin \theta = e^{-j\theta}. \tag{2}$$

$$\tfrac{1}{2}(1) + \tfrac{1}{2}(2) \cdots \cos \theta = \tfrac{1}{2}(e^{j\theta} + e^{-j\theta}), \tag{3}$$

$$\frac{1}{2j}(1) - \frac{1}{2j}(2) \cdots \sin \theta = \frac{1}{2j}(e^{j\theta} - e^{-j\theta}). \tag{4}$$

Students are sometimes worried by the fact that j appears in the denominator in equation (4); they think this means that $\sin \theta$ will not be a real number. But the quantity $e^{j\theta} - e^{-j\theta}$ in the bracket contains a factor j, as may be seen by substituting in the series for e^x. If we did not have j in the denominator to cancel this factor we would indeed be confronted by a paradox.

Equations (3) and (4) fulfil our promise; they define cosine and sine purely in terms of exponential functions.

An abbreviation will prove convenient. For this section only, we introduce a convention. Given quantities $A, B, C, D, \ldots$ denoted by capital letters we shall understand by the corresponding small letters $a, b, c, d, \ldots$ the following:

$$a = e^{jA}, \qquad b = e^{jB}, \qquad c = e^{jC}, \qquad d = e^{jD}.$$

Then we have $e^{-jA} = 1/a$ so we can obtain purely algebraic expressions for $\cos A$ and $\sin A$, namely

$$\cos A = \frac{1}{2}\left(a + \frac{1}{a}\right), \qquad \sin A = \frac{1}{2j}\left(a - \frac{1}{a}\right).$$

Thus the trigonometric identity $\cos^2 A + \sin^2 A = 1$ may be regarded as a consequence of the algebraic identity

$$\frac{1}{4}\left(a + \frac{1}{a}\right)^2 - \frac{1}{4}\left(a - \frac{1}{a}\right)^2 = 1.$$

In the exercises below, students are asked to translate into algebra trigonometric expressions of gradually increasing complexity. It will be found that even an expression such as $\cos(A - 2B + 3C + 4D)$ has a fairly simple algebraic equivalent.

In the nineteenth century, students were expected to achieve great virtuosity in the handling of trigonometric expressions. Today there is rightly less stress on such manipulative skill, but there are still occasions where it is useful for an engineer to understand some argument involving trigonometric formulas. This may be because engineers deal with objects that have actual shapes, so that trigonometry arises in its primitive role; it may also be in some situation where waves or vibrations are involved; again it may be that trigonometric functions arise in the middle of some mathematical process. The exercises below lead to a sort of dictionary, allowing any statement about sines and cosines to be translated into a purely algebraic equation. Algebra is usually easier to handle than trigonometry; it is more systematic. However translating trigonometry into algebra does not automatically remove all difficulties; sometimes the resulting algebraic equation is quite hard to verify or prove.

Exercises

1 By using the convention explained above, translate the expressions below into algebraic expressions involving only the small letters a, b, c, d.

(a) $\cos(A + B)$ (b) $\cos(A - B)$ (c) $\sin(A + B)$
(d) $\sin(A - B)$ (e) $\cos 2A$ (f) $\sin 2A$
(g) $\cos 3A$ (h) $\sin 3A$ (i) $\cos 4A$
(j) $\sin 4A$ (k) $\cos(A + 2B)$ (l) $\sin(A + 2B)$
(m) $\cos(A - 2B)$ (n) $\sin(A - 2B)$ (o) $\cos(A - 2B + 3C)$
(p) $\sin(A - 2B + 3C)$ (q) $\cos(A - 2B + 3C - 4D)$ (r) $\sin(A - 2B + 3C - 4D)$

2 By means of the results found in Exercise 1 above, translate the following equations into algebra, investigate whether they are correct or incorrect.

(a) $1 + \cos 2A = 2\cos^2 A$. (b) $2\cos A \cos B = \cos(A + B) + \cos(A - B)$.
(c) $\cos(A + B) = \cos A \cos B + \sin A \sin B$. (d) $\cos(A - B) = \cos A \cos B + \sin A \sin B$.
(e) $\cos 3A = 3\cos^3 A - 4\cos A$. (f) $\cos 3A = 4\cos^3 A - 3\cos A$.

17 Complex numbers and convergence

One of the arguments used in §16 depended on the use of infinite series. The development of calculus 300 years ago led rapidly to the discovery that very many functions could be represented as infinite series and this has remained a very powerful method for dealing with them. However, an infinite series, as one might guess, is a dangerous thing to use; if you do not understand it properly it can lead to entirely incorrect results. In this section we shall state (but not prove) a very simple result that enables us to distinguish correct from incorrect usage of infinite series. This result depends on complex numbers and the Argand diagram, and could not be stated or understood without the use of complex numbers.

The dangers of infinite series can be seen from the simplest example, one that occurs already in secondary school work. The series $1 + x + x^2 + x^3 + \cdots$ is a geometrical progression; its sum to infinity, calculated from the formula $a/(1 - r)$ is $1/(1 - x)$. This series can also be reached by starting with $1/(1 - x)$ and doing long division, or using the binomial theorem with exponent -1.

We will now compare the behaviour of the series and the fraction $1/(1 - x)$. In Fig. 48, the graph of $y = 1/(1 - x)$ is shown by a curve, the rectangular hyperbola. The value

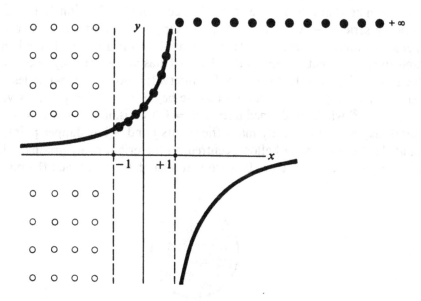

Fig. 48

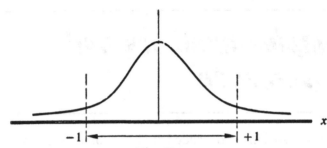

Fig. 49

of the series is indicated by large dots. For $-1 < x < 1$, as is well known, the series agrees perfectly with $1/(1 - x)$; the dots lie on the curve. When $x > 1$, this agreement totally disappears. For example, when $x = 2$, the series gives $1 + 2 + 4 + 8 + \cdots$ with sum $+\infty$. The fraction gives -1. Any value of x larger than 1 gives $+\infty$ for the series. We indicate this, rather symbolically, by dots along the 'ceiling' of the graph. For $x < -1$, the series is unsatisfactory in another way; if we put, for example, $x = -2$, the series gives $1 - 2 + 4 - 8 + 16 - 32 \cdots$; it oscillates more and more wildly and does not settle down to any number at all. The diagram tries to indicate this by dots scattered at random. Thus the series is a true and faithful servant so long as x lies between -1 and $+1$; outside this region it gives results such as $-1 = +\infty$ and is totally unreliable. In this its behaviour is typical for power series.

If we examine the graph, we can see why something unusual should be expected as x approaches $+1$, for here the graph goes off to infinity. But why should the series begin to misbehave when we pass -1? When $x = -1$, the fraction has the value $\frac{1}{2}$, and the graph is perfectly smooth and well behaved.

This question becomes even more acute if we consider the fraction $1/(1 + x^2)$ and the corresponding series $1 - x^2 + x^4 - x^6 + \cdots$ One could hardly ask for a tamer and better behaved graph than that of $1/(1 + x^2)$, as shown in Fig. 49. There is nothing in this graph to suggest that the interval -1 to $+1$ has any special significance, yet here again the series behaves perfectly within that interval; it becomes meaningless outside it.

The situation changes entirely as soon as we begin to consider complex values. The fraction $1/(1 + x^2)$ will be undefined if $x^2 + 1 = 0$, that is, if $x = j$ or $x = -j$. Thus on the Argand diagram we naturally mark the points j and $-j$ as danger points. Imagine pins put at these points and a balloon centred on the origin being gradually blown up, (Fig. 50). Suppose we blow it up, but are careful not to let it reach the pins. Then in

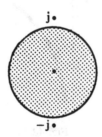

Fig. 50

the part of the Argand diagram covered by the balloon, all will be well; the series will sum to its correct value. But this is not all; within this region, you can apply any formal procedure that any sane person would think of; you can integrate the series, differentiate it, rearrange the terms, multiply it by itself or by another well behaved series, and in each case arrive at the result you want.

The danger points are known as *singularities*. The series for a complex function, $f(z)$, will behave well in any circle, with centre O, that does not reach as far as the nearest singularity. The functions e^z, sin z, cos z do not have a singularity for any finite value of z. Thus the series for them can be used with confidence for every value of z. As we have seen, $1/(1 - z)$ and $1/(1 + z^2)$ have singularities at unit distance from the origin. They are well behaved in any circle with centre O and radius less than 1.

A function can have a singularity without itself becoming infinite. For example $\sqrt{1 - z}$ has a singularity at $z = 1$ although its value there is zero. However, its derivative is infinite when $z = 1$.

For most engineering applications, it is sufficient to know that a function has a singularity where it, or any derivative, becomes infinite or undefined. The circle, with centre O, that goes through the nearest singularity, is called the *circle of convergence*. Inside this circle, all is well. On it, or outside it, anything may happen.

18   Complex numbers: terminology

This section is quite brief. Its aim is simply to explain certain names and symbols the student may meet in books or lectures.

When we were dealing with multiplication of complex numbers, a great role was played by r, the distance of a point from the origin, and θ, the angle at which the point occurs. It is not surprising that names have been coined for these.

For any real number x, the sign $|x|$ denotes its distance from 0. For complex numbers the same symbol is used for the distance from O in the Argand diagram. Thus if z is a complex number at distance r from O, we write $|z| = r$. This quantity is referred to as the *modulus* or *absolute value* of z.

The angle θ at which z occurs is called the *argument* of z, or arg z for short.

It can often happen that we are dealing with a complex number, $z = x + jy$, but we are only interested in one of the numbers x, y. There are names which allow us to indicate this; x is called the *real part* of $x + jy$, and y the *imaginary part*. Abbreviations are used, such as $x = \text{Re}(x + jy)$ or $y = \text{Im}(x + jy)$. These may vary a little from author to author. Note that the imaginary part of $7 + 13j$ indicates the real number 13; it does *not* indicate 13j. This is obviously more convenient; in engineering, any measurement we make involves *real numbers*. For purposes of calculation it may be convenient to know that the number we want turns up as the *coefficient* of j in some complicated expression. If the opposite convention were used, we would have to keep pointing out that we did not want jy, the part of $x + jy$ that contains $\sqrt{-1}$, but only y, the number that appears multiplying j in this part of the expression.

Routine exercises

1 Write down $|z|$ for the following values of z:

(*a*) j   (*b*) −j   (*c*) 4 + 3j

(*d*) 4 − 3j   (*e*) 1 + j   (*f*) − 3j

(*g*) $a + bj$ (*h*) $\cos\theta + j\sin\theta$ (*i*) $e^{j\theta}$

2 Give arg z for:

(*a*) j   (*b*) − 3   (*c*) − j

(*d*) 1 + j   (*e*) − 2 + 2j   (*f*) $(\sqrt{3}) + j$

(*g*) $1 - j\sqrt{3}$ (*h*) $\cos\theta + j\sin\theta$ (*i*) $e^{j\theta}$

3 Re and Im indicate real and imaginary parts. Give the following:

(*a*) Re(2 + 3j)   (*b*) Im(2 + 3j)   (*c*) Re($a + bj$)   (*d*) Im($a + bj$)

(*e*) Re $1/(1 + j)$ (*f*) Im $1/(1 + j)$ (*g*) Re $e^{j\theta}$ (*h*) Im $e^{j\theta}$

(*i*) Re $e^{-j\theta}$ (*j*) Im $e^{-j\theta}$

19   *The logic of complex numbers*

We have now followed the development of complex numbers through two stages. The first stage lasted about two centuries; in it mathematicians applied the rules of algebra to complex numbers and obtained correct and useful results, but always with some uncertainty as to what complex numbers were and why they worked so well. In the second stage, it was found that the Argand diagram gave a useful geometrical method for calculating with complex numbers.

We now come to the third stage, in which a logical account of complex numbers is sought, and a fourth stage in which mathematicians pass beyond complex numbers and see if there are yet other systems ('hypercomplex numbers') that can be profitably studied.

The Argand diagram is a kind of analogue computer. If we are given a complex number, we know where to put a dot to represent it. Conversely, given a dot on the diagram, we know how to read off the complex number it represents. Given two dots, p and q, we know geometrical constructions for marking the dots $p + q$ and pq. By continuing these constructions we can arrive at points corresponding to longer algebraic expressions, such as for example $(p + q)^2$ and $p^2 + 2pq + q^2$. (A little care is necessary here to specify the order in which the operations are done.) In the algebra of real numbers, the two expressions just mentioned would always be equal. But in building up a logical theory of complex numbers we are not allowed to make use of the guess that complex numbers obey the same laws as real ones. Rather, we define the geometrical operations that give sum and product. Starting from any two points p and q in the plane, a sequence of additions and multiplications leads us to $(p + q)^2$. So we can construct the point $(p + q)^2$. Similarly, the constructions corresponding to a sequence of additions and multiplications will lead us to $p^2 + 2pq + q^2$. When these constructions are carried out, it may be that $(p + q)^2$ and $p^2 + 2pq + q^2$ will always coincide, or they may not – so far as we know at present. But the matter is outside our control. We have defined the process, and have to carry it out according to the definitions. We may experiment and find in many particular cases that the points do coincide, or we may be able to prove that they always will.

Thus the Argand diagram becomes something like a machine we have set up, into which complex numbers can be fed and operations carried out. A statement such as 'For all p and q, $(p + q)^2 = p^2 + 2pq + q^2$' is to be regarded as true if the dots that correspond to the two constructions always coincide. On the basis of long experience, we expect that every formula valid in elementary algebra will be valid for this machine. Can we prove this?

Here we find ourselves up against a difficulty. There are infinitely many equations that hold in elementary algebra. We cannot deal with each one individually. How shall we know when we have proved enough to make sure that all the rest hold?

Mathematicians put this question to themselves in the years 1800–40, and it was in these years that the terms 'commutative', 'associative' and 'distributive' (which so obsess the writers of recent school textbooks) were coined. The idea was that every formula in elementary algebra is a logical consequence of these properties. Accordingly, if we can demonstrate that our Argand diagram 'machine' obeys $p + q = q + p$, $p(qr) = (pq)r$, $p(q + r) = pq + pr$ and so forth, then we know that operations with polynomials will work just the same for complex numbers as for real numbers. There are other properties that ought to be checked, if the full analogy with arithmetic and algebra is to be established – that subtraction and division (except by zero) are always possible; that complex numbers playing the roles of 1 and 0 exist; that $pq = 0$ only if $p = 0$ or $q = 0$. In all, not more than a dozen properties have to be checked. These properties are known as the axioms of a field. A field may informally be defined as a system in which there are operations called addition and multiplication, which can be handled just as if we were dealing with our usual arithmetic.

In the nineteenth century it was proved that complex numbers were such a system, and that we had logical justification for applying the usual procedures of arithmetic and algebra to them.

In many ways, for an engineer and even for a mathematician, the conclusion reached is more important than the details of the proof. There are many occasions on which all we need to remember is that we can apply the usual methods of arithmetic, algebra and calculus to complex numbers. It can hinder you, in the course of such a calculation, if you start thinking of every complex number as a point in a plane, that is as a vector. The Argand diagram should be used when it is appropriate. For some calculations, as we have seen, it is extremely useful. It also can be used as one way of showing that work with complex numbers has a logical justification. But there are also occasions when the wisest thing is to forget all about it and return to the simple faith that expressions like $(2 + 3j)/(4 + 5j)$ and $e^{j\theta}$ do have some meaning and can be handled just as if they were real numbers.

Exercises

1 Investigate the properties of commutativity, associativity and distributivity for operations with complex numbers.

2 Discuss subtraction for complex numbers.

3 Investigate when division by $a + jb$ is possible.

20   The algebra of transformations

One of the most persistent habits of mathematicians is generalizing. A particular device has proved useful for solving a certain type of problem. Almost the first question a mathematician asks is whether this device can be amended in some way so that it can be used in a wider class of problems.

In the Argand diagram multiplication by a complex number, $z \to cz$, appears as a particular kind of transformation, involving rotation and change of scale. If p and q are two complex numbers, we find ourselves considering the transformation $z \to (p + q)z$ and the transformation $z \to (pq)z$. Thus, for these very special transformations addition and multiplication are defined. It will only be a matter of time before some mathematician asks, 'Using our experience with complex numbers as a guide, can we define addition and multiplication for *any* transformations?'

A definition of multiplication is easily found. To multiply by pq, we can first multiply by q and then multiply the result by p. Thus, if $qz = z^*$ and $pz^* = z^{**}$, we have $z^{**} = (pq)z$. Now this last statement remains meaningful if p and q are thought of as representing *any* transformation of some space.

Thus, if S and T are transformations of any space, we define ST as having the effect of the transformation T followed by the transformation S. This definition is quite in keeping with the usual way of writing functions of a real variable; for instance, log sin x could be found by first looking up sin x in a table of sines, and then finding the logarithm of that number. Similarly, given tables of sines and logarithms, we could construct a table for sin log x, by doing the operations in the reverse order.

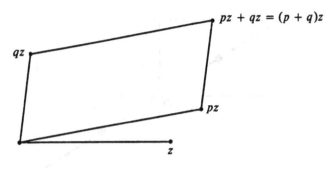

Fig. 51

Addition calls for a little more thought. In the Argand diagram we know all the laws of algebra hold. So, if we want to find $(p + q)z$ we know we can write this as $pz + qz$. The addition here will be seen in the Argand diagram as vector addition. We shall have the diagram of Fig. 51.

This suggests that we should define the transformation $S + T$ as follows; to find $(S + T)z$, first find Sz and Tz, and then add these to give $Sz + Tz$. This assumes we are dealing with a space in which addition has been defined. As we are going to be concerned only with transformations of vector spaces, in all our work this condition will be met.

Students sometimes raise the objection – we set out to define $S + T$ but we have not succeeded in doing that. We have only defined $(S + T)z$. How do we get rid of this z?

In fact there is no need to get rid of z. A transformation sends a vector z to a vector z^*. The transformation is completely defined when we know where it sends each vector. Now in the diagram and construction above, z represents *any* vector. So we can use the construction to find where each vector goes; that is to say, the transformation $S + T$ is completely defined by the construction above.

For example, suppose we are dealing with a plane. Let I denote the identity transformation, M_1 reflection in the horizontal axis and M_2 reflection in the vertical axis.

What is $I + M_1$? Consider any point z. By definition, $(I + M_1)z$ means $Iz + M_1z$. Now Iz is simply z, while M_1z is its reflection as shown in Fig. 52. We can find

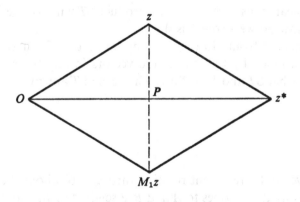

Fig. 52

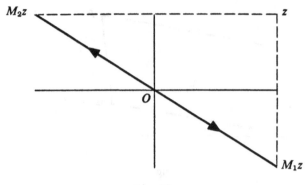

Fig. 53

$z + M_1 z$ by drawing a parallelogram. This gives us z^*, the point to which $I + M_1$ sends z. If P is the projection of z on the horizontal axis, z^* is twice as far from O as P. We now know exactly what the transformation $I + M_1$ does to each point of the plane. The transformation is fully defined; we have a complete specification of it.

In the same way, we can find $M_1 + M_2$. Fig. 53 shows any point z, and its reflections in the axes. It will be seen that $M_1 z$ and $M_2 z$ are equal and opposite vectors; their sum is thus the zero vector, 0. Thus the transformation $M_1 + M_2$ sucks every point of the plane into the origin. This transformation is usually thought of as the zero transformation; we may denote it by 0, and write $M_1 + M_2 = 0$.

Note that we have three zeros in our work, the number 0, the vector 0 and the transformation 0. It will usually be clear which of these we have in mind; for instance, in the equation $M_1 + M_2 = 0$, as the sum of two transformations is a transformation, not a vector or a number, it is clear that the 0 in this equation must stand for the zero transformation.

If we write $0v$, where v is a vector, it may not be clear whether we mean the vector v multiplied by the *number* 0, or the result of applying the *transformation* 0 to the vector v. Fortunately both interpretations give the same result – the zero vector, 0.

You may meet some books in which different signs are used to distinguish the three zeros; the sign θ (Greek theta) is sometimes used for one of them.

If S and T are linear transformations, the product ST will also be linear, and so will the sum $S + T$. (How do we know this?)

It will be remembered from §10 that for any linear transformation, the fate of the entire space is determined by the fate of the vectors in any basis. This is a possible method for finding what ST and $S + T$ are, when S and T are given. It is not always the best method.

Worked example

Let M_1 be as above, and J represent rotation through 90° about the origin. Find $M_1 J$. In Fig. 54, consider what $M_1 J$ does to A and B. J sends A to B, and M_1 then sends B to

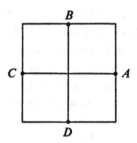

Fig. 54

D. Thus $(M_1J)A = D$. J sends B to C and M_1 leaves C where it was. So $(M_1J)B = C$. Thus for M_1J we see $A \to D$, $B \to C$.

Thus M_1J is reflection in the line with equation $x + y = 0$.

Multiplying a transformation by a number

We have explained how transformations are added. There is nothing to stop us adding a transformation to itself. Thus, given T, we can find $T + T$. If we add T again we can find $T + T + T$. Naturally we abbreviate these to $2T$ and $3T$. Let v be any vector, and let $Tv = v^*$. Then, by definition, $(T + T)v = Tv + Tv = v^* + v^* = 2v^*$. Thus $(2T)v = 2v^*$. In the same way, $(3T)v = 3v^*$.

These results suggest that, if k is any number (not necessarily a natural number), we should define kT so that $(kT)v = kv^*$. Now v^* is short for Tv, so our equation means $(kT)v = k(Tv)$.

We accordingly *define* kT by the equation just given. To anyone used to elementary algebra, it may be hard to see that this equation says anything: the two sides seem to be exactly the same thing.

Suppose however this equation had not been given but that instead students suddenly found an expression such as $3.5T$ occurring in a problem. Surely someone would ask, 'What do you mean by multiplying a transformation by 3.5?' And the student would be quite right to ask this: it is not obvious what, for instance, 3.5 times a reflection means.

The right-hand side of the equation $(kT)v = k(Tv)$ is meaningful. The transformation T sends the vector v to a vector represented by Tv. Thus Tv is a vector, and we know what multiplying a vector by k means. So $k(Tv)$ has a clear meaning. On the left, we have kT, a symbol that has not been used before. It is up to us to explain what we mean by it. A person who introduces a new symbol is entitled to say what he uses it to stand for. This seems to give him a lot of freedom; he could say kT stood for a pound of cheese. But in practice, his freedom is not so great. The symbol kT looks like multiplication in elementary algebra. If we defined it in such a way that it did not behave, more or less, like the multiplications we did at school, it would be misleading. People would be influenced by their old habits and would continually be making mistakes and getting wrong results. This would certainly happen if we defined our new symbol kT in such a way that the equation $(kT)v = k(Tv)$ did *not* hold. So we are practically forced to try this definition. Now of course it might turn out that even this definition did not work out well; it might lead to complications because of some other definition we wanted to adopt. This would mean that the behaviour of transformations was essentially different from that of numbers. If this were so, we would be wise to use a symbolism for transformations utterly unlike that for numbers; this would be the best way to avoid confusion. However, this possibility is not realized in fact. There is a very strong analogy between the behaviour of transformations and that of numbers. It does pay us to use the symbols of algebra for transformations; more often than not, the results we expect will be true. We can learn, in a reasonably short time, what processes in elementary algebra we must *not* apply to transformations. It is really remarkable that transformations and numbers should have properties as similar as they do.

The equation $(kT)v = k(Tv)$ that defines the transformation kT can be put into words. It says that the transformation kT can be carried out as follows: apply the transformation T, then change the scale in the ratio k:1.

Exercises

I, M_1, M_2, and J have the meanings already defined in this section.

1 Find a geometrical description, as simple as circumstances permit, for each of the transformations listed below. In each case, say what are the invariant lines (if any); state the eigenvectors and eigenvalues (if any).

(a) $M_2 J$ (b) $J M_2$ (c) $J M_1$
(d) $M_1 M_2$ (e) $M_2 M_1$ (f) M_1^2, that is, $M_1 M_1$
(g) M_2^2 (h) J^2 (i) $M_2 J^2 M_1$
(j) $I + M_2$ (k) $I + J$ (l) $I + M_1 + M_2$
(m) $M_2 J + J M_2$ (Use your answers to (a) and (b) above.)
(n) $J M_1 + J M_2$

2 Which of the following statements are true and which false?

(a) $J M_1 = M_1 J$ (b) $J M_2 = M_2 J$ (c) $M_1 M_2 = M_2 M_1$
(d) $J M_1 = M_2 J$ (e) $J M_2 = M_1 J$

3 In the light of your work with the earlier questions, would you expect the statement $ST = TS$, where S and T are any two transformations to be true always, sometimes or never? What about the statement $S + T = T + S$?

4 $J M_1 + J M_2$ was found in question 1(n). Would we get the same result if we found $J(M_1 + M_2)$?

5 (a) In the text $I + M_1$ was found. How would you describe $\frac{1}{2}(I + M_1)$ most simply as a geometrical operation?
(b) Similarly describe $\frac{1}{2}(I + M_2)$.
(c) What is $(1/\sqrt{2})(I + J)$? Have we met essentially this result in any earlier section?

21 Subtraction of transformations

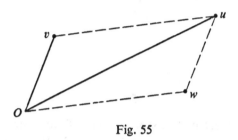

Fig. 55

In §20 we defined addition and multiplication of transformations, and by repeated use of these definitions we can build expressions such as $STS + 2S + 3T$ or $S^2 + 5S + 6$. However, we cannot yet attach a meaning to an expression involving subtraction such as $S - T$ or even $-T$. Such expressions do arise naturally. For instance, in §20 we found $M_1 + M_2 = 0$ for the reflections in the axes. This suggests $M_2 = -M_1$, if we can find some way of explaining what the minus sign means in this connection.

We deal with this question in much the same way as a teacher might in Grade 1. If a child does not know $7 - 4$ the teacher may ask '4 and what make 7?' Thus a child does not have to learn a 'subtraction table'. Knowing the addition table gives the subtraction results. 'What is $7 - 4$?' and '4 and what makes 7?' are equivalent questions. If we replace 'what?' by an algebraic symbol, we may say that $x = 7 - 4$ and $4 + x = 7$ are equivalent.

We know that $7 - 4$ indicates a definite number, 3. This means that there is some number denoted by $7 - 4$, and that there is only one such number – the equation $4 + x = 7$ has a solution, and (unlike, say, a quadratic) it has only one solution.

When we are dealing with transformations, or even with vectors, subtraction is defined along essentially the same lines.

If u and v are vectors, by $u - v$ we mean that vector which, added to v, will give u.

Suppose $w = u - v$, that is, suppose $w + v = u$. Then u must be the corner of a parallelogram formed when w and v are added. Given u and v, we can construct w, as shown in Fig. 55. There is only one possible position for w. It will be noticed that the path from O to w is parallel to the path from v to u, and of equal length. Thus the magnitude and direction of the vector $u - v$ can be seen, by considering the path 'from v to u'.

Usually $O - v$, the difference between v and O, is shortened to $-v$. Thus $-v$ means (in accordance with our definition) the vector that has to be added to v to give O. This of course, is the vector equal and opposite to v, as in Fig. 56.

Now of course $u - v$ and $u + (-v)$ ought to mean the same thing, and you can check that in fact they do. In Fig. 55, you may notice that from u to w is a displacement equal and opposite to that from O to v.

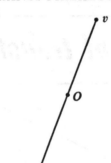

Fig. 56

Now we return to the subject of transformations. If we write $W = S - T$, then W means that transformation which, added to T, gives S. We have to check that some transformation does this, and that only one does.

If $S = W + T$, for any vector v, $Sv = Wv + Tv$, from the definition of addition in §20. Thus $Wv = Sv - Tv$. Sv and Tv are both vectors – they are the points to which v is sent under the transformations S and T. $Sv - Tv$ is the difference of these vectors. As we have just seen, you can subtract any vector from any vector and a definite vector results. Thus there will always be one, and only one, Wv that satisfies the required condition.

If we like, we can define $-T$, and then get $S - T$ by taking $S + (-T)$. This is often the most convenient way to do things. By $-T$ we understand of course $0 - T$, the transformation that, added to T, gives 0. This means $T + (-T) = 0$. So for any vector v, $Tv + (-T)v = 0v$. Now 0 is the transformation that sucks everything into the origin. Whatever v, $0v = 0$. (Notice how this looks right, for someone used to elementary algebra. What kinds of things are the two zeros here – transformations, vectors, numbers?)

So $Tv + (-T)v = 0$, that is $(-T)v$ is the vector equal and opposite to Tv. Again this looks right; we could write it $(-T)v = -(Tv)$.

So we can find the effect of $-T$ as follows; see where T sends each point, then reverse all the vectors.

For example, consider $-J$, where J as before denotes rotation through 90°. J sends P to Q (Fig. 57). Reversing OQ, we find the point P^*. Thus $-J$ is a rotation through $-90°$ for $-J$ sends P to P^*.

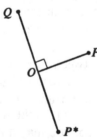

Fig. 57

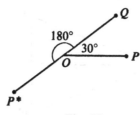

Fig. 58

Do not jump to the conclusion that, if R is a rotation through θ, $-R$ is a rotation through $-\theta$. This is *not* true; 90° is a very special angle. Suppose for instance R denotes rotation through 30°. Then R sends P to Q as shown in Fig. 58, and $-Q$ gives P^*. We do not get from P to P^* by a rotation of $-30°$.

It will be useful to consider $-I$. The identity transformation I leaves any point P unchanged. Thus $-I$ sends P to P^* as in Fig. 59. Thus $-I$ denotes a rotation through 180°.

If we now look back at the figure where R denoted rotation through 30°, we see that $-R$ denoted rotation through 30° followed by rotation through 180°. As rotation through 30° is denoted by R and through 180° by $-I$, the previous sentence can be boiled down into the equation $-R = (-I)R$, which again looks very plausible as a piece of algebra.

In §20, we noticed $M_1 + M_2 = 0$ which we may now write as $M_2 = -M_1$. It is useful at this stage to look back at the diagram showing M_1z and M_2z, and to see how this particular case illustrates what we have just been considering. We see that indeed $M_2 = (-I)M_1$; the effect of reflection in the horizontal followed by a rotation of 180° is a reflection in the vertical.

Now that addition, subtraction and multiplication have been defined we can build expressions that look very much like those at the beginning of an elementary algebra book, such as $(S - T)^2$; $(S - T)(S + T)$; $(S + T)(S^2 - ST + T^2)$. We have seen above that many true statements about transformations look exactly like statements in elementary algebra. *However, as the exercises below will show, equations that are true for numbers are not always true for transformations.* So, for the present, we have to check each time, by going back to the meaning of the statement, whether it is true for transformations or not. Later we shall try to find a more systematic way of deciding which equations work for transformations and which do not.

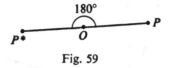

Fig. 59

Co-ordinate Methods

So far, our discussion of transformations has been based entirely on *geometrical diagrams*. This has been done because drawing is one of the main ways in which an engineer

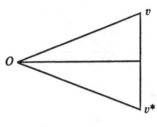

Fig. 60

thinks and communicates. It is not enough for an engineer to calculate; he must be able to *see* what he is doing. In North American education this aspect of mathematics tends to be neglected. However, it is often helpful to calculate; though, whenever possible, calculation should be illustrated by a diagram. Many results that we have found earlier by purely geometrical thinking can be found or checked by co-ordinate methods. The following worked examples illustrate the procedure.

(1) Where does I send the point (x, y)? I leaves it unchanged; so I makes $(x, y) \rightarrow (x, y)$.

(2) Where does M_1 send (x, y)? M_1 is a reflection in the horizontal: it sends v to v^* in Fig. 60. Thus the x co-ordinate is unchanged, but the y component changes sign. So we have $M_1: (x, y) \rightarrow (x, -y)$.

(3) Where does $I + M_1$ send (x, y)? By the definition of addition,

$$(I + M_1)v = Iv + M_1v = v + v^*$$

in the figure just used. Now $v = (x, y)$ and $v^* = (x, -y)$ as we found above. Superposing these, we find $(x, y) \rightarrow (2x, 0)$ when $I + M_1$ acts. This agrees with the conclusion reached in §20, and gives a more compact way of recording the result. In words we can express the effect of $I + M_1$ as follows; the x co-ordinate is doubled, the y co-ordinate becomes zero.

(4) Where does M_1J send (x, y)? By the definition of the product of transformations, we must let J act, and then see what M_1 does to the result. Fig. 61 shows v, and we have

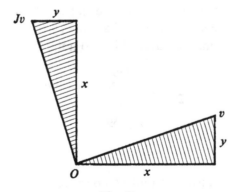

Fig. 61

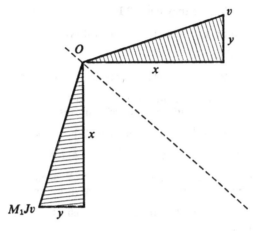

Fig. 62

shaded in a certain triangle. J rotates everything through $90°$. If we imagine it picking up the shaded triangle, of which v is one corner, it will take it to the shaded triangle of which Jv is a corner. We see that the co-ordinates of Jv are $(-y, x)$. This of course is a standard result and the argument just used should be familiar to the student.

M_1 changes the sign of the second co-ordinate. So M_1 acting on Jv gives $(-y, -x)$, as in Fig. 62. This again agrees with our conclusion found in §20, that M_1J is a reflection in the line with slope -1.

The co-ordinate approach may be found helpful in some of the following exercises.

It is important that students should work right through the following questions. There may seem to be a lot of questions here, but each is quite short, and each is really a small part of an investigation. For instance, the first five questions form a single topic. The questions are not merely exercises on the definitions, but are designed to show what the algebra of transformations is like. A student who worked Exercises (1) to (12) only and then stopped would be left with an entirely misleading impression. These questions show an effect that happens *sometimes*; the later questions show that it does *not* happen always.

Exercises

1 Where does $2M_1$ send (x, y)?

2 Where does M_1^2 send (x, y)? What is the simplest symbol for M_1^2?

3 Where does $I + 2M_1 + M_1^2$ send (x, y)?

4 Where does $(I + M_1)^2$ send (x, y)? (The effect of $I + M_1$ has been discussed in the text.)

5 Elementary algebra suggests that $(I + M_1)^2$ and $I + 2M_1 + M_1^2$ might be the same transformation. Do the answers to questions 3 and 4 confirm this or not?

6 Where does $-I$ send (x, y)?

7 Where does M_2 send (x, y)?

8   Investigate $M_1 M_2$ and $M_2 M_1$. Are they equal? Is there any simpler way of writing them?

9   Where does $M_1 + M_2$ send (x, y)? What is the simplest symbol for $M_1 + M_2$?

10   Where does $(I + M_2)$ send (x, y)?

11   Where does $(I + M_1)(I + M_2)$ send (x, y)? What is the simplest symbol for this transformation?

12   Elementary algebra suggests that $(I + M_1)(I + M_2)$ and $I + M_1 + M_2 + M_1 M_2$ might be the same transformation. Investigate this with the help of your answers to questions 8, 9 and 11.

13   The effects of J and M_1 have been given in the text. Where does $J + M_1$ send (x, y)?

14   Where does $J + M_1$ send $(7, 3)$? Where does it send $(4, 4)$? Where does $(J + M_1)^2$ send $(7, 3)$?

15   Choose any point, e.g. $(3, 2)$, and see where $(J + M_1)^2$ sends it. Experiment a little with different points. Investigate by algebra the effect of $(J + M_1)^2$ on any point (x, y).

16   Where does JM_1 send (x, y)? Does JM_1 have the same effect as $M_1 J$, which was considered in the Worked example 4 above?

17   Simplify $J^2 + 2JM_1 + M_1^2$, or find its effects on (x, y), whichever you prefer. Is

$$J^2 + 2JM_1 + M_1^2$$

the same transformation as $(J + M_1)^2$?

18   Is $(J + M_1)^2$ the same as either of the following?
   (a) $J^2 + 2M_1 J + M_1^2$ (b) $J^2 + M_1 J + JM_1 + M_1^2$

19   In elementary algebra, a quadratic has only two solutions. For transformations, the quadratic $X^2 = I$ has the solutions $X = I$ and $X = -I$. Has it any other solutions? If so, how many?

20   In elementary algebra, $ab = 0$ only if $a = 0$ or $b = 0$. Is it true for transformations that $ST = 0$ only if $S = 0$ or $T = 0$? (Some of your answers to earlier questions provide evidence relevant to this.)

22   *Matrix notation*

In working the exercises of §21, you probably found that the use of the co-ordinates to specify points made the work easier. This idea can be carried further. By a very simple device, we can specify linear transformations in terms of numbers. This allows us to specify a transformation in a convenient form, and it also allows us to calculate the numbers that specify transformations such as $S + T$ and ST, when we are given the numbers that specify S and T.

In §10, 'Specification of a linear mapping' we met the doctrine that the fate of a basis determines the fate of the whole space. In the plane, the points $A = (1, 0)$ and $B = (0, 1)$ constitute a basis. If a linear transformation T sends these points, say, to $A^* = (4, 1)$

and $B^* = (2, 3)$, this fixes what happens to every point of the plane. As A^* has 2 co-ordinates and B^* also has 2 co-ordinates, it follows that any linear transformation of the plane can be specified by giving 4 numbers, the co-ordinates of these two points. Thus we could specify T by writing $(1, 0) \rightarrow (4, 1)$, $(0, 1) \rightarrow (2, 3)$. However, a shorter and more convenient symbolism can be found.

In this section, we are concerned to translate into algebra definitions which have already been given in geometrical form. Much of this work is quite straightforward. Students will find it improves their confidence and understanding of the definitions if they carry out the details of this work for themselves, rather than simply following an exposition of it. Accordingly this section is presented as a series of questions. After each question, there will be a row of stars. Such a row of stars is an invitation to the student to turn away from the book and find the answer for himself. Below the stars, the answer to the question will be explained.

For the particular transformation T considered above we have the information $(1, 0) \rightarrow (4, 1)$, $(0, 1) \rightarrow (2, 3)$, and we know that this fixes where every point of the plane goes. It ought to be possible, then, to derive a formula showing where any given point, (x, y), goes. If T sends (x, y) to (x^*, y^*), what equations give x^* and y^* in terms of (x, y)? We know T is linear, so we can use the principles of proportionality and superposition.

* * *

We use proportionality first. If the input $(1, 0)$ is multiplied by any number x, the output $(4, 1)$ will be multiplied by that same number. So $(x, 0) \rightarrow (4x, x)$. By the same principle, applying multiplication by y to the statement $(0, 1) \rightarrow (2, 3)$, we find $(0, y) \rightarrow (2y, 3y)$ for any number y. We now know where any point on either axis goes. Superposing $(x, 0)$ and $(0, y)$ gives (x, y). So we conclude $(x, y) \rightarrow (4x + 2y, x + 3y)$ and we may write

$$x^* = 4x + 2y,$$
$$y^* = x + 3y.$$

So, appropriately enough, a linear transformation, T, is expressed by a pair of linear equations. We would obtain much the same kind of result if we took any other numbers in place of $(4, 1)$ and $(2, 3)$. If we replace the particular numbers by algebraic symbols and carry through exactly the same argument, we find that $(1, 0) \rightarrow (a, c)$, $(0, 1) \rightarrow (b, d)$ lead to the equations

$$\left. \begin{array}{l} x^* = ax + by, \\ y^* = cx + dy. \end{array} \right\} \tag{1}$$

These equations show why we made the apparently strange choice of symbols in $(1, 0) \rightarrow (a, c)$. By having $(1, 0) \rightarrow (a, c)$ rather than $(1, 0) \rightarrow (a, b)$ we get the letters a, b, c, d in their natural order in the equations (1).

Here again four numbers are used to specify a transformation. Matrix notation is based on one very simple idea, that of arranging the four numbers in the positions they occupy in the equations above. Thus we would specify the transformation T by writing

$$\begin{pmatrix} 4 & 2 \\ 1 & 3 \end{pmatrix}$$

while the general linear transformation, corresponding to equations (1), would be written

$$\begin{pmatrix} a & b \\ c & d \end{pmatrix}.$$

In many of the transformations we have considered in our examples, some of the terms seem to be missing. For instance M_1, reflection in the horizontal, has $x^* = x$, $y^* = -y$. In order to exhibit this in the pattern of equations (1), we insert the coefficients 0 and 1 implied by these equations. Thus, for M_1

$$x^* = 1x + 0y,$$
$$y^* = 0x - 1y.$$

and so, in matrix notation

$$M_1 = \begin{pmatrix} 1 & 0 \\ 0 & -1 \end{pmatrix}.$$

In the same way, if we meet the symbol

$$\begin{pmatrix} 0 & -1 \\ 1 & 0 \end{pmatrix}$$

this tells us that in equations (1) we are to substitute $a = 0$, $b = -1$, $c = 1$, $d = 0$. We then find $x^* = -y$, $y^* = x$, that is $(x, y) \rightarrow (-y, x)$, which near the end of §21 we saw to be the transformation J, rotation through 90°.

One advantage of using the matrix symbolism is that we can easily write the effect of successive transformations. Thus M_1J is immediately written as

$$\begin{pmatrix} 1 & 0 \\ 0 & -1 \end{pmatrix} \begin{pmatrix} 0 & -1 \\ 1 & 0 \end{pmatrix}.$$

Much more writing would be needed if we had to express this combined transformation by means of equations.

Successive transformations frequently arise in engineering. There are obvious examples such as transistor amplifiers in which each stage takes a signal and operates on it. The final output is the result of all these operations in turn applied to the original input. But in any branch of engineering less obvious examples arise. In a computing process, certain data may be taken, a calculation made with them giving another set of numbers, which then become the starting point for a further calculation or subroutine. In numerical analysis, frequent use is made of iteration, the process mentioned in §11, by which the same transformation is applied again and again. In structural analysis, we find, if not successive transformations, at any rate successive mappings, which can be represented by matrices. Thus in the direct stiffness method, we start with the loads applied to a structure, use a matrix to deduce the displacements, and from these displacements, with the help of another matrix, we find the stresses.

Suppose we have two transformations, H and N, specified in matrix form by

$$N = \begin{pmatrix} p & q \\ r & s \end{pmatrix}, \qquad H = \begin{pmatrix} a & b \\ c & d \end{pmatrix}.$$

The combined transformation NH has been defined in §20; it is found by applying H and then applying N to the result. Thus if H sends (x, y) to (x^*, y^*) and N then sends (x^*, y^*) to (x^{**}, y^{**}), the transformation NH sends (x, y) to (x^{**}, y^{**}). We have the equations

$$\left. \begin{array}{l} x^* = ax + by, \\ y^* = cx + dy, \end{array} \right\} \quad (1) \qquad \left. \begin{array}{l} x^{**} = px^* + qy^*, \\ y^{**} = rx^* + sy^*. \end{array} \right\} \quad (2)$$

These correspond to the matrices given for H and N.

NH is the transformation $(x, y) \rightarrow (x^{**}, y^{**})$. To find the matrix specification for NH, we have to answer two questions, neither of which calls for any special trick or ingenuity. What equations give (x^{**}, y^{**}) in terms of (x, y)? What is the matrix corresponding to these equations?

*   *   *

The first question can be answered immediately by substituting in equations (2) the values of (x^*, y^*) given in equations (1). For example, for x^{**} we find

$$p(ax + by) + q(cx + dy).$$

Collecting together the x and y terms, we find the first equation shown in (3) below. The second equation below is found by applying a similar process to y^{**}. We obtain

$$\left. \begin{array}{l} x^{**} = (pa + qc)x + (pb + qd)y, \\ y^{**} = (ra + sc)x + (rb + sd)y. \end{array} \right\} \quad (3)$$

The matrix form is obtained by writing the coefficients here in the appropriate places, as in the matrix for NH below. We may show the result of combining N and H to obtain NH by the scheme below.

$$\begin{pmatrix} p & q \\ r & s \end{pmatrix} \begin{pmatrix} a & b \\ c & d \end{pmatrix} = \begin{pmatrix} pa + qc & pb + qd \\ ra + sc & rb + sd \end{pmatrix}. \qquad (4)$$

There is a mechanical process for obtaining this result. It we examine, for example, $pb + qd$, the number that occurs in the first row and second column of NH, we see that it involves the numbers p, q in the first row of N, and the numbers b, d in the second column of H. Corresponding numbers in these two sets are multiplied (giving pb and

qd) and the results added (giving $pb + qd$). It is the custom to perform these calculations at first with the help of finger movements. A finger of the left-hand moves across the first row of N; a finger of the right-hand moves down the second column in H. Our attention is thus concentrated on the entries we have to deal with.

Similar considerations apply to the other entries in NH.

In writing out a matrix such as that for NH care should be taken to keep the four expressions involved clearly separated. One should remember that $pa + qc$ represents a single number, and that it is not added or multiplied to any of the other numbers appearing in the scheme. It simply records the coefficient of x in the equation for x^{**}.

It may be helpful to consider a numerical example. If we take $p = 10, q = 3, r = 40,$ $s = 2, a = 20, b = 30, c = 1, d = 3$, equations (1) and (2) become

$$\left.\begin{aligned} x^* &= 20x + 30y \\ y^* &= x + 3y \end{aligned}\right\} \quad (1a) \qquad \left.\begin{aligned} x^{**} &= 10x^* + 3y^*, \\ y^{**} &= 40x^* + 2y^*. \end{aligned}\right\} \quad (2a)$$

From these we find

$$\left.\begin{aligned} x^{**} &= 203x + 309y, \\ y^{**} &= 802x + 1206y. \end{aligned}\right\} \tag{3a}$$

In matrix shorthand this is expressed by

$$\begin{pmatrix} 10 & 3 \\ 40 & 2 \end{pmatrix} \begin{pmatrix} 20 & 30 \\ 1 & 3 \end{pmatrix} = \begin{pmatrix} 203 & 309 \\ 802 & 1206 \end{pmatrix}.$$

The numbers used here have been chosen so that the algebraic formula behind each can be seen. Thus 1206 corresponds to $rb + sd$ in the general formula. If 1206 is read as 12 hundred and 6, the '12 hundred' corresponds to rb, namely 40×30, and the 6 to sd, namely 2×3. In the other entries similarly the 'hundreds' correspond to one term in the formula, the 'units' to another term.

This comment may seem unnecessarily elementary. However, experience shows that errors arise in students' work, owing to algebraic symbols becoming detached from their correct position, and getting mixed up with symbols in some other column of the matrix.

The columns in a matrix

In §10 we saw that the fate of a basis under a linear mapping determines the fate of the whole space. As we mentioned at the beginning of this section, in the plane an obvious basis consists of the vectors A and B, where A is $(1, 0)$ and B is $(0, 1)$. What happens to these under the transformation specified by

$$\begin{pmatrix} a & b \\ c & d \end{pmatrix}?$$

To answer this question, we go back to the equations indicated by this matrix, namely $x^* = ax + by, y^* = cx + dy$.

To find the effect of the transformation on A, we substitute the co-ordinates of A, that is, $x = 1, y = 0$. We find $x^* = a, y^* = c$.

To find the effect on B we substitute $x = 0, y = 1$ and find $x^* = b, y^* = d$.

Thus, for this transformation, $(1, 0) \rightarrow (a, c)$ and $(0, 1) \rightarrow (b, d)$.

If we look at the matrix above, we notice that a and c, the co-ordinates of A^*, occur in the first column, while b and d, the co-ordinates of B^*, occur in the second column.

This is a very useful result. It means that, given a matrix, we can see from it immediately, without any calculation, where it sends the points A and B. We simply read off the columns of the matrix.

For instance, early in this section we showed that a certain transformation T was given by the matrix

$$\begin{pmatrix} 4 & 2 \\ 1 & 3 \end{pmatrix}.$$

Reading off the figures in the columns we see that, for T, we have $A^* = (4, 1)$ and $B^* = (2, 3)$. In this particular case, this is not new information, since T was originally defined by specifying A^* and B^*. But the general doctrine is valuable: if we meet any matrix, we can get some idea of its geometrical effect simply by reading off the columns of the matrix.

Conversely, if for some transformation we know A^* and B^*, we can rapidly write down the matrix for this transformation. For instance, suppose we need the matrix for J, rotation through $90°$. J sends A to $(0, 1)$, so in the first column we must write 0 and 1, in that order, starting at the top. J sends B to $(-1, 0)$ so -1 and 0 must be written, in that order, in the second column. Thus we find

$$J = \begin{pmatrix} 0 & -1 \\ 1 & 0 \end{pmatrix}.$$

We assume here that conventional graph paper is being used, with perpendicular axes and a grid composed of squares.

This procedure, while simple, is useful, and to fix it in mind a student should take some transformations known to him, – reflections, rotations, shears, etc. – and write down the matrices for them by this method. In particular, the matrix for rotation through θ, which occurs in many engineering problems, should be found by this method.

Students with some skill in manipulation may find the matrix for reflection in the line $y = mx$ (or the line $y = m \tan A$) by this method.

Vectors as columns

When we are specifying a transformation we may want to show, not only the coefficients a, b, c, d in the equations

$$x^* = ax + by,$$
$$y^* = cx + dy,$$

but also the fact that this transformation has input (x, y) and output (x^*, y^*). Now x and y occur on the same level, while x^* occurs above y^* in these equations. Yet (x, y) and (x^*, y^*) represent the same kind of object; they both specify points (or vectors) in the plane. Should we show them horizontally or vertically?

There are two arguments to show why, in matrix work, it is convenient to write vectors as columns $\begin{pmatrix} x \\ y \end{pmatrix}$ rather than as rows (x, y).

The first argument comes from our paragraphs above on the columns of a matrix. These columns show the co-ordinates of A^* and B^*. If we form the habit of writing vectors as columns, it will be easy for us to split the matrix $\begin{pmatrix} a & b \\ c & d \end{pmatrix}$ and read off

$$A^* = \begin{pmatrix} a \\ c \end{pmatrix}, \qquad B^* = \begin{pmatrix} b \\ d \end{pmatrix}.$$

If we do not form this habit, we are liable to make the mistake of reading off the rows (a, b) and (c, d) which of course do *not* tell us where A^* and B^* are.

A second reason comes from the mechanical procedure for matrix multiplication. We saw by studying equation (4) that the entries in the matrix product involved numbers from a row in the first matrix and a column in the second:

$$(\rightarrow) \quad (\downarrow).$$

If we write a vector as a column we can use the same mechanical procedure to work out Mv, a *matrix* acting on a *vector*. Thus we find

$$\begin{pmatrix} a & b \\ c & d \end{pmatrix} \begin{pmatrix} x \\ y \end{pmatrix} = \begin{pmatrix} ax + by \\ cx + dy \end{pmatrix} = \begin{pmatrix} x^* \\ y^* \end{pmatrix}.$$

The product has 2 rows, because the first factor, the matrix, has 2 rows; it has 1 column because the second factor, the vector, has 1 column. It will be seen that both the input and the output vectors now appear as columns.

It will be found that writing vectors as columns proves a very convenient convention for matrix work. It is unfortunate that in co-ordinate geometry, invented long before matrices, we use the arrangement (x, y) rather than the column. Some effort is needed if one is not to become confused.

Exercises

1 Write in the usual notation of elementary algebra the equations expressed by

$$\begin{pmatrix} x^* \\ y^* \end{pmatrix} = \begin{pmatrix} 2 & 5 \\ 8 & 3 \end{pmatrix} \begin{pmatrix} x \\ y \end{pmatrix}.$$

2 Write in matrix notation (as in question 1)

$$x^* = 4x + 9y,$$
$$y^* = 16x + 25y.$$

3 Vectors u and v are specified by $u = \begin{pmatrix} 1 \\ 3 \end{pmatrix}$, $v = \begin{pmatrix} 20 \\ 10 \end{pmatrix}$. Write, in column form, the following vectors:

(a) $u + v$ (b) $2u$ (c) $3v$ (d) $2u + 3v$

(e) ku (f) mv (g) $ku + mv$ (h) $u - v$.

4 Work out the following matrix products.

(a) $\begin{pmatrix} 2 & 1 \\ 5 & 3 \end{pmatrix} \begin{pmatrix} 3 & -1 \\ -5 & 2 \end{pmatrix}$

(b) $\begin{pmatrix} 1 & 2 \\ 2 & 4 \end{pmatrix} \begin{pmatrix} 2 & -4 \\ -1 & 2 \end{pmatrix}$

(c) $\begin{pmatrix} \cos\theta & -\sin\theta \\ \sin\theta & \cos\theta \end{pmatrix} \begin{pmatrix} \cos\theta & \sin\theta \\ -\sin\theta & \cos\theta \end{pmatrix}$

(d) $\begin{pmatrix} 1 & 1 \\ 0 & 1 \end{pmatrix} \begin{pmatrix} a & b \\ c & d \end{pmatrix}$

(e) $\begin{pmatrix} a & b \\ c & d \end{pmatrix} \begin{pmatrix} 1 & 1 \\ 0 & 1 \end{pmatrix}$

(f) $\begin{pmatrix} a & b \\ c & d \end{pmatrix} \begin{pmatrix} d & -b \\ -c & a \end{pmatrix}$

(g) $\begin{pmatrix} 1 & 2 \\ 3 & 4 \end{pmatrix} \begin{pmatrix} a & b \\ c & d \end{pmatrix}$

(h) $\begin{pmatrix} a & b \\ c & d \end{pmatrix} \begin{pmatrix} 1 & 2 \\ 3 & 4 \end{pmatrix}$

Optional puzzle; when are the answers to (g) and (h) equal?

5 Work out the squares of the following matrices.

(a) $\begin{pmatrix} 0 & 1 \\ 1 & 0 \end{pmatrix}$

(b) $\begin{pmatrix} 2 & 0 \\ 0 & 3 \end{pmatrix}$

(c) $\begin{pmatrix} 0 & -1 \\ 1 & 0 \end{pmatrix}$

(d) $\begin{pmatrix} 1 & 0 \\ 0 & -1 \end{pmatrix}$

(e) $\begin{pmatrix} 1 & -2 \\ 0 & -1 \end{pmatrix}$

(f) $\begin{pmatrix} 1 & b \\ c & -1 \end{pmatrix}$

(g) $\begin{pmatrix} 1 & 1 \\ -1 & -1 \end{pmatrix}$

(h) $\begin{pmatrix} 0 & 1 \\ 0 & 0 \end{pmatrix}$

(i) $\begin{pmatrix} 0 & 2 \\ 0 & 0 \end{pmatrix}$

6 The transformation I leaves every point where it was; the transformation 0 maps every point to the origin. Write the matrices for I and 0.

7 Write the matrices for M_1, M_2 and J, the transformations involved in the exercises at the end of §20. Use matrix multiplication to find the products in (a) to (i) of question 1, and all parts of question 2 in those exercises.

8 *An investigation.* In elementary algebra, the equation $x^2 = 1$ has only 2 solutions, and $x^2 = 0$ only 1 solution. The calculations in question 5 above suggest that the matrix equations $M^2 = I$ and $M^2 = 0$ may behave very differently. Find all 2 × 2 matrices that satisfy $M^2 = I$; try to interpret your results geometrically. Do the same for $M^2 = 0$.

Worked example

Find the matrix that represents a rotation of 45° about the origin, O. It is understood that the co-ordinates refer to conventional squared graph paper.

Solution The columns of the required matrix represent the points to which $(1, 0)$ and $(0, 1)$ are mapped by this rotation. A rotation of 45° sends $(1, 0)$ to $(1/\sqrt{2}, 1/\sqrt{2})$ and $(0, 1)$ to $(-1/\sqrt{2}, 1/\sqrt{2})$. We write this information, using column vectors:

$$\begin{pmatrix} 1 \\ 0 \end{pmatrix} \rightarrow \begin{pmatrix} 1/\sqrt{2} \\ 1/\sqrt{2} \end{pmatrix} \qquad \begin{pmatrix} 0 \\ 1 \end{pmatrix} \rightarrow \begin{pmatrix} -1/\sqrt{2} \\ 1/\sqrt{2} \end{pmatrix}.$$

The required matrix is now obtained by putting the two output columns side by side:

$$\begin{pmatrix} 1/\sqrt{2} & -1/\sqrt{2} \\ 1/\sqrt{2} & 1/\sqrt{2} \end{pmatrix}.$$

9 By the method of the worked example, or otherwise, find matrices that represent the following transformations of the plane.

(a) reflection in the horizontal axis
(b) reflection in the vertical axis
(c) reflection in the line $y = x$
(d) rotation of 180° about O
(e) rotation of 90° about O
(f) rotation of $-90°$ about O
(g) rotation of $-45°$ about O
(h) rotation of 30° about O
(i) rotation through any given angle α about O
(j) reflection in the line $y = x \tan \alpha$

Addition and subtraction of matrices

In §20 and §21 we saw that the sum and difference of two transformations, say H and N, were defined in such a way that for any vector v we would have the equations,

$$(H + N)v = Hv + Nv \quad \text{and} \quad (H - N)v = Hv - Nv.$$

Suppose we have the matrix specification of H and N, namely

$$H = \begin{pmatrix} a & b \\ c & d \end{pmatrix}, \qquad M = \begin{pmatrix} p & q \\ r & s \end{pmatrix}.$$

The statements in the first paragraph show us the procedure by which we can find the matrix for $H + N$. Here again, it is good if the student can derive this result for himself, by following the strategy now to be indicated.

First, we suppose the vector v to have arbitrary co-ordinates x and y. The student may work with the traditional co-ordinate form (x, y) or with the column vector form $\begin{pmatrix} x \\ y \end{pmatrix}$. If he can work with the latter, he will be forming a useful habit; if, however, he finds this unfamiliar format inhibits his thinking, he may be wise to use the more familiar arrangement.

In order to find $H + N$, we have to answer the following questions:

(1) What is the vector Hv in terms of x and y?

(2) What is Nv?

(3) What is $Hv + Nv$, found by adding our answers to (1) and (2)?

By definition, our answer to (3) gives $(H + N)v$. This answer involves x and y. To write the matrix for $H + N$ we have to pick out the coefficients in this answer, and write them in appropriate positions, as explained earlier in this section, when matrix notation was first introduced.

No ingenuity is required to carry this programme through. The work is left to the student; it is essential that the two exercises below should be worked, as the results to which they lead are essential for all later work.

Exercises

1 Carry out the instructions above for finding the matrix $H + N$.

2 By a similar procedure, find the matrix $H - N$.

Multiplying a matrix by a number

In §20, we defined kT, the product of a number k and a transformation T, by the equation $(kT)v = k(Tv)$. This told us what the transformation kT did to any vector v.

We now wish to express this definition in matrix notation.

Let us consider the transformation H, specified by the matrix $\begin{pmatrix} a & b \\ c & d \end{pmatrix}$. What matrix will specify kH ?

By definition, $(kH)v = k(Hv)$. We will suppose v to be $\begin{pmatrix} x \\ y \end{pmatrix}$, and work out the expression $k(Hv)$ on the right-hand side of the equation:

$$Hv = \begin{pmatrix} a & b \\ c & d \end{pmatrix} \begin{pmatrix} x \\ y \end{pmatrix} = \begin{pmatrix} ax + by \\ cx + dy \end{pmatrix}.$$

The last expression represents a vector in the plane. It has two numbers in it, $ax + by$ and $cx + dy$, which are the two co-ordinates of the vector Hv.

To obtain $k(Hv)$ we must multiply the vector Hv by the number k. This means that each co-ordinate must be multiplied by k. Accordingly

$$k(Hv) = \begin{pmatrix} kax + kby \\ kcx + kdy \end{pmatrix}.$$

This last vector gives $(kH)v$, the output of the transformation kH when the input is v. Let x^* and y^* be the co-ordinates of this output vector. Then the effect of the transformation kH is shown by the equations,

$$x^* = kax + kby,$$
$$y^* = kcx + kdy.$$

To specify the transformation kH in vector form we need only to pick out the coefficients in these equations. We find

$$kH = \begin{pmatrix} ka & kb \\ kc & kd \end{pmatrix}.$$

Thus the rule for passing from the matrix H to the matrix kH is very simple – multiply each entry by k.

Summary

We now have routines by which matrices may be added, subtracted and multiplied. Division by a matrix is not always possible; this question will be discussed later.

An electronic computer is capable of dealing with a calculation that has been reduced to routine. In this section we have written only 2×2 matrices – matrices with 2 rows and 2 columns. For these the input vector, v, is specified by 2 numbers, and the output vector v^* also by 2 numbers. One convenience of matrix notation is that we can use the same symbolism $v^* = Mv$ when the input and the output each consist of 1000 numbers.

The routines for $n \times n$ matrices, where n is any natural number, are essentially the same as for 2×2 matrices. Thus we can instruct a computer to carry out a certain sequence of operations, and these instructions will apply to a whole class of problems. In one problem there may be 50 unknowns, in another 500. Of course, in each problem it will be necessary to feed into the computer data for that particular problem, so the computer knows how many unknowns are involved and what transformations have to be applied to these. But the general procedure, the strategy to be followed, can be laid down once and for all, and matrix notation gives a very convenient way of doing this.

In this section we have introduced matrix notation by considering transformations. Transformations map a space to itself; if the space is of n dimensions, the input and output each involve n numbers. This is in many ways the most fruitful and interesting case. However, other mappings can be of engineering importance. We have met examples in which the input consisted of 3 forces and the output of 2 reactions, or the input of 2 voltages and the output 3 current measurements. Such mappings lead to rectangular rather than square matrices, and we do not then have the same freedom to add, subtract and multiply as in this section. However, when such operations are possible, the formalisms are very similar to those just considered, and we shall come to them in due course.

Exercises on addition, subtraction and multiplication of matrices

1 Find $\begin{pmatrix} 1 & 2 \\ 3 & 4 \end{pmatrix} + \begin{pmatrix} 80 & 60 \\ 70 & 50 \end{pmatrix}$.

2 Find $\begin{pmatrix} 10 & 20 \\ 30 & 40 \end{pmatrix} - \begin{pmatrix} 5 & 6 \\ 7 & 8 \end{pmatrix}$.

Questions 1 and 2 serve simply to check that the student has understood the procedure for addition and subtraction of matrices. If further routine examples of this kind are needed, they can be produced simply by varying the numbers involved.

For questions 3 to 11, the following symbols are used:

$$A = \begin{pmatrix} 1 & 1 \\ 0 & 1 \end{pmatrix}, \qquad B = \begin{pmatrix} 1 & 0 \\ 1 & 1 \end{pmatrix}, \qquad C = \begin{pmatrix} 0 & 1 \\ 1 & 0 \end{pmatrix}, \qquad I = \begin{pmatrix} 1 & 0 \\ 0 & 1 \end{pmatrix}.$$

3 Find $A + B - C$.

4 Find $A^2 - 2A + I$ and $(A - I)^2$. Are they equal?

5 Find $(A + B)(A - B)$; $(A - B)(A + B)$; $A^2 - B^2$. Are any two of these equal?

6 Find $B^2 - 2B + I$ and $(B - I)^2$. Are they equal?

7 If, for some matrix M we find $M^2 = 0$, does it follow that $M = 0$? Justify your answer.

8 Are $(A + B)(I + C)$ and $(I + C)(A + B)$ equal?

9 Find $(I + C)(I - C)$.

10 Find $AC - B$ and C^2.

11 Elementary algebra suggests $(AC - B)C = AC^2 - BC$. With the help of your answers to question 10 check whether this equation is correct or not.

12 Let $D = \begin{pmatrix} 2 & 1 \\ 3 & 2 \end{pmatrix}$. Write $2D$, $3D$, $4D$, $5D$.

13 Is there any number k for which $D^2 + I = kD$?

14 Let $E = \begin{pmatrix} 1 & 2 \\ 3 & 4 \end{pmatrix}$. Calculate $E^2 - 5E - 2I$.

15 Let $F = \begin{pmatrix} 0 & 1 \\ 2 & 3 \end{pmatrix}$. Is it possible to find numbers p, q, for which $F^2 - pF + qI = 0$?

16 We have seen that some 2×2 matrices satisfy quadratic equations. Do they all do so? Given the matrix $M = \begin{pmatrix} a & b \\ c & d \end{pmatrix}$, can we be sure there will exist numbers p and q such that

$$M^2 - pM + qI = 0?$$

23 An application of matrix multiplication

The remarks at the end of §22 about the engineering applications of matrix algebra can be illustrated by an electrical example that does not call for any specialized knowledge.

We consider boxes, each having 4 terminals. On the input side, we suppose a current i flows in at one terminal and out at the other. Similarly on the output side a current I flows as shown in Fig. 63. Let v be the voltage difference between the input terminals,

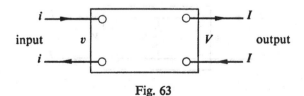

Fig. 63

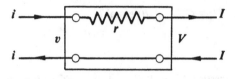

Fig. 64

V between the output terminals. The vector $w = \begin{pmatrix} v \\ i \end{pmatrix}$ measures the input, while $W = \begin{pmatrix} V \\ I \end{pmatrix}$ measures the output. In the cases we shall consider $w \to W$ will be a linear transformation.

We now consider certain special boxes, which ultimately will be put together to form more complicated circuits. Box (1) is very simple. It contains a resistance of r ohms, as shown in Fig. 64.

Clearly $I = i$ since the current has no chance to do anything but flow straight through. The current i flowing through resistance r produces a potential drop ri, so $V = v - ri$. Thus

$$V = v - ri,$$
$$I = i.$$

Picking out the coefficients of v and i, we see that the output of this box is related to the input by the equation $W = Aw$, where

$$A = \begin{pmatrix} 1 & -r \\ 0 & 1 \end{pmatrix}.$$

Box (2) contains a resistance R connected as shown in Fig. 65. Since the input and output terminals are directly connected, there can be no change of voltage. We have $V = v$. However, by Ohm's Law, a current v/R flows down the resistance, and there will be this much less current to pass to the output. So $I = i - (v/R)$.

Accordingly, being careful to write the equations and symbols in the correct order, we obtain the equations

$$V = v,$$
$$I = -(1/R)v + i.$$

So, for this box, $W = Bw$ where

$$B = \begin{pmatrix} 1 & 0 \\ -1/R & 1 \end{pmatrix}.$$

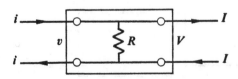

Fig. 65

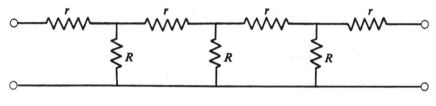

Fig. 66

Now consider two stations connected by a cable which is imperfectly insulated. After all, no insulation is perfect. We might approximate the behaviour of such a cable by a circuit such as that shown in Fig. 66.

The current flows along the cable through the resistances r, but leaks away to earth through the resistances R. Now this circuit can be constructed by connecting a number of boxes (1) and (2), as in Fig. 67.

Let w_1 represent the input to the first box; w_2, the output of that box, is also the input of the second box; w_3 the output of the second box is the input of the third box, and so we continue until we reach the final output w_8. We have $w_2 = Aw_1$, $w_3 = Bw_2$, $w_4 = Aw_3$, $w_5 = Bw_4$, $w_6 = Aw_5$, $w_7 = Bw_6$, $w_8 = Aw_7$.

It follows that $w_8 = ABABABAw_1$. Thus the effect of such a leaky cable can be found by a routine process of matrix multiplication. It is clear that with the help of an electronic computer we could find the effect of such a circuit with thousands of boxes so connected.

Alternating current circuits involving resistance, capacity and inductance can be dealt with in a somewhat similar way, by using matrices in which complex numbers occur.

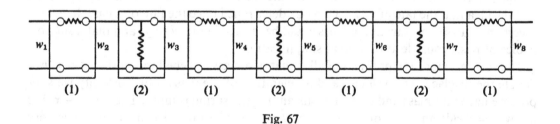

Fig. 67

Exercises

1 Calculate the matrix $ABABABA$ for the case $r = 1$, $R = 1$.

2 The circuit shown is symmetrical, so that interchanging input and output should not make any difference. Check that this is so for the matrix you found in question 1. (*Note.* The condition that a matrix represents a symmetrical circuit needs to be formulated with some care, as there is a small point that is easily overlooked. It may help to consider a simpler matrix such as ABA or even A; if the circuits corresponding to these are drawn it will be seen that they too are symmetrical.)

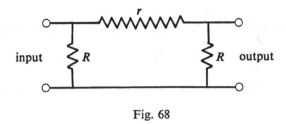

Fig. 68

3 Find the matrix corresponding to the circuit shown in Fig. 68.

4 Draw the circuits corresponding to the matrices AB and BA. Are they electrically equivalent?

24 An application of linearity

In §23, a fairly complex circuit was built up from very simple components and its behaviour was deduced by a routine procedure. The processes to be described in this section differ greatly from those used in §23, but they have the same general purpose – to cope with a complex situation by relating it to simple ones.

In this section, we will use certain vibration problems as illustrations, but it should be realized that the methods are perfectly general; they can be used in many situations that have nothing to do with vibration.

Our present considerations are built on the ideas of two earlier sections. In §10 we met the idea that the fate of any vector was determined once the fate of a basis had been ascertained. In §12 we saw that a basis could be chosen in many ways and that a suitable choice of basis could lead to a great simplification.

The simplest vibration problem of all is that of a mass attached to a spring. Let x, v, a denote the displacement, velocity and acceleration of the mass. For simplicity we suppose we have unit mass and that the constant for the spring is unity. Then $a = -x$. We know $v = \mathrm{d}x/\mathrm{d}t$ and $a = \mathrm{d}v/\mathrm{d}t$, since $\mathrm{d}/\mathrm{d}t$ means 'the rate of change of'. It is very convenient to represent differentiation by a dot, so $\mathrm{d}x/\mathrm{d}t$ is written $\dot{x}$ and $\mathrm{d}^2x/\mathrm{d}t^2$ becomes $\ddot{x}$. We then may write $v = \dot{x}$, $a = \dot{v} = \ddot{x}$ and $a = -x$ takes the form $\ddot{x} = -x$.

If we consider the graph drawn by a vibrating body on a surface that moves steadily past it, we are immediately reminded of the graphs of sine or cosine. If we try $x = \sin t$, on differentiating twice we find $\dot{x} = \cos t$ and $\ddot{x} = -\sin t$, which means $\ddot{x} = -x$ for every time t. Thus $x = \sin t$ is a possible motion for our mass on a spring. Similarly, we can check that $x = \cos t$ is a possible motion.

Now if we have a mass on the end of the spring, we can displace it any small distance we like and send it off with a speed chosen by ourselves. Once that is done, the process develops automatically. At every moment, the amount the spring is stretched determines the acceleration of the mass. So the initial position and initial velocity determine the subsequent motion.

The initial position and initial velocity are specified by a pair of numbers. This is our input. The formula for the subsequent motion contains a function of time, t. This is our output.

For the motion $x = \sin t$ we have, when $t = 0$, $x = 0$ and $v = 1$. Our input is $(0, 1)$. We may write $(0, 1) \to \sin t$.

For the motion $x = \cos t$, when $t = 0$ we have $x = 1$, $v = 0$. Our input here is $(1, 0)$ and so $(1, 0) \to \cos t$.

It can be shown that the mapping, input $\to$ output, is linear. As $(1, 0)$ and $(0, 1)$ form a basis for the input, it follows from the doctrine of §10 that the output for any input is now known.

By proportionality, since $(1, 0) \to \cos t$, multiplying by a constant A, we find $(A, 0) \to A \cos t$. (The A here is simply a number. It has nothing to do with the matrix A in §23.)

Similarly, since $(0, 1) \to \sin t$, multiplying by a constant B, we find

$$(0, B) \to B \sin t.$$

Superposing these two results, we find that $(A, B) \to A \cos t + B \sin t$, so now we know that $x = A \cos t + B \sin t$ if the mass starts from position A with velocity B. This covers every possible motion.

In a calculus course, $x = A \cos t + B \sin t$ is shown to be the most general solution of the differential equation $\ddot{x} + x = 0$. The language may differ somewhat from that used here, but the essential content is the same.

We will now consider an arrangement, as shown in Fig. 69, with two unit masses and three springs. The central spring is stronger than the others; it has constant 4 while they each have constant 1. The ends are fixed and the arrangement is symmetrical. If we imagine the masses displaced distances x and y, as shown, with $y > x$, the central spring will be in tension and exerting forces of magnitude $4(y - x)$ at each end. The left spring will exert a tension of x, and the right-hand spring will be compressed and exerting a force y. The forces on the masses will thus be $-x + 4(y - x)$ and $-4(y - x) - y$ respectively. Thus the equations of motion (found from force = mass × acceleration) will be $\ddot{x} = -5x + 4y$ and $\ddot{y} = 4x - 5y$.

Now we can start the system moving by giving an arbitrary displacement and velocity to each of the particles, so we have 4 constants at our disposal. We do not want to work

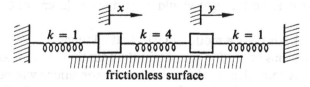

frictionless surface

Fig. 69

in 4 dimensions, so let us agree that both masses start from the equilibrium position. From that position we give each mass a bang to start it moving. This reduces the number of constants to 2; we can choose an initial velocity for each mass. Thus by an input of, say, (3, 8), we understand that we have given the left-hand mass an initial velocity of 3 units, and the right-hand mass an initial velocity of 8 units, both in the positive direction.

Our first example might suggest that we try to see the motions that result from inputs (1, 0) and (0, 1). But it is not easy to see what happens in these cases. Input (1, 0) for instance means that we leave the right-hand mass alone, but start the left-hand mass moving towards the right with velocity 1. The central spring will become compressed; it will bring the right-hand mass into motion, and it is not simple to predict how the later movements will develop.

We are dealing with a symmetrical system. This suggests that, instead of considering unsymmetrical inputs like (1, 0) and (0, 1), we might do better to put the masses on an equal footing and consider the input (1, 1). Input (1, 1) means that both masses begin moving to the right with velocity 1. Since they have the same velocity, at the beginning of the motion the distance between them is not changing, and the middle spring is being neither stretched nor compressed. This may suggest to us the question – can this state of affairs continue? Can the masses continue to move exactly in step, so the middle spring exerts no force? It is fairly easy to see that this can happen. Imagine the middle spring removed. The two end springs have the same stiffness constants, so the masses can perform identical oscillations. If they are sent off, each with unit velocity, we shall have after time t the equations $x = \sin t$, $y = \sin t$, as we saw earlier. If, now we replace the central spring, it will make no difference, since $y - x = 0$ at all times. This spring will exert no force. (We are assuming the masses of the springs to be negligible, so replacing the middle spring does not affect the inertia of the system.)

Accordingly we have identified a possible motion:

$$\text{input } (1, 1) \rightarrow \text{output } (\sin t, \sin t). \tag{1}$$

The two entries in the output are the formula for x and the formula for y, at time t.

A rather more formal way of seeing that this motion is possible is the following. We pose the question, is it possible to have a motion in which at all times $x = y$? If we substitute $x = y$ in the equations of motion $\ddot{x} = -5x + 4y$, $\ddot{y} = 4x - 5y$ *both* equations lead to $\ddot{x} = -x$. Thus $x = y = \sin t$ satisfies all requirements.

There is another symmetrical motion. Our system has mirror symmetry about its centre. If we started the masses off with unit velocity towards the centre, we would expect this kind of symmetry to continue. The system would appear balanced; the displacement of one mass to the right would equal the displacement of the other to the left.

This leads us to ask – is there a motion in which at all times $y = -x$? If we substitute $-x$ for y in the equations of motion we obtain the equations $\ddot{x} = -9x$ and $-\ddot{x} = 9x$. The second of these is equivalent to the first. Thus all conditions will be met if $y = -x$ and $\ddot{x} = -9x$. Now $\ddot{x} = -9x$ has the solution $x = \sin 3t$. Differentiating gives

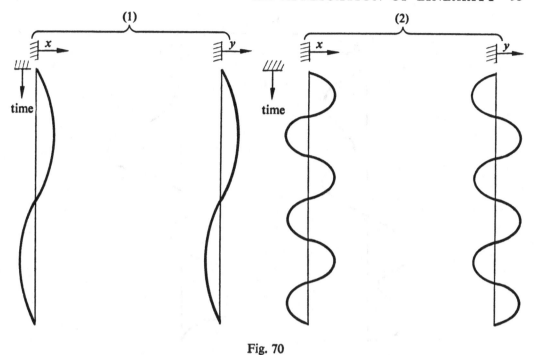

Fig. 70

$\dot{x} = 3 \cos 3t$, so when $t = 0$, $\dot{x}$, the velocity of the first mass, has the value 3. Since $y = -x$, we know $y = -\sin 3t$ at all times, and the initial velocity of the second mass is -3.

Thus

$$(3, -3) \rightarrow (\sin 3t, -\sin 3t). \tag{2}$$

If the vibrating masses were made to record their positions on a moving strip, the results would be as in Fig. 70, in which (1) shows the motion $(\sin t, \sin t)$ and (2) shows $(\sin 3t, -\sin 3t)$, in which the vibration is 3 times as rapid as in (1). Both motions are easy to imagine.

It can be proved that for this system the mapping input $\rightarrow$ output is linear. Accordingly we may use the principles of superposition and proportionality. Since

$$(1, 1) \rightarrow (\sin t, \sin t),$$

for any constant number A, $(A, A) \rightarrow (A \sin t, A \sin t)$. Similarly, for any constant number B, $(3B, -3B) \rightarrow (B \sin 3t, -B \sin 3t)$. Superposing we see

$$(A + 3B, A - 3B) \rightarrow (A \sin t + B \sin 3t, A \sin t - B \sin 3t).$$

Thus, for any numbers A and B, a possible vibration is given by the equations

$$x = A \sin t + B \sin 3t, \, y = A \sin t - B \sin 3t.$$

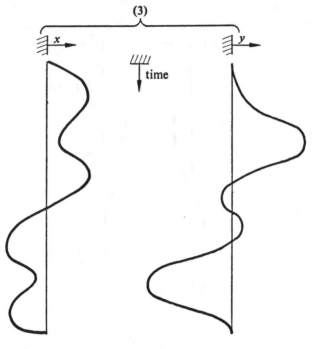

Fig. 71

For example, we may take $A = 2$, $B = 1$. This gives

$$x = 2 \sin t + \sin 3t, \; y = 2 \sin t - \sin 3t. \tag{3}$$

The corresponding motion is indicated by the graphs in Fig. 71. This motion is anything but simple to imagine. The two simple vibrations with different periods have combined to produce an unsymmetrical and rather irregular motion, which would still, however, be periodic. The graphs above would repeat indefinitely, at any rate in the idealized situation where frictional damping is ignored.

If the time scale were suitably chosen, these vibrations would produce audible

sounds. If vibration 1 gave ♮♪ vibration 2 would give ♮♪

A person with a good musical ear, when exposed to vibration 3, would hear both the notes, bass C and treble G. It is interesting that our ears (or brains) spontaneously analyse the complex vibration 3 into the two simple vibrations 1 and 2, of which it is composed.

The simple vibrations, in which only one frequency or musical note occurs, are known as the *normal modes* of vibration. The simple system of masses and springs that we have considered is not exceptional. In a subject known as general dynamics a theory of vibrations is developed which shows that similar features are to be expected in the vibrations of any structure.

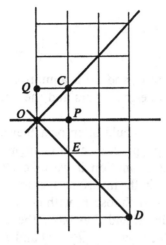

Fig. 72

Inputs, such as (1, 0), (0, 1), (1, 1), (3, −3), involve only 2 numbers each and can be plotted on graph paper. In Fig. 72 these four cases are represented by the points P, Q, C and D. As we saw, the basis A, B was not helpful in our problem, but the vectors C and D, which also form a basis, gave us a simple treatment. In this particular problem we were led to consider C and D by considerations of symmetry. Now symmetry is an important idea in mathematical work but we cannot expect always to find it. For instance, in our vibrating system if the masses had differed in magnitude by even 1 per cent, symmetry would have been destroyed. Yet surely physically this would have made very little difference to the system; surely it would still produce two notes, and by setting it suitably in motion we could make it produce just one of these notes. Can we then find some way of showing the significance of the vectors C and D without appealing to symmetry? If we look at our equations of motion

$$\ddot{x} = -5x + 4y$$
$$\ddot{y} = 4x - 5y$$

we see that they embody the matrix

$$\begin{pmatrix} -5 & 4 \\ 4 & -5 \end{pmatrix} = M, \quad \text{say.}$$

Are the vectors C and D specially related to this matrix? Let us see what it does to them. If we let this transformation act on C we get

$$\begin{pmatrix} -5 & 4 \\ 4 & -5 \end{pmatrix} \begin{pmatrix} 1 \\ 1 \end{pmatrix} = \begin{pmatrix} -1 \\ -1 \end{pmatrix}.$$

When it acts on D we get

$$\begin{pmatrix} -5 & 4 \\ 4 & -5 \end{pmatrix} \begin{pmatrix} 3 \\ -3 \end{pmatrix} = \begin{pmatrix} -27 \\ 27 \end{pmatrix}.$$

We notice that C^* is a multiple of C, and D^* is a multiple of D. In fact $MC = -C$ and $MD = -9D$. Thus C and D are eigenvectors and the lines OC and OD are invariant lines.

In §12 we saw that simplification could often be achieved by choosing invariant lines as axes, and in §9 we had a simple procedure for changing axes. Accordingly let us see what happens to our equations of motion if we take OC and OD as axes. As base vectors we can choose any points in the invariant lines (one point in each line, of course). In the line OD it will be simpler to choose E, with co-ordinates $(1, -1)$ rather than D with co-ordinates $(3, -3)$. Taking C and E as basis, the point with co-ordinates (X, Y) is given by the vector $XC + YE$. Now $C = P + Q$ and $E = P - Q$ so

$$XC + YE = X(P + Q) + Y(P - Q) = (X + Y)P + (X - Y)Q.$$

We are thus led to put $x = X + Y, y = X - Y$. Substituting these values in the equations of motion we find $\ddot{X} + \ddot{Y} = -X - 9Y$ and $\ddot{X} - \ddot{Y} = -X + 9Y$. Solving for $\ddot{X}$ and $\ddot{Y}$ we obtain $\ddot{X} = -X$, $\ddot{Y} = -9Y$.

At the beginning we had equations such as $\ddot{x} = -5x + 4y$ in which the x and y were mixed up together, and a student might well not know what to do. By going over to the invariant lines as axes, we obtain the equation $\ddot{X} = -X$ in which X alone appears, and which we recognize as being just like the equation for a single mass on the end of a spring. Similarly, we obtain the equation $\ddot{Y} = -9Y$, another simple vibration equation, involving Y alone.

It is thus suggested that, if *we choose the invariant lines as axes, the problem will split into simpler problems.*

We can see why this should be so. If (x, y) are the co-ordinates of a moving point, then $(\ddot{x}, \ddot{y})$ are the co-ordinates of the vector that represents its acceleration. The point (x, y) is of course only a graphical device to show the displacements of the masses on the spring. As the system vibrates, the point (x, y) will move around in the plane. The vector $(\ddot{x}, \ddot{y})$ shows how this point is accelerated at any time.

Fig. 73 shows the direction of the acceleration for various positions of the point (x, y). It will be seen that if the point (x, y) starts on one of the invariant lines, its acceleration will be such as to keep it in that line. If, however, the point starts not on either of these lines, the acceleration will cause it to move on some curve, the nature of which we cannot readily predict. Students may be familiar with the Lissajous curves that can be produced by combining simple oscillations. In general the path followed by the moving point (x, y) will be such a curve.

What was said earlier in this section may now be repeated. The method used here is not restricted to vibration problems; it would produce a simplification in any problem where the matrix $\begin{pmatrix} -5 & 4 \\ 4 & -5 \end{pmatrix}$, that we have been considering, occurred. And of course it would be equally useful for many other matrices.

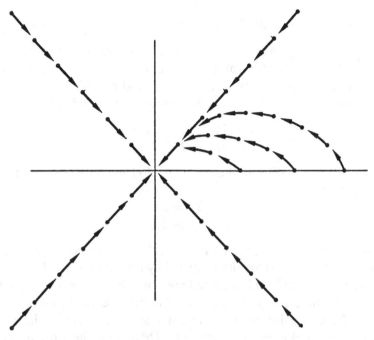

Fig. 73 In this diagram the dot represents a position (x, y). The arrow gives the direction of the acceleration $(\ddot{x}, \ddot{y})$ corresponding to that position.

Having seen that problems are simplified when we change our axes to the invariant lines, the question naturally arises – how do we find the invariant lines? This question will be considered in the next section.

25 Procedure for finding invariant lines, eigenvectors and eigenvalues

The procedure for finding invariant lines, eigenvectors and eigenvalues falls into two stages. In Stage 1, the eigenvalues are found from a certain equation. In Stage 2 the eigenvectors belonging to each eigenvalue are found; this immediately gives the invariant lines.

For understanding the procedure, it is best to work backwards. We will suppose that, somehow, Stage 1 has been completed. Someone, say, has told us what the eigenvalues are. How would we then find the eigenvectors?

Suppose, for instance, we are dealing with the matrix $\begin{pmatrix} 1 & 2 \\ -1 & 4 \end{pmatrix}$, and we have received the information that the eigenvalues are 2 and 3. This means that there is one vector which gets doubled when the transformation acts on it; there is another vector that gets tripled. Our task is to find these vectors. It is not at all difficult.

Which vector gets doubled? Suppose its co-ordinates are x and y. It gets doubled if

$$\begin{pmatrix} 1 & 2 \\ -1 & 4 \end{pmatrix} \begin{pmatrix} x \\ y \end{pmatrix} = 2 \begin{pmatrix} x \\ y \end{pmatrix}. \tag{1}$$

We interpret this matrix equation in the language of elementary algebra. This gives the simultaneous equations

$$x + 2y = 2x,$$
$$-x + 4y = 2y.$$

If we simplify these equations, we find that each of them boils down to the same equation $2y = x$. Students are sometimes worried by this; they ask, 'How can I find two unknowns from one equation?' In fact something would be seriously wrong if they could find a definite solution for x and y. We saw in §12 that, if v is an eigenvector, so is kv for any number $k \neq 0$. Any point on an invariant line (except the origin, which is not acceptable for this purpose) gives an eigenvector. There must therefore be an infinity of solutions; the equations cannot possibly fix x and y.

In our example, the algebra leads us to the single equation $2y = x$. This is the only condition the vector has to meet. It does not matter what y is (barring always $y = 0$) provided x is twice as large. You can check that equation (1) is satisfied if $x = 2$, $y = 1$ or if $x = 100$, $y = 50$ or if $x = -0.3$, $y = -0.15$. We can if we like give the general solution, $\begin{pmatrix} 2t \\ t \end{pmatrix}$ where t can be any number except 0, or we may pick out a particular solution $\begin{pmatrix} 2 \\ 1 \end{pmatrix}$ on the understanding that any non-zero multiple of this will do equally well (see the paragraph in §12 immediately before the questions at the end of that section).

As was mentioned in §12, the symbol λ is widely used for eigenvalues. Thus our result may be recorded as $\begin{pmatrix} 2 \\ 1 \end{pmatrix}$, $\lambda = 2$. This means that the vector $\begin{pmatrix} 2 \\ 1 \end{pmatrix}$ gets multiplied by 2 when the transformation acts on it.

We were also told that 3 was an eigenvalue. So, searching for a vector that gets tripled, we write the equation

$$\begin{pmatrix} 1 & 2 \\ -1 & 4 \end{pmatrix} \begin{pmatrix} x \\ y \end{pmatrix} = 3 \begin{pmatrix} x \\ y \end{pmatrix}. \tag{2}$$

This gives us $x + 2y = 3x$ and $-x + 4y = 3y$, and these are equivalent to the single condition $x = y$. Thus any vector $\begin{pmatrix} t \\ t \end{pmatrix}$ is tripled by the transformation, and has $\lambda = 3$.

The invariant lines are $2y = x$ for $\lambda = 2$, and $y = x$ for $\lambda = 3$.

If we introduce new co-ordinates (X, Y) with the invariant lines as axes the transformation will take the form $(X, Y) \rightarrow (X^*, Y^*)$ with $X^* = 2X$, $Y^* = 3Y$, since the vectors in the first line are doubled and in the second line are tripled. The matrix expressing the transformation relative to the new axes is thus (by reading off the coefficients in the two equations) $\begin{pmatrix} 2 & 0 \\ 0 & 3 \end{pmatrix}$.

Note that we are still dealing with the same transformation. What happens to each point is exactly the same throughout. But this transformation is expressed by a different matrix when we introduce a new system of axes.

The matrix $\begin{pmatrix} 2 & 0 \\ 0 & 3 \end{pmatrix}$ is said to be in *diagonal form* since the entries not on the main diagonal are zero. Note that the entries in the main diagonal are the eigenvalues. This always happens. The order in which the eigenvalues occur is arbitrary. If we had taken $y = x$ for our first axis and $2y = x$ for our second axis, the transformation would have been specified by the matrix $\begin{pmatrix} 3 & 0 \\ 0 & 2 \end{pmatrix}$.

Exercises

In the following questions, students should check that they have found the correct eigenvectors, by applying the transformation to the vectors they have found and seeing if in fact these get multiplied by the numbers specified.

1 Find the eigenvectors, invariant lines, and matrix resulting when the invariant lines are taken as axes, for each of the following transformations.

(a) $\begin{pmatrix} 2 & -3 \\ 1 & -2 \end{pmatrix}$ which has eigenvalues 1 and -1

(b) $\begin{pmatrix} -4 & 2 \\ -15 & 7 \end{pmatrix}$ with eigenvalues 1 and 2

(c) $\begin{pmatrix} 3 & -1 \\ 0 & 2 \end{pmatrix}$ with eigenvalues 2 and 3

(d) $\begin{pmatrix} 2 & 1 \\ 1 & 2 \end{pmatrix}$ with eigenvalues 1 and 3

(e) $\begin{pmatrix} 2 & 2 \\ 2 & -1 \end{pmatrix}$ with eigenvalues 3 and -2

(f) $\begin{pmatrix} 1 & 1 \\ 1 & 1 \end{pmatrix}$ with eigenvalues 2 and 0

(g) $\begin{pmatrix} 1 & -1 \\ 0 & 0 \end{pmatrix}$ with eigenvalues 0 and 1

2 The matrix $\begin{pmatrix} a & b \\ c & d \end{pmatrix}$ is called *symmetric* when $b = c$, as in questions (d), (e), (f) above. What do you notice about the invariant lines in these questions?

Finding eigenvalues

In practice, of course, we are rarely in the position of being told the eigenvalues of a matrix. Before we can do the type of work in the exercises just given, we have to find out for ourselves what the eigenvalues are.

Equations (1) and (2) at the beginning of this section were formed when we looked for vectors that were multiplied by 2 and by 3 when the transformation $\begin{pmatrix} 1 & 2 \\ -1 & 4 \end{pmatrix}$ acted. Equation (1) was equivalent to the pair of equations $x + 2y = 2x$, $-x + 4y = 2y$ and these turned out to be equivalent to the single equation $2y = x$. It followed that $x = 2$, $y = 1$ would do as a solution.

What would have happened if our informant had been mistaken and had asked us to search for a vector that got multiplied by 5 when the transformation was applied? We would have written

$$\left. \begin{array}{l} x + 2y = 5x, \\ -x + 4y = 5x. \end{array} \right\} \tag{3}$$

The first equation simplifies to $y = 2x$, the second to $2y = 3x$. The only solution is $x = 0$, $y = 0$. Now under *any* linear transformation, the origin goes to the origin. Knowing this fact does not in any way help us to find new axes in which the transformation will appear simpler. Thus looking for a vector that gets enlarged 5 times does not help us.

If we are looking for a vector that gets multiplied by any number λ, we write the equations

$$\begin{array}{l} x + 2y = \lambda x, \\ -x + 4y = \lambda y, \end{array}$$

which may be put in the form

$$\left. \begin{array}{l} (1 - \lambda)x + 2y = 0, \\ -x + (4 - \lambda)y = 0. \end{array} \right\} \tag{4}$$

Both equations represent straight lines through the origin. If we choose an unsuitable value for λ, such as $\lambda = 5$, we get two different lines through the origin as happened for equations (3). If we take a suitable value of λ, such as $\lambda = 3$ used in equations (2), we find we get the same line twice. In fact $\lambda = 3$ gives $-2x + 2y = 0$ and $-x + y = 0$. These equations do not appear identical, but their graphs coincide. For $\lambda = 5$ there is only one point that lies on both lines, and so satisfies both equations (3). It is the origin which, as we have seen, is no use to us. For $\lambda = 3$, any point of the line $y = x$ satisfies both equations, and thus qualifies as an eigenvector.

Thus to find a suitable value of λ, we have to arrange for the two lines in equations (4) to coincide.

Let us look at the general question of when two lines through the origin coincide. Suppose we have two lines with equations $px + qy = 0$, $rx + sy = 0$. What is the condition for them to coincide? Students often produce the following solution. The first line has slope $-p/q$, the second $-r/s$. For the lines to coincide these slopes must be equal. After clearing fractions the condition $ps - qr = 0$ is found.

Now this result is in fact perfectly correct. The lines will coincide if $ps - qr = 0$, and will not coincide if $ps - qr \neq 0$. But the manner of proof is open to some objections. We have divided by q and by s; what if q or s or both should happen to be zero?

We can tidy up the argument so as to avoid any question of dividing by zero. We use a procedure that is sometimes used to solve equations. We can get rid of y by taking s times the first equation minus q times the second. This gives us $(ps - qr)x = 0$. Similarly, we can get an equation not involving x if we take p times the second equation minus r times the first. This gives $(ps - qr)y = 0$. Thus, if $ps - qr \neq 0$, we must have $x = 0$ and $y = 0$. The lines in this case intersect only at the origin.

We still have to show that if $ps - qr$ is zero, then the equations have a solution other than $x = 0$, $y = 0$. This is easy. The equation $px + qy = 0$ is automatically satisfied by $x = -q$, $y = p$. These values make $rx + sy$ equal to $ps - qr$, which is given to be zero. So we have a point that satisfies both equations. But there is one last snag. If $p = q = 0$, this solution will not be other than $x = 0$, $y = 0$. However, we can clear this up. If $p = q = 0$, the first equation is $0x + 0y = 0$ and *any* values of x and y will satisfy this. We have only to pick a point, other than the origin, to satisfy the second equation $rx + sy = 0$ and we have done what is required.

A special sign exists for $ps - qr$. It is called the *determinant* and is written $\begin{vmatrix} p & q \\ r & s \end{vmatrix}$

The determinant is a single number – not a vector or a matrix. Thus we have a theorem. *The equations*

$$px + qy = 0,$$
$$rx + sy = 0,$$

have a solution other than the trivial solution $x = 0$, $y = 0$ when, and only when, the determinant $\begin{vmatrix} p & q \\ r & s \end{vmatrix} = 0$.

We now return to our particular problem. If we did not already know the eigenvalues, how would we determine λ to make the equations (4) have a solution other than $x = 0$, $y = 0$? We write the determinant, replacing p, q, r, s by the coefficients that occur in equations (4).

This gives us

$$\begin{vmatrix} 1 - \lambda & 2 \\ -1 & 4 - \lambda \end{vmatrix} = 0$$

which is the same as saying

$$(1 - \lambda)(4 - \lambda) - (-1)(2) = 0.$$

This reduces to $\lambda^2 - 5\lambda + 6 = 0$ with solutions $\lambda = 2$ and $\lambda = 3$.

We have required a fairly long explanation to arrive at this procedure, but once the idea has been understood, the actual calculation is quite short. This constitutes Stage 1, mentioned at the beginning of the section.

Worked example

Find the eigenvalues of $\begin{pmatrix} 1 & 4 \\ 2 & 3 \end{pmatrix}$.

Solution If λ is an eigenvalue,

$$\begin{pmatrix} 1 & 4 \\ 2 & 3 \end{pmatrix}\begin{pmatrix} x \\ y \end{pmatrix} = \lambda \begin{pmatrix} x \\ y \end{pmatrix}$$

will have a non-trivial solution.

This leads to the equations

$$\left. \begin{array}{r} (1 - \lambda)x + ry = 0, \\ 2x + (3 - \lambda)y = 0. \end{array} \right\} \tag{5}$$

These will have a non-trivial solution if and only if

$$\begin{vmatrix} 1 - \lambda & 4 \\ 2 & 3 - \lambda \end{vmatrix} = 0,$$

that is

$$(1 - \lambda)(3 - \lambda) - (2)(4) = 0.$$

Whence $\lambda^2 - 4\lambda - 5 = 0$, with solutions $\lambda = -1$, $\lambda = 5$. The correctness of this calculation is automatically checked at Stage 2, when we come to find the eigenvectors. Substituting $\lambda = -1$ in equations (5) above, we obtain $2x + 4y = 0$, $2x + 4y = 0$, so $x = 2$, $y = -1$ for instance will do as an eigenvector. Similarly $\lambda = 5$ leads to $-4x + 4y = 0$, $2x - 2y = 0$ with solution $x = 1$, $y = 1$ for an eigenvector.

If, when we reach Stage 2, we find things going wrong – that is to say, we find $x = 0$, $y = 0$ is the only solution of our equations – this means that we must have made a mistake in Stage 1.

Exercises

Find eigenvalues and eigenvectors for the following matrices:

(a) $\begin{pmatrix} 3 & 1 \\ 1 & 3 \end{pmatrix}$ (b) $\begin{pmatrix} 4 & 1 \\ 1 & 4 \end{pmatrix}$ (c) $\begin{pmatrix} 0 & 2 \\ -2 & 5 \end{pmatrix}$

(d) $\begin{pmatrix} -2 & 2 \\ -3 & 5 \end{pmatrix}$ (e) $\begin{pmatrix} 1 & 1 \\ 2 & 2 \end{pmatrix}$ (f) $\begin{pmatrix} 2 & 1 \\ 0 & 3 \end{pmatrix}$

(g) $\begin{pmatrix} a & b \\ b & a \end{pmatrix}$ where a, b are any pair of numbers.

(h) What difficulties arise in trying to find eigenvalues and eigenvectors for $\begin{pmatrix} a & -b \\ b & a \end{pmatrix}$? Are there any values a, b for which this matrix has eigenvectors? (We are concerned only with real numbers.)

26 Determinant and inverse

We have now met two expressions that look somewhat similar but have very different meanings. They are M and Δ where

$$M = \begin{pmatrix} p & q \\ r & s \end{pmatrix} \quad \text{and} \quad \Delta = \begin{vmatrix} p & q \\ r & s \end{vmatrix}.$$

In each we see the same four numbers, p, q, r, s; in M these numbers are enclosed in curved brackets, in Δ in straight lines. It is important to observe carefully this small distinction in notation. It indicates that M is a matrix, which specifies a transformation, while Δ is a determinant, a single number. We speak of Δ as 'the determinant of the matrix M', and we may write $\Delta = \det M$. Thus Δ is a single number associated with the transformation M. It is natural to ask what is the meaning of the number Δ; what does it tell us about the transformation?

If

$$A = \begin{pmatrix} 1 \\ 0 \end{pmatrix} \quad \text{and} \quad B = \begin{pmatrix} 0 \\ 1 \end{pmatrix},$$

we know that A^* and B^* can be read off from the columns of M (see §22). We have

$$A^* = \begin{pmatrix} p \\ r \end{pmatrix}, \qquad B^* = \begin{pmatrix} q \\ s \end{pmatrix}.$$

Let us consider the area of the parallelogram $O^*A^*C^*B^*$ in Fig. 74.

It will help us if we drop perpendiculars from A^* and C^* onto the x-axis and from B^* and C^* onto the y-axis as in Fig. 75. The parallelogram is now inside a rectangle, and its area can be found by subtracting from the area of the whole rectangle the area of the shaded parts. This gives us $(p + q)(r + s) - pr - qs - 2qr$ which simplifies to $ps - qr$, the determinant Δ.

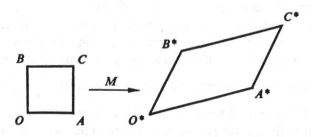

Fig. 74

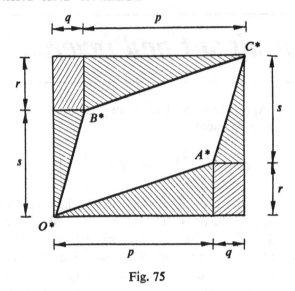

Fig. 75

The vectors A^*, B^* are the columns of the matrix M. We could indicate this by writing $M = (A^*, B^*)$, and our result could then be stated – the area of the parallelogram determined by the vectors A^* and B^* is $\det(A^*, B^*)$. Incidentally we have a result here that gives a method for finding the area of a parallelogram; it could be used by someone who did not know anything about the idea of a linear transformation.

This result, however, is very significant in connection with the transformation M. Imagine any region in the original plane that can be broken up into a number of unit squares such as $OACB$. When the transformation M acts, this region will be changed into a number of parallelograms, each congruent to $O^*A^*C^*B^*$. Thus, if the original region contained N unit squares, the transformed region would have area $N\Delta$. From this we can be led to the conclusion that, when the transformation M acts on the plane, all areas are changed in the ratio $1:\Delta$. *The determinant of M thus measures the ratio in which areas change when M acts on the plane.*

This gives us a way of generalizing the idea of determinant. Transformations in 3 dimensions are specified by a matrix with 3 rows and 3 columns. The transformation will change the unit·cube into a region with parallel plane faces, and all volumes will be changed in the same ratio. We *define* the determinant as being the number that specifies this ratio. In 4, 5 or more dimensions similar results apply: there are 'boxes' which we cannot easily imagine that correspond to squares in 2 dimensions or cubes in 3 dimensions. We can specify the 'volume' of a region by breaking it up into boxes. Mathematicians have carried out the details of this in a subject known as Measure Theory.

One point should be noted. Suppose we consider the matrix $\begin{pmatrix} 0 & 1 \\ 1 & 0 \end{pmatrix}$ which represents a reflection in the line $y = x$. If we substitute these values in $\Delta = ps - qr$, we get the value -1. The size of the answer is correct – in a reflection areas are unchanged; they change in the ratio $1:1$, so we might have expected to find Δ equal to 1. The negative

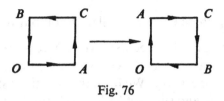

Fig. 76

sign is an indication that the plane has been turned over; anticlockwise has been changed to clockwise.

So the determinant gives us an extra piece of information. By its sign it tells us whether such a reversal has taken place or not.

If we have two 2×2 matrices S and T, the product ST indicates that we first apply T to the plane, and then apply S to the result. When T is applied, all areas will be multiplied by det T. When S acts, they will again be multiplied by det S. Thus altogether areas get multiplied by (det S)·(det T). Now det(ST) indicates the ratio in which areas change under the transformation ST. Accordingly we must have

$$\det(ST) = (\det S) \cdot (\det T).$$

The determinant of a product equals the product of the determinants.

For example, we have by matrix multiplication

$$\begin{pmatrix} 2 & 3 \\ 1 & 4 \end{pmatrix} \begin{pmatrix} 20 & 10 \\ 3 & 2 \end{pmatrix} = \begin{pmatrix} 49 & 26 \\ 32 & 18 \end{pmatrix}.$$

The determinants of the matrices on the left-hand side are

$$\begin{vmatrix} 2 & 3 \\ 1 & 4 \end{vmatrix} = 5 \quad \text{and} \quad \begin{vmatrix} 20 & 10 \\ 3 & 2 \end{vmatrix} = 10.$$

Accordingly we expect the determinant of the matrix on the right-hand side to be 50. And in fact

$$\begin{vmatrix} 49 & 26 \\ 32 & 18 \end{vmatrix} = 882 - 832 = 50.$$

This last result holds in any number of dimensions. To prove it in 3 dimensions we replace the word 'area' by 'volume'; in higher dimensions we use the corresponding measure of what a box holds.

Inverse matrix

The matrix M introduced at the beginning of this section corresponds to the equations

$$x^* = px + qy, \tag{1}$$

$$y^* = rx + sy. \tag{2}$$

If we know the input, $\begin{pmatrix} x \\ y \end{pmatrix}$, these equations tell us the output vector $\begin{pmatrix} x^* \\ y^* \end{pmatrix}$. Occasionally we need to proceed in the opposite direction. We require a certain output; what input will give it? In this case, x^* and y^* are given; we want to find x and y. So we have to solve equations (1) and (2) for x and y. We form the equations $s(1) - q(2)$. (This means, we multiply (1) by s, (2) by q, and subtract.) We also form the equation $p(2) - r(1)$. We thus obtain

$$sx^* - qy^* = (ps - qr)x, \tag{3}$$

$$-rx^* + py^* = (ps - qr)y. \tag{4}$$

We notice that both x and y have the coefficient Δ, the determinant of M, and we want to divide by this. As a rule, we shall be able to do so, but there is the exceptional case, in which $\Delta = 0$. These cases we must consider separately.

(1) General case, $\Delta \neq 0$ In this case, we can divide equations (3) and (4) by Δ and obtain

$$x = (s/\Delta)x^* - (q/\Delta)y^*, \tag{5}$$

$$y = -(r/\Delta)x^* + (p/\Delta)y^*. \tag{6}$$

Equations (1) and (2) tell us that, for any input, there is just one possible output. Equations (5) and (6) tell us that, for any output, there is just one possible input. We have what is known as a one-to-one mapping. We may represent it diagrammatically as in Fig. 77.

We began with $v \to v^*$, where $v^* = Mv$. If we start with v^* and want to get back to v, it is natural to write $v = M^{-1}v^*$. M^{-1} is called the *inverse matrix*: it specifies the inverse transformation. Some authors prefer M^I or ^IM, with I here indicating 'inverse'.

The product $M^{-1}M$ indicates that we start with (say) v, cross to v^*, and return to v. Thus $M^{-1}Mv = v$ for any vector v. Thus $M^{-1}M$ is the identity transformation; we may write $M^{-1}M = I$. Products usually depend on order; ST and TS are not always the same. So we had better look at MM^{-1}. We start with v^* in the output; M^{-1} sends it back to v; then M sends it forward to v^*. Thus MM^{-1} gives $v^* \to v^*$. It also is an identity transformation. Accordingly we have $MM^{-1} = M^{-1}M = I$.

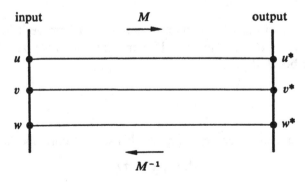

Fig. 77

Definition If we have two matrices, M and M^{-1}, such that $MM^{-1} = M^{-1}M = I$, then M^{-1} is called the inverse of M.

(2) Special case, $\Delta = 0$ Our algebra above showed that this case led to difficulties, since in equations (3) and (4) the coefficients of x and y become zero. It is impossible to divide and obtain equations (5) and (6).

Geometrically, $\Delta = 0$ means that the parallelogram $O^*A^*C^*B^*$, used earlier in this section, has area zero. How can this arise? There are two possibilities. One of the sides, O^*A^* or O^*B^*, may be of zero length, or the angle $B^*O^*A^*$ may be zero.

The first possibility would arise, for example, with the matrix $\begin{pmatrix} 1 & 0 \\ 0 & 0 \end{pmatrix}$. For this matrix $x^* = x$, $y^* = 0$. Every point is projected onto the x-axis.

The second possibility arises with $\begin{pmatrix} 1 & 2 \\ 1 & 2 \end{pmatrix}$. For this matrix,

$$A^* = \begin{pmatrix} 1 \\ 1 \end{pmatrix}, \qquad B^* = \begin{pmatrix} 2 \\ 2 \end{pmatrix}.$$

Both points lie on the line through the origin from south-west to north-east.

In both cases, the plane is crushed into a line. Thus every area becomes zero. There is an even more extreme possibility, represented by the matrix 0, for which every point maps to the origin.

The equation $\Delta = 0$ indicates that the space has been transformed into a space of lower dimension. We have been considering the example of the plane, but this remark applies equally well to spaces of any dimension. For instance, in 3 dimensions, if every volume is multiplied by 0, this can only mean that space has been crushed down into a plane, a line or a point.

We can now see why there is difficulty in forming the inverse transformation. This point is sufficiently illustrated by the projection $x^* = x$, $y^* = 0$ mentioned above. In this transformation, every point P goes to the point P^* on the x-axis immediately below it (Fig. 78).

If we now try to prescribe an output (x^*, y^*) and go back from it to an input (x, y) we run into two kinds of difficulty – one a famine and one a glut. If we want an output

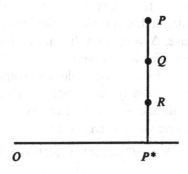

Fig. 78

such as (3, 2), a point that does not lie on the x-axis, we cannot get it. No point goes to this position. On the other hand if we take a point P^* on the x-axis, as in Fig. 78, it is true $P \to P^*$, but also $Q \to P^*$ and $R \to P^*$ and so do an infinity of other points. We do not know which to take for $M^{-1}P^*$. Someone might suggest: 'Choose one of them at random'. But this will not work out right. Suppose we say, let $M^{-1}P^* = P$. Now $M^{-1}M = I$, when an inverse exists. Apply this equation to Q. $M^{-1}MQ = IQ = Q$. Now $MQ = P^*$, so $M^{-1}MQ = M^{-1}P^* = P$ by what we agreed just now. We have obtained $Q = P$. But this is wrong, P and Q are not the same point. The glut of candidates for the position $M^{-1}P^*$ when P^* is on the x-axis is just as embarrassing as the famine, the absence of any candidate, when P^* is off that axis.

These things work out in essentially the same way in any number of dimensions. For convenience of language, we will discuss this question in terms of 3-dimensional space, on the understanding that essentially the same arguments and results apply in any number of dimensions.

When $\Delta = 0$, the unit cube transforms to a region whose volume is zero. This can happen only if A^*, B^*, C^* (corresponding to the unit vectors in the axes, A, B, C) lie in a plane through the origin, O^*. (In the more extreme cases, when A^*, B^*, C^* lie in a line through O^*, or even coincide with O^*, this condition still holds.) Then M^{-1} cannot exist, for we cannot find an input that gives a point outside this plane.

When $\Delta = 0$, there is a complete line, every point of which transforms into the origin. For the three vectors, A^*, B^*, C^* lie in a plane. By this we mean that the lines O^*A^*, O^*B^*, O^*C^* lie in a plane. As we agreed in §7, three vectors in a plane cannot be linearly independent; one of them, say C^*, must be a combination of the other two. Suppose then $C^* = hA^* + kB^*$. This means $hA^* + kB^* - C^* = 0$. Now in a linear mapping, $xA + yB + zC \to xA^* + yB^* + zC^*$. Putting $x = h$, $y = k$, $z = -1$, it follows $hA + kB - C \to hA^* + kB^* - C^* = 0$. So $hA + kB - C$ is a vector that maps to 0. This vector cannot itself be 0, since it has $z = -1$. Call this vector v_1. Then $v_1 \to 0$ and, by proportionality, $tv_1 \to 0$ for any number t. The points of the form tv_1, the multiples of v_1, fill a line. Q.E.D.

Let u be any vector. Since $u \to u^*$ and $tv_1 \to 0$, $u + tv_1 \to u^*$, by superposition. Since t can be any number, we see that there are infinitely many points that map to u^*.

This shows that quite generally when $\Delta = 0$ we have the alternative of famine or glut. Either an output is unobtainable from any input, or it arises from infinitely many different inputs. On both counts, such a transformation cannot possess an inverse.

Let us look at the general case, $\Delta \neq 0$, for 3-dimensional space. In this case, A^*, B^*, C^* cannot be linearly dependent; for if they were, the vectors would lie in a plane, and volumes would be sent to zero. It looks reasonable, on geometrical grounds, and it can in fact be proved, that any three linearly independent vectors in 3 dimensions form a basis for the space – we can build a co-ordinate framework on them. Thus every point P^* can be represented in just one way as $xA^* + yB^* + zC^*$. Now

$$xA + yB + zC \to xA^* + yB^* + zC^*,$$

so we certainly can find a point P that maps to P^*. And there can only be one such point.

For suppose $x_1A + y_1B + z_1C$ and $x_2A + y_2B + z_2C$ both mapped to P^*. We would have $P^* = x_1A^* + y_1B^* + z_1C^* = x_2A^* + y_2B^* + z_2C^*$, so P^* would have co-ordinates (x_1, y_1, z_1) and (x_2, y_2, z_2) in the A^*, B^*, C^* system. But a point can only have one set of co-ordinates, so $x_1 = x_2$, $y_1 = y_2$, $z_1 = z_2$; this means the points mapping to P^* must coincide. So when $\Delta \neq 0$, every point P^* can be got by choosing a suitable input, and only one such input exists. We have a one-to-one correspondence, and the inverse transformation M^{-1} exists.

It will be a linear transformation, since $u + v \rightarrow u^* + v^*$ and $ku \rightarrow ku^*$ for M. For M^{-1}, we reverse the arrows.

We may sum this up as follows.

Theorem   When det $M \neq 0$, there exists an inverse matrix, M^{-1}, such that

$$MM^{-1} = M^{-1}M = I.$$

The equation $Mv = w$ then has the unique solution $v = M^{-1}w$.

When det $M = 0$, no inverse exists. The equation $Mv = w$ will have either 0 or ∞ solutions. There will always be some non-zero vector v_1 such that $Mv_1 = 0$.

Note on the role of determinants

In the nineteenth century great attention was paid to determinants and their properties, and determinants were used to solve systems of equations. In this century, computers have been used to solve problems in which hundreds of variables may be involved. For such work determinants are completely *un*-suitable. It is useful to recognize determinants when we are working in 2 or 3 dimensions. For work involving several variables, determinants occur in theoretical arguments, such as those just used, and in the procedure for finding eigenvalues (§25).

This procedure gives interest to the theorem above, which states that $Mv = 0$ has a non-zero solution when det $M = 0$. The first part of the theorem shows that, when det $M \neq 0$, the only solution of $Mv = 0$ is $v = 0$; for $v = 0$ obviously is a solution, and the transformation being one-to-one, there cannot be any other solution.

In the eigenvector–eigenvalue search, we are trying to find λ and non-zero v so that $Mv = \lambda v$. This equation may be written $Mv = \lambda Iv$ or $(M - \lambda I)v = 0$. By the theorem, a non-zero solution will exist when $\det(M - \lambda I) = 0$. The equations used in the final stages of §25 were of this form. For instance, in the worked example, we had

$$M = \begin{pmatrix} 1 & 4 \\ 2 & 3 \end{pmatrix}.$$

Since

$$\lambda I = \begin{pmatrix} \lambda & 0 \\ 0 & \lambda \end{pmatrix}, \qquad M - \lambda I = \begin{pmatrix} 1 - \lambda & 4 \\ 2 & 3 - \lambda \end{pmatrix}.$$

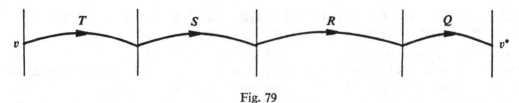

Fig. 79

In the worked example, it will be seen that the determinant of this matrix was equated to zero, so that suitable values of λ could be found. When these values of λ are substituted, $\det(M - \lambda I) = 0$ and so it is certain that we shall be able to satisfy $(M - \lambda I)v = 0$ by some non-zero vector v, that is, a vector that is acceptable as an eigenvector.

Terminology   A matrix that has no inverse is said to be *singular*.

The equation $\det(M - \lambda I) = 0$ is called the *characteristic equation* of M. The roots of the characteristic equation, as we have seen, are the *eigenvalues* of M. Some authors refer to the eigenvalues as *characteristic* numbers or *latent roots*.

A matrix that has an inverse is sometimes called *invertible*, sometimes *non-singular*.

Inverse of a product

It is sometimes necessary to consider the inverse of a product of matrices, such as $QRST$ say. Here we suppose that each of the matrices Q, R, S and T has an inverse. We may represent the product transformation graphically by Fig. 79, where

$$v^* = QRSTv.$$

The inverse transformation brings us back from v^* to v. Moving back, we find we perform Q^{-1}, R^{-1}, S^{-1}, T^{-1} in that order – the reverse of the order we followed when we went from v to v^*. Thus the inverse of $QRST$ is $T^{-1}S^{-1}R^{-1}Q^{-1}$.

We can verify this formally, for we have

$$(QRST)(T^{-1}S^{-1}R^{-1}Q^{-1}) = QRS(TT^{-1})S^{-1}R^{-1}Q^{-1}$$
$$= QRS \cdot I \cdot S^{-1}R^{-1}Q^{-1} = QR(SS^{-1})R^{-1}Q^{-1} = QR \cdot I \cdot R^{-1}Q^{-1}$$
$$= Q(RR^{-1})Q^{-1} = QIQ^{-1} = QQ^{-1} = I.$$

At several stages of this argument we have omitted a factor I. This is legitimate, since I means 'leave things as they were'. In the same way in arithmetic we can omit the factors 1 in a product such as $3 \times 7 \times 1 \times 2 \times 1 \times 9$.

If we wish, we can use the same procedure to check formally that the product is still I when we put $T^{-1}S^{-1}R^{-1}Q^{-1}$ to the left of $QRST$ instead of the right; that is, we can verify that $(T^{-1}S^{-1}R^{-1}Q^{-1})(QRST) = I$.

Questions

1   For each matrix listed below, draw a diagram to show its effect on the square $OACB$, where $O = (0, 0)$, $A = (1, 0)$, $B = (0, 1)$ and $C = (1, 1)$. (As explained in §22, the columns of the matrix provide a quick way of doing this, in cases where the geometrical meaning of the transformation is not immediately obvious.) Find the area of the resulting region, and compare it with the determinant of the matrix. In which cases does the plane experience an effect of turning over, so that clockwise and counterclockwise circuits are interchanged?

(a) $\begin{pmatrix} 2 & 0 \\ 0 & 1 \end{pmatrix}$ (b) $\begin{pmatrix} 1 & 0 \\ 0 & 3 \end{pmatrix}$ (c) $\begin{pmatrix} 2 & 0 \\ 0 & 3 \end{pmatrix}$ (d) $\begin{pmatrix} 2 & 0 \\ 0 & \frac{1}{2} \end{pmatrix}$ (e) $\begin{pmatrix} 1 & 0 \\ 0 & -1 \end{pmatrix}$ (f) $\begin{pmatrix} 1 & -1 \\ 1 & 1 \end{pmatrix}$

(g) $\begin{pmatrix} 1 & 1 \\ 1 & -1 \end{pmatrix}$ (h) $\begin{pmatrix} \cos \alpha & -\sin \alpha \\ \sin \alpha & \cos \alpha \end{pmatrix}$ (i) $\begin{pmatrix} \cos \alpha & \sin \alpha \\ \sin \alpha & -\cos \alpha \end{pmatrix}$.

2 For complex numbers $(4 + j3)(x + jy) = x^* + jy^*$ with $x^* = 4x - 3y$ and $y^* = 3x + 4y$, so multiplication by $4 + j3$ is associated with the matrix

$$M = \begin{pmatrix} 4 & -3 \\ 3 & 4 \end{pmatrix}.$$

If every complex number in the Argand diagram is multiplied by $4 + j3$, in what way does this transformation affect (a) lengths, (b) areas? Show that your answers agree with those that would be obtained by considering the value of det M.

Replace $4 + j3$ by the general number $a + jb$ and carry out the same investigation.

Does the value of the determinant in this general case throw any light on the question – could a complex number exist, such that multiplication by it produced a reflection in the Argand diagram?

3 Let

$$S = \begin{pmatrix} a & -b \\ b & a \end{pmatrix}, \qquad T = \begin{pmatrix} x & -y \\ y & x \end{pmatrix}.$$

Calculate the product matrix ST. Find the determinants of S, T and ST. Verify that

$$\det S \cdot \det T = \det(ST).$$

4 Which of the following matrices do *not* possess inverses?

(a) $\begin{pmatrix} 1 & 0 \\ 0 & 0 \end{pmatrix}$ (b) $\begin{pmatrix} 1 & 2 \\ 3 & 4 \end{pmatrix}$ (c) $\begin{pmatrix} 1 & 2 \\ 3 & 6 \end{pmatrix}$ (d) $\begin{pmatrix} 0 & 2 \\ 3 & 0 \end{pmatrix}$

(e) $\begin{pmatrix} 1 & 1 \\ 0 & 1 \end{pmatrix}$ (f) $\begin{pmatrix} 1 & 1 \\ 1 & 1 \end{pmatrix}$ (g) $\begin{pmatrix} 8 & 10 \\ 12 & 15 \end{pmatrix}$.

5 Let

$$S = \begin{pmatrix} 2 & 1 \\ 3 & 2 \end{pmatrix}, \qquad T = \begin{pmatrix} 1 & 2 \\ 3 & 7 \end{pmatrix}.$$

Calculate ST. Find det S, det T and det(ST). Find S^{-1}, T^{-1} and $(ST)^{-1}$. Of the matrices $(ST)^{-1}$ $S^{-1}T^{-1}$ and $T^{-1}S^{-1}$ which, if any, would you expect to be equal? Check your answer by direct calculation of these matrices.

6 (a) Make a sketch to show the effect of the matrix $\begin{pmatrix} 2 & 1 \\ 1 & 2 \end{pmatrix}$ on the square with corners $O = (0, 0)$, $C = (1, 1)$, $D = (0, 2)$ and $E = (-1, 1)$. In what ratios are areas changed? What is the determinant of the matrix? What are the eigenvectors and eigenvalues of this matrix? (These can be seen from the diagram; it is not necessary to use the routine of §25.)

(b) Answer the same questions as in (a), but with the matrix $\begin{pmatrix} 2.5 & 0.5 \\ 0.5 & 2.5 \end{pmatrix}$.

(c) Do the same, but with the matrix $\begin{pmatrix} a & b \\ b & a \end{pmatrix}$.

(d) Does there appear to be any connection between the eigenvalues and the value of the determinant? Is there any logical reason for expecting such a connection?

7 Each matrix below has determinant zero. According to §26, it should map the whole plane into a certain line. There should be another line, every point of which maps to the origin. For each matrix, find these two lines and indicate by a sketch how the plane is transformed by the matrix. Each matrix has two invariant lines; show these on your sketch and label each of them with the appropriate eigenvalue, λ.

(a) $\begin{pmatrix} 1 & 0 \\ 0 & 0 \end{pmatrix}$ (b) $\begin{pmatrix} 1 & 0 \\ 1 & 0 \end{pmatrix}$ (c) $\begin{pmatrix} \frac{1}{2} & \frac{1}{2} \\ \frac{1}{2} & \frac{1}{2} \end{pmatrix}$ (d) $\begin{pmatrix} 1 & 1 \\ 1 & 1 \end{pmatrix}$ (e) $\begin{pmatrix} 0.2 & 0.4 \\ 0.4 & 0.8 \end{pmatrix}$.

8 Let

$$S = \begin{pmatrix} 1 & a \\ 0 & 1 \end{pmatrix}, \qquad T = \begin{pmatrix} 1 & 0 \\ b & 1 \end{pmatrix}.$$

Verify that the inverses of S and T are

$$S^{-1} = \begin{pmatrix} 1 & -a \\ 0 & 1 \end{pmatrix}, \qquad T^{-1} = \begin{pmatrix} 1 & 0 \\ -b & 1 \end{pmatrix}.$$

Calculate the matrices ST and $S^{-1}T^{-1}$. Would you expect $S^{-1}T^{-1}$ to be the inverse of ST? Would you expect the product of ST and $S^{-1}T^{-1}$ to be a matrix of determinant 1? Check your answers by calculation.

Would the answer to either question be altered if we replaced $S^{-1}T^{-1}$ throughout by $T^{-1}S^{-1}$?

27 Properties of determinants

We have defined determinants geometrically. If u and v are the columns of a matrix in 2 dimensions, the determinant of that matrix is the area of the parallelogram with sides u and v. For a 3×3 matrix, with columns u, v, w, the determinant equals the volume for the cell* with sides u, v, w, and for higher dimensions there are corresponding definitions. This leaves determinants a little up in the air. We would like to know the explicit formulas for these quantities, and their properties for occasions when we need to calculate with them.

In §26 we found an explicit formula for $\det(u, v)$, the area of the parallelogram with sides u and v. For $u = \begin{pmatrix} p \\ r \end{pmatrix}$, $v = \begin{pmatrix} q \\ s \end{pmatrix}$ we found

$$\det(u, v) = \begin{vmatrix} p & q \\ r & s \end{vmatrix} = ps - qr.$$

---

* The traditional name for the figure in 3 dimensions corresponding to the parallelogram is 'parallelepiped', an awful word both to spell and to pronounce. I use 'cell' instead, and understand by this word the figure in any number of dimensions that corresponds to the parallelogram in 2 dimensions.

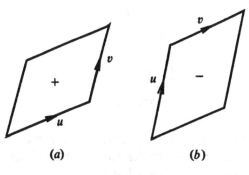

Fig. 80

It was pointed out in that section that this quantity may be positive or negative. It is useful to indicate u and v as in Fig. 80, with the second vector drawn from the end of the first. In case (a), where the arrows run round the parallelogram in a counterclockwise direction, $\det(u, v)$ is positive; in case (b), where the arrows run in a clockwise sense, $\det(u, v)$ is negative.

This might seem to be a complication, but in fact it makes the work easier. This feature appears already in one dimension. Suppose we have points on a line, and we know B is 10 units from A, and C is 3 units from B. Then the distance AC may be either 7 or 13 units, depending on the directions of AB and BC. But if we use directed distances, and write $AB = +10$, $BC = +3$ to mean that B is 10 units to the right of A and C is 3 to the right of B, we are sure that $AC = +13$. With this system, if A, B and C have co-ordinates a, b, c we write $AB = b - a$, $BC = c - b$, $AC = c - a$ and we can be sure $AB + BC = AC$. Now of course AB, BC and AC may be positive or negative depending on the relative positions of the points A, B, C.

With areas it helps in many ways if we allow $+$ and $-$ signs. Fig. 81 illustrates $\det(u, v)$ and $\det(3u, v)$. Obviously, $\det(3u, v)$ is 3 times $\det(u, v)$. This suggests that, for any number k we ought to have $\det(ku, v) = k\det(u, v)$. But what about $k = -1$?

If the formula is to hold, we must have $\det(-u, v) = -\det(u, v)$, so the parallelogram of Fig. 82 should have area equal to that of (a) in size, but opposite in sign.

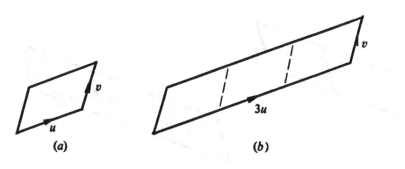

(a)   (b)

Fig. 81   Fig. 82

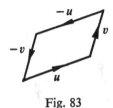

Fig. 83

It does no harm if we put arrows on all the sides of the parallelogram, as in Fig. 83. Any two consecutive sides can be used to describe the area. Thus we could call the area $\det(v, -u)$ or $\det(-u, -v)$ or $\det(-v, u)$. All of these are equal to $\det(u, v)$ as you can check by using the formula $ps - qr$.

The sign convention is particularly handy when we want to join regions together. Thus, if we take two counterclockwise parallelograms such as (a) and (b) in Fig. 81 we can put them together to make the parallelogram in Fig. 84. You will notice that the arrows on the common boundary are in opposite directions. We can regard this part as cancelling out, and we are simply left with the parallelogram involving $4u$ and v. This agrees with the result $\det(3u, v) + \det(u, v) = \det(4u, v)$.

If we want to put together a clockwise parallelogram as in Fig. 82 and an anticlockwise one like (b), to get this cancelling of common boundaries we have to arrange them as in Fig. 85.

After cancelling the parts with opposing arrows, we are left with the shaded parallelogram, counterclockwise, in agreement with the result

$$\det(3u, v) + \det(-u, v) = \det(2u, v),$$

given by the determinant formula.

Note that if we superposed Fig. 81 (a) and Fig. 82, everything would cancel. This agrees with $\det(u, v) + \det(-u, v) = 0$.

A familiar idea is that the area of a parallelogram, such as $OCED$ in Fig. 86 is not altered if we chop off a triangle ODF from one side and replace it by a congruent triangle CEG on the other side.

The vector $CG = CE + EG$. Now EG has the same direction as u, and so must be ku for some k. Thus $CG = v + ku$. The statement area $OCED$ = area $OCGF$ may

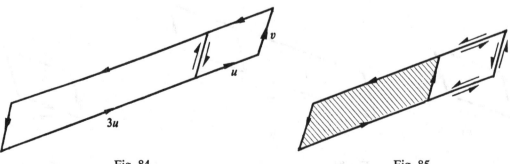

Fig. 84   Fig. 85

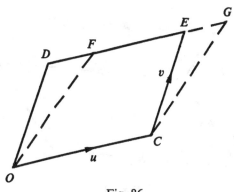

Fig. 86

accordingly be written $\det(u, v) = \det(u, v + ku)$ for any number k. Thus the determinant is unaltered in value if we add a multiple of one column to another.

This works in any number of dimensions. For 3 dimensions we can draw the picture in Fig. 87, which shows $\det(u, v, w) = \det(u, v + ku, w)$. For dimensions higher than 3 we cannot draw pictures, but we can explain by means of co-ordinate geometry or vector notation what set of points is replaced by a congruent set.

Of course the three vectors u, v, w are on an equal footing. We have, for instance, $\det(u, v, w) = \det(u + kw, v, w)$. We may add a multiple of any column to any other column.

So far we have two principles.

(I) Enlarging one vector k times enlarges the volume of the cell k times. For instance, $\det(ku, v, w) = k \det(u, v, w)$.

(II) Adding a multiple of one vector to another makes no difference to the determinant. For instance $\det(u, v, w) = \det(u, v, w + kv)$.

From these two principles we can deduce a useful result:

(III) Interchanging two vectors changes the sign of the determinant.

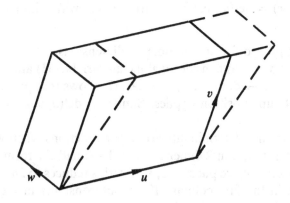

Fig. 87

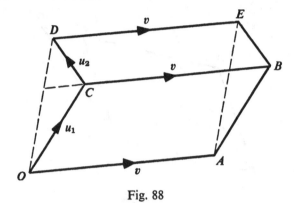

Fig. 88

Proof; $\det(u, v) = \det(u, u + v)$ by (II) $= \det(-v, u + v)$ by (II); we have added -1 times second vector to first.

Use (II) again, add first vector to second; this gives $\det(-v, u)$ which by (I), with $k = -1$, is $-\det(v, u)$ as required.

If we are in three or more dimensions, we carry out the same steps. For example $\det(u, v, w) = \det(u, u + v, w) = \det(-v, u + v, w) = \det(-v, u, w) = -\det(v, u, w)$. The presence of w makes no difference to the argument.

We can of course apply (II) more than once to obtain results such as

$$\det(u, v, w) = \det(u, v, w + au) = \det(u, v, w + au + bv)$$

for any two numbers a, b. Geometrically, this is equivalent to a statement about two cells with the same base (sides u, v) and the same height.

In Fig. 88 we have two parallelograms $OABC$ and $CBED$ lying in a plane. It is easily seen that area $OABC$ + area $CBED$ = area $OAED$. As $OD = u_1 + u_2$, we may write this result as $\det(u_1 + u_2, v) = \det(u_1, v) + \det(u_2, v)$. In 3 dimensions we have a similar result, namely $\det(u_1 + u_2, v, w) = \det(u_1, v, w) + \det(u_2, v, w)$. This follows from the fact that cells with the same base and same height have equal volumes. So we have:

(IV) $\det(u_1 + u_2, v, w) = \det(u_1, v, w) + \det(u_2, v, w)$, with similar results involving the vectors v and w.

This again can be proved for any number of dimensions.

Suppose we fix v and w, and consider u alone as variable. (I) and (IV) taken together show that the mapping $u \to \det(u, v, w)$ is linear. (I) shows that proportionality applies, and (IV) shows that superposition applies. Similarly, $\det(u, v, w)$ is linear in v and is linear in w.

This does not mean that determinants are linear functions when we consider all the vectors varying. For instance, in 2 dimensions we have had the formula $ps - qr$ for the determinant. If, for example, we put $q = 3$, $s = 2$, this reduces to $2p - 3r$, an expression linear in the entries in the first column. If we put constants in the first column, say $p = 5$, $r = 6$, we get $5s - 6q$, again a linear expression. But if we consider p, q, r, s as all variable the determinant has to be regarded as of the second degree, just as for instance in co-ordinate geometry the hyperbola $xy = 1$ is a conic, and is classified with the ellipse and parabola, equations of the second degree, although it is linear in x alone (y fixed) or in y alone (x fixed).

A fifth principle is sometimes referred to as 'skinning a determinant'. Consider the determinant

$$\begin{vmatrix} 1 & 0 & 0 \\ 0 & p & q \\ 0 & r & s \end{vmatrix}$$

If we call the three columns here u, v and w, u is the unit vector along the x-axis, while v and w are in the plane perpendicular to the x-axis (Fig. 89). The volume of the cell they determine is thus the area of the base (the parallelogram with sides v, w) multiplied by the height (which is 1). The area of the base is $\begin{vmatrix} p & q \\ r & s \end{vmatrix}$. We know this from our study of areas in the plane. Accordingly we have

(V)

$$\begin{vmatrix} 1 & 0 & 0 \\ 0 & p & q \\ 0 & r & s \end{vmatrix} = \begin{vmatrix} p & q \\ r & s \end{vmatrix}.$$

The name 'skinning the determinant' is given because we here strip off the top row and the first column.

Now of course the x-axis is no better than the y-axis or the z-axis, and there must be corresponding results when the first column is

$$\begin{pmatrix} 0 \\ 1 \\ 0 \end{pmatrix} \quad \text{or} \quad \begin{pmatrix} 0 \\ 0 \\ 1 \end{pmatrix}.$$

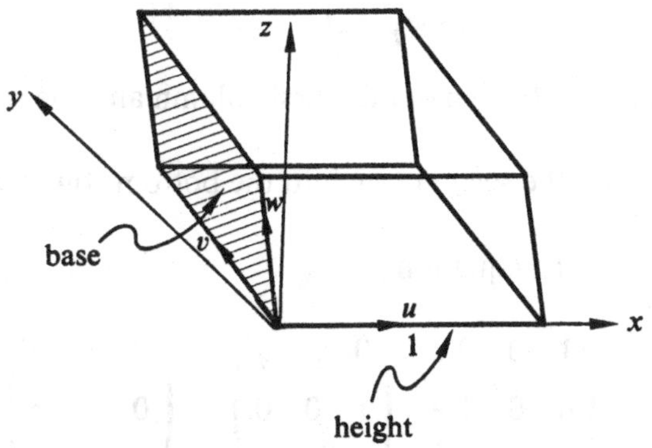

Fig. 89

In these cases, some care is needed to see whether the sign is $+$ or $-$. There is a rule covering this, and perhaps the following is as good a way as any to obtain it.

Matrix multiplication gives the following equation:

$$\begin{pmatrix} 0 & 1 & 0 \\ 1 & 0 & 0 \\ 0 & 0 & 1 \end{pmatrix} \begin{pmatrix} 1 & 0 & 0 \\ 0 & p & q \\ 0 & r & s \end{pmatrix} = \begin{pmatrix} 0 & p & q \\ 1 & 0 & 0 \\ 0 & r & s \end{pmatrix}.$$

We saw in §26 that, for any two matrices S, T, we have $\det(ST) = (\det S) \cdot (\det T)$.

What is the determinant of the first matrix in the equation above? If we interchanged the first two columns, the matrix would become

$$\begin{pmatrix} 1 & 0 & 0 \\ 0 & 1 & 0 \\ 0 & 0 & 1 \end{pmatrix},$$

the identity matrix, I. The identity transformation leaves all volumes unchanged; it multiplies them by 1. So $\det I = 1$. By (III), interchanging two columns changes the sign of the determinant. So the determinant of the first matrix is -1. (We might have guessed this. The matrix represents a reflection in the plane $y = x$. This is why we brought it in – to relate the unit vector on the y-axis to the unit vector on the x-axis.)

Accordingly, taking the determinants of the matrices in the equation we find

$$-\begin{vmatrix} 1 & 0 & 0 \\ 0 & p & q \\ 0 & r & s \end{vmatrix} = \begin{vmatrix} 0 & p & q \\ 1 & 0 & 0 \\ 0 & r & s \end{vmatrix}.$$

Thus

$$\begin{vmatrix} 0 & p & q \\ 1 & 0 & 0 \\ 0 & r & s \end{vmatrix} = -\begin{vmatrix} p & q \\ r & s \end{vmatrix}.$$

In this case, we pay for striking out the first column and the second row by introducing a minus sign.

We can get the 1 from the second row into the bottom row if we apply a reflection in the plane $y = z$.

Accordingly we write the equation

$$\begin{pmatrix} 1 & 0 & 0 \\ 0 & 0 & 1 \\ 0 & 1 & 0 \end{pmatrix} \begin{pmatrix} 0 & p & q \\ 1 & 0 & 0 \\ 0 & r & s \end{pmatrix} = \begin{pmatrix} 0 & p & q \\ 0 & r & s \\ 1 & 0 & 0 \end{pmatrix}$$

which again comes by matrix multiplication. Here again, the first matrix has determinant -1, and, taking determinants we find

$$\begin{vmatrix} 0 & p & q \\ 0 & r & s \\ 1 & 0 & 0 \end{vmatrix} = - \begin{vmatrix} 0 & p & q \\ 1 & 0 & 0 \\ 0 & r & s \end{vmatrix} = + \begin{vmatrix} p & q \\ r & s \end{vmatrix}.$$

Thus, in the three cases considered, the required signs are $+$, $-$, $+$. In higher dimensions the signs alternate as we continue in this way. Thus, for example, in a 6×6 determinant, if 1 occurs somewhere in the first column, and zeros fill the rest of that column and the rest of the row in which 1 appears, we may strike out that column and row, and affix a sign which is $+$, $-$, $+$, $-$, $+$, $-$ according as the row deleted is the 1st, 2nd, 3rd, 4th, 5th or 6th.

We now have the machinery needed for giving the explicit formula for a determinant. We will carry it through for 3 dimensions; a similar method can be used for any number of dimensions.

Suppose then we wish to find the determinant

$$\begin{vmatrix} a & d & g \\ b & e & h \\ c & f & i \end{vmatrix}.$$

The determinant, as we saw, is linear in each of the vectors involved. The vector in the first column is

$$\begin{pmatrix} a \\ b \\ c \end{pmatrix} \quad \text{which is} \quad \begin{pmatrix} a \\ 0 \\ 0 \end{pmatrix} + \begin{pmatrix} 0 \\ b \\ 0 \end{pmatrix} + \begin{pmatrix} 0 \\ 0 \\ c \end{pmatrix}.$$

By superposition then

$$\begin{vmatrix} a & d & g \\ b & e & h \\ c & f & i \end{vmatrix} = \begin{vmatrix} a & d & g \\ 0 & e & h \\ 0 & f & i \end{vmatrix} + \begin{vmatrix} 0 & d & g \\ b & e & h \\ 0 & f & i \end{vmatrix} + \begin{vmatrix} 0 & d & g \\ 0 & e & h \\ c & f & i \end{vmatrix}.$$

We consider the three parts separately. By (I), the first determinant equals

$$a \begin{vmatrix} 1 & d & g \\ 0 & e & h \\ 0 & f & i \end{vmatrix}.$$

We apply (II) twice. Add $-d$ times the first column to the second; then add $-g$ times the first column to the third column. This gives

$$a \begin{vmatrix} 1 & 0 & 0 \\ 0 & e & h \\ 0 & f & i \end{vmatrix} = a \begin{vmatrix} e & h \\ f & i \end{vmatrix}, \quad \text{by skinning.}$$

Apply similar procedures to the other two determinants, with care to remember the signs required. We find the original determinant equals

$$a \begin{vmatrix} e & h \\ f & i \end{vmatrix} - b \begin{vmatrix} d & g \\ f & i \end{vmatrix} + c \begin{vmatrix} d & g \\ e & h \end{vmatrix}.$$

This result is easily remembered. The coefficient of a is the result of crossing out the row and column containing a. For the coefficient of b we cross out the row and column containing b, but require a minus sign. Similar considerations apply to the coefficient of c, but the sign returns to plus.

The 2×2 determinants are given by the formula introduced in §25. We use this and by multiplying out obtain the result

$$\begin{vmatrix} a & d & g \\ b & e & h \\ c & f & i \end{vmatrix} = a(ei - fh) - b(di - fg) + c(dh - eg)$$

$$= aei + bfg + cdh - afh - bdi - ceg.$$

It will be noticed that there are 6 terms. We never find two letters in the same row or the same column appearing in a product. The numerical coefficients are all either $+1$ or -1.

It is possible to check the sign of any term as follows. Suppose we wish to check for cdh. We put $c = d = h = 1$, and all the other entries zero. This gives

$$\begin{vmatrix} 0 & 1 & 0 \\ 0 & 0 & 1 \\ 1 & 0 & 0 \end{vmatrix} = +1.$$

We then see how many interchanges of columns are needed to bring the determinant to the form det I. We find, by (III):

$$\begin{vmatrix} 0 & 1 & 0 \\ 0 & 0 & 1 \\ 1 & 0 & 0 \end{vmatrix} = - \begin{vmatrix} 1 & 0 & 0 \\ 0 & 0 & 1 \\ 0 & 1 & 0 \end{vmatrix} = + \begin{vmatrix} 1 & 0 & 0 \\ 0 & 1 & 0 \\ 0 & 0 & 1 \end{vmatrix} = +1.$$

This is in agreement with our earlier equation.

A 2×2 determinant has 2 terms when multiplied out in the form $ps - qr$. When we calculate a 3×3 determinant, as we did above, each entry in the first column has a

2×2 determinant as its coefficient. Thus there are 3×2 terms in the complete expression. If we apply the same procedure to a 4×4 determinant, there are 4 entries in the first column, and each of these has for its coefficient a 3×3 determinant with 3×2 terms. Thus the complete expression for a 4×4 determinant has $4 \times 3 \times 2$ terms, that is 24 terms. A 5×5 determinant has $5 \times 24 = 120$ terms and quite generally an $n \times n$ determinant has $n!$ terms. So a 10×10 determinant already has 3 628 800 terms.

As engineers may need to handle problems with hundreds or thousands of variables, it is easily understood why determinants are usually avoided as a means of computation. Thus given a matrix M there is a formula for its inverse, M^{-1}, in terms of determinants but it is rarely wise to use it. It is better to solve the equations $Mv = w$ directly, since, when an inverse exists, $v = M^{-1}w$.

Determinants can be useful, both for theoretical results and for particular calculations, when the number of dimensions is small, as for instance in geometrical, mechanical and electrical problems about our everyday physical space of 3 dimensions.

The eigenvalues of a matrix are usually of physical and engineering significance. Whole conferences are held on how to calculate them, but here again the methods used do not involve determinants. Yet it is useful to know that the eigenvalues of a matrix are given by an equation $\det(M - \lambda I) = 0$. If M is an $n \times n$ matrix, this equation is of the nth degree and we can classify the possible situations, by considering whether the roots of this equation are real or complex, distinct or repeated. In the same way, it may be useful to know that there is a single condition, $\det M = 0$, for a matrix M to possess no inverse.

The present position seems to be that an engineer does not need the expertise in manipulating determinants that was required of nineteenth century schoolboys, but that the properties of determinants should be part of his general knowledge. He may encounter them at any time in his reading.

In the exercises below, Questions 2 and 3 bring out properties of determinants that should be known to the student. These questions are in effect part of the theory covered in this section.

In the exercises below, questions $1(c)$, $1(d)$, $1(e)$, 4, 5, 6 and 7 could be slogged out by multiplying the determinants out completely. This, however, would be unwise and in some cases extremely laborious. These questions are intended to illustrate the properties of determinants (such as (I), (II), (III), (IV) in §27). The worked examples below show how such questions can be answered without heavy calculations.

Worked example 1

Calculate

$$\begin{vmatrix} 10 & 4 & 7 \\ 20 & 5 & 8 \\ 30 & 6 & 9 \end{vmatrix}.$$

Solution   The first column contains the factor 10. By (1) the determinant equals

$$10\begin{vmatrix} 1 & 4 & 7 \\ 2 & 5 & 8 \\ 3 & 6 & 9 \end{vmatrix}.$$

To the second column add -1 times the first (by (II)). This gives

$$10\begin{vmatrix} 1 & 3 & 7 \\ 2 & 3 & 8 \\ 3 & 3 & 9 \end{vmatrix}.$$

Now add -1 times the first column to the third. We obtain

$$10\begin{vmatrix} 1 & 3 & 6 \\ 2 & 3 & 6 \\ 3 & 3 & 6 \end{vmatrix}.$$

Finally, add -2 times the second column to the third, to give

$$10\begin{vmatrix} 1 & 3 & 0 \\ 2 & 3 & 0 \\ 3 & 3 & 0 \end{vmatrix}.$$

The third column now contains the zero vector; hence the determinant is zero.

Worked example 2

Simplify $\det(v + w, u + w, u + v)$ where u, v, w are any 3 vectors in 3 dimensions.

$$
\begin{aligned}
\det(v + w, u + w, u + v) &= \det(v + w, u - v, u + v) && \text{by (column 2)} - \text{(column 1)}, \\
&= \det(v + w, u - v, 2u) && \text{by (column 3)} + \text{(column 2)}, \\
&= 2 \det (v + w, u - v, u) && \text{from factor 2 in column 3}, \\
&= 2 \det(v + w, -v, u) && \text{by (column 2)} - \text{(column 3)}, \\
&= 2 \det(w, -v, u) && \text{by (column 1)} + \text{(column 2)}, \\
&= -2 \det(w, v, u) && \text{from factor} -1 \text{ in column 2}, \\
&= 2 \det(u, v, w) && \text{on interchanging columns 1 and} \\
& && \text{3, with resulting change of sign.}
\end{aligned}
$$

Note   Fallacies can result from unwise attempts to carry out several steps at once, as is shown by the following example.

A fallacious procedure   Let u, v, w be any 3 vectors in 3 dimensions. Then

$$\det(u, v, w) = \det(u - v, v - w, w - u)$$

by subtracting column 2 from column 1, column 3 from column 2 and column 1 from column 3. Add column 3 to column 2; the result is $\det(u - v, v - u, w - u)$. Add column 2 to column 1. This gives $\det(0, v - u, w - u)$. As the first column now contains the zero vector, the determinant must be zero.

The argument above purports to show that every 3×3 determinant is zero. The fallacy lies in the first step. If the steps are carried out one at a time, it will be found that no combination of (I), (II), (III), (IV) will lead from $\det(u, v, w)$ to

$$\det(u - v, v - w, w - u).$$

In order to avoid fallacies, it is wise to perform one step at a time, and to indicate the principle involved. This helps considerably if you need at a later time to check the correctness of the work. A detailed explanation, as in Worked example 1 above, is very wearisome to write. In Worked example 2, the commentary is much more concise, and it is clear that further abbreviations are possible.

Needless to say, the numbering (I), (II), (III), (IV) is purely for reference within this book, and this numbering need not be memorized.

Exercises on Determinants

1 Calculate the following:

$$(a)\ \begin{vmatrix} 0 & 0 & 1 \\ 0 & 1 & 0 \\ 1 & 0 & 0 \end{vmatrix} \qquad (b)\ \begin{vmatrix} 0 & 0 & 1 \\ 1 & 0 & 0 \\ 0 & 1 & 0 \end{vmatrix} \qquad (c)\ \begin{vmatrix} 1 & 0 & 0 \\ 0 & 1 & 1 \\ 0 & 0 & 0 \end{vmatrix}$$

$$(d)\ \begin{vmatrix} a & d & d \\ b & e & e \\ c & f & f \end{vmatrix} \qquad (e)\ \begin{vmatrix} a & d & a+d \\ b & e & b+e \\ c & f & c+f \end{vmatrix}.$$

2 Let

$$A = \begin{pmatrix} k & 0 & 0 \\ 0 & 1 & 0 \\ 0 & 0 & 1 \end{pmatrix}, \qquad B = \begin{pmatrix} 1 & k & 0 \\ 0 & 1 & 0 \\ 0 & 0 & 1 \end{pmatrix}, \qquad C = \begin{pmatrix} 1 & 0 & 0 \\ k & 1 & 0 \\ 0 & 0 & 1 \end{pmatrix}, \qquad M = \begin{pmatrix} a & d & g \\ b & e & h \\ c & f & i \end{pmatrix}.$$

Some of the calculations below provide alternative proof of principles already stated; identify the principles. Others lead to new principles; which should be noted.

First work out determinant A, determinant B and determinant C. Substitute these values in the equations below, after working out any matrix multiplications appearing in these equations.

(a) $\det(MA) = (\det M)(\det A)$
(b) $\det(AM) = (\det A)(\det M)$
(c) $\det(MB) = (\det M)(\det B)$
(d) $\det(MC) = (\det M)(\det C)$
(e) $\det(BM) = (\det B)(\det M)$
(f) $\det(CM) = (\det C)(\det M).$

3 By M^T we understand the transpose of M, that is to say, the matrix obtained when the columns of M are changed into rows, that is

$$M^T = \begin{pmatrix} a & b & c \\ d & e & f \\ g & h & i \end{pmatrix}.$$

Work out det M^T and compare it with the expression for det M given in the text. What do you notice? What general conclusion is suggested?

4 (a) Simplify

$$\begin{vmatrix} a & d & 5a + 7d \\ b & e & 5b + 7e \\ c & f & 5c + 7f \end{vmatrix}.$$

(b) Find the value of

$$\begin{vmatrix} 1 & 4 & 4001 \\ 2 & 5 & 5002 \\ 3 & 6 & 6003 \end{vmatrix}.$$

5 For 3 particular vectors, u, v, w, it is known that $\det(u, v, w) = 10$. Find the values of the following determinants:

(a) $\det(2u, 3v, 4w)$ (b) $\det(v, u, w)$ (c) $\det(w, u, v)$
(d) $\det(u, v, 3u + 2v)$ (e) $\det(u + v + w, v + w, w)$.

6 Calculate

$$(a) \begin{vmatrix} 1 & 17 & 1017 \\ 2 & 18 & 2018 \\ 3 & 19 & 3019 \end{vmatrix} \quad (b) \begin{vmatrix} 108 & 107 \\ 106 & 105 \end{vmatrix}.$$

7 Simplify
(a) $\det(2u + 3w, 3w, 2u + 5v) \div \det(u, v, w)$
(b) $\det(u + w, v + w, -u - v - 2w)$
(c) $\det(u - 4v + 6w, -3u + 5v - 3w, 2u - v - 3w)$.

8 The following matrices have determinant zero, so each matrix must map some non-zero vector to the origin. For each matrix, find some non-zero vector it maps to the origin.

$$(a) \begin{pmatrix} 0 & 1 & -1 \\ -1 & 0 & 1 \\ 1 & -1 & 0 \end{pmatrix} \quad (b) \begin{pmatrix} 1 & -2 & 1 \\ 2 & -1 & 0 \\ 3 & 0 & -1 \end{pmatrix} \quad (c) \begin{pmatrix} 1 & 2 & 3 \\ -2 & -1 & 0 \\ 1 & 0 & -1 \end{pmatrix}$$

$$(d) \begin{pmatrix} 1 & 2 & 3 \\ 4 & 5 & 6 \\ 7 & 8 & 9 \end{pmatrix} \quad (e) \begin{pmatrix} 1 & 4 & 7 \\ 2 & 5 & 8 \\ 3 & 6 & 9 \end{pmatrix}.$$

9 The following matrices have determinant zero and each maps a plane to the origin. For each matrix find any two independent vectors that it maps to the origin.

$$(a) \begin{pmatrix} 1 & 0 & 0 \\ 0 & 0 & 0 \\ 0 & 0 & 0 \end{pmatrix} \quad (b) \begin{pmatrix} 1 & 1 & 1 \\ 1 & 1 & 1 \\ 1 & 1 & 1 \end{pmatrix} \quad (c) \begin{pmatrix} 1 & 1 & 1 \\ 2 & 2 & 2 \\ 3 & 3 & 3 \end{pmatrix} \quad (d) \begin{pmatrix} 1 & 2 & 3 \\ 1 & 2 & 3 \\ 1 & 2 & 3 \end{pmatrix} \quad (e) \begin{pmatrix} 1 & 2 & 3 \\ 2 & 4 & 6 \\ 3 & 6 & 9 \end{pmatrix}.$$

10 Let

$$M = \begin{pmatrix} 7 & 2 & 0 \\ 2 & 6 & -2 \\ 0 & -2 & 5 \end{pmatrix}.$$

Verify that $\det(M - \lambda I) = 0$ for $\lambda = 3$, 6 or 9. Find non-zero vectors u, v, w such that

$$(M - 3I)u = 0, \quad (M - 6I)v = 0, \quad (M - 9I)w = 0.$$

(The vectors so found are eigenvectors of M, corresponding to eigenvalues 3, 6 and 9 respectively.)

11 Verify that, for

$$M = \begin{pmatrix} 0 & \sqrt{2} & 0 \\ \sqrt{2} & 0 & \sqrt{2} \\ 0 & \sqrt{2} & 0 \end{pmatrix}$$

$\det(M - \lambda I) = -\lambda^3 + 4\lambda$. For what values of λ is this determinant zero? Find the corresponding eigenvectors. (*Note.* An eigenvalue can be zero; an eigenvector cannot.)

12 Show that for

$$M = \begin{pmatrix} -1 & -2 & 0 \\ -2 & 0 & 2 \\ 0 & 2 & 1 \end{pmatrix}$$

we have $\det(M - \lambda I) = -\lambda^3 + 9\lambda$. Find the eigenvalues and eigenvectors of M.

13 Show that for

$$M = \begin{pmatrix} 2 & 2 & -3 \\ 2 & 2 & 3 \\ -3 & 3 & 3 \end{pmatrix}$$

the characteristic equation $\det(M - \lambda I) = 0$ is equivalent to $\lambda^3 - 7\lambda^2 - 6\lambda + 72 = 0$. Solve this equation, it being given that one solution is $\lambda = -3$. Find the eigenvalues and eigenvectors of M.

14 Find the eigenvalues and eigenvectors of the matrix

$$\begin{pmatrix} 3 & 4 & -4 \\ 4 & 5 & 0 \\ -4 & 0 & 1 \end{pmatrix}.$$

15 Let

$$M = \begin{pmatrix} \frac{1}{3} & \frac{1}{3} & \frac{1}{3} \\ \frac{1}{3} & \frac{1}{3} & \frac{1}{3} \\ \frac{1}{3} & \frac{1}{3} & \frac{1}{3} \end{pmatrix}.$$

For this matrix both $\det M$ and $\det(M - I)$ are zero. Show that $M - I$ maps a non-zero vector to the origin (as in question 8) while M itself maps an entire plane to the origin. Describe the eigenvalues and corresponding eigenvectors of the matrix M.

16 Show that the matrix

$$\begin{pmatrix} \frac{1}{6} & \frac{1}{6} & \frac{1}{3} \\ \frac{1}{6} & \frac{1}{6} & \frac{1}{3} \\ \frac{1}{3} & \frac{1}{3} & \frac{2}{3} \end{pmatrix}$$

has eigenvalues 0 and 1. Investigate the eigenvectors belonging to these values.

17 Investigate the eigenvalues and corresponding eigenvectors of the matrix

$$\begin{pmatrix} \frac{2}{3} & -\frac{1}{3} & -\frac{1}{3} \\ -\frac{1}{3} & \frac{2}{3} & -\frac{1}{3} \\ -\frac{1}{3} & -\frac{1}{3} & \frac{2}{3} \end{pmatrix}.$$

In what way does the answer resemble, and in what way does it differ from, the answer to question 15?

18 The matrix

$$\begin{pmatrix} \frac{5}{6} & -\frac{1}{6} & -\frac{1}{3} \\ -\frac{1}{6} & \frac{5}{6} & -\frac{1}{3} \\ -\frac{1}{3} & -\frac{1}{3} & \frac{1}{3} \end{pmatrix}$$

has eigenvalues 0 and 1. Investigate the eigenvectors corresponding to these values. Compare and contrast the answer to this question with the answer to question 16.

Cramer's rule

Suppose we have two equations

$$ax + by + cz = 0, \tag{1}$$

$$dx + ey + fz = 0. \tag{2}$$

Here we have 2 equations for 3 unknowns, so we cannot expect to fix x, y, z completely. In fact, we can see that if x, y, z, is any solution, so is kx, ky, kz. The most we can hope to do is to determine the ratios $x:y:z$.

Geometrically, equations (1) and (2) represent two planes through the origin, in space of 3 dimensions; their intersection will be a line. At least, this will be so in general. There will be exceptional cases; for instance we may have $a = b = c = 0$, in which case the first equation tells us nothing at all; or the two planes may coincide. We will suppose we are dealing with the general situation.

If we take $f(1) - c(2)$ we get rid of z and find $x(af - cd) + y(bf - ce) = 0$. This will be satisfied if $x = k(bf - ce)$ and $y = -k(af - cd)$, where k is any number.

If we get rid of x by taking $-d(1) + a(2)$ we obtain $y(ae - bd) + z(af - cd) = 0$. This will be satisfied if we use the value already found for y and take $z = k(ae - bd)$.

It may be checked that the values of x, y, z just found do in fact satisfy equations (1) and (2).

It will be noticed that the expressions found have the form of determinants. We may write our solution in the form

$$x = k \begin{vmatrix} b & c \\ e & f \end{vmatrix}, \qquad y = -k \begin{vmatrix} a & c \\ d & f \end{vmatrix}, \qquad z = k \begin{vmatrix} a & b \\ d & e \end{vmatrix}.$$

The equations (1) and (2) had the coefficient scheme

$$\begin{array}{ccc} x & y & z \\ a & b & c \\ d & e & f \end{array}$$

In the solution, the determinant associated with x can be formed by striking out the symbols underneath x in the coefficient scheme. Similar rules apply for y and z.

With these determinants we must put the signs $+$, $-$, $+$, alternating just as they did when we found the formula for 3×3 determinant. There is also of course the arbitrary multiplier k.

This procedure, known as Cramer's rule, can be extended to the general case of n linear equations in $(n + 1)$ unknowns, but its practical value is so small that it will not be discussed here

28 Matrices other than square; partitions

So far we have used only square matrices. These correspond to transformations in which a space is mapped onto itself, or possibly to cases in which a space is mapped onto another space having the same number of dimensions.

Students sometimes seem to feel that it is wrong to have a map involving spaces with different dimensions, but in fact this is perfectly in order. It is indeed an everyday experience. If we buy some pencils and notebooks, the price of a pencil and of a notebook being fixed, the mapping, purchase $\rightarrow$ payment, is from 2 dimensions to 1 dimension. For we need two numbers, x, y, to specify the purchase, x pencils and y notebooks, but only one number to specify the money paid. In an electrical circuit the voltages supplied by 2 batteries may determine the currents in 3 resistances, or in 32, or in 3007. In a pin jointed framework, the small displacements of the pins determine the tensions in the bars. There is no reason to expect that the number of co-ordinates needed to specify the positions of the pins will equal the number of the bars. Thus it is in no way unusual for an input, specified by m numbers to lead to an output specified by n numbers, with $m \neq n$. Thus a mapping from m dimensions to n dimensions is in no way strange or impermissible.

A linear mapping $(x, y, z) \rightarrow (s, t)$ would be represented by equations

$$\left. \begin{array}{l} ax + by + cz = s, \\ dx + ey + fz = t. \end{array} \right\} \tag{1}$$

In matrix form this would appear as

$$\begin{pmatrix} a & b & c \\ d & e & f \end{pmatrix} \begin{pmatrix} x \\ y \\ z \end{pmatrix} = \begin{pmatrix} s \\ t \end{pmatrix}. \tag{2}$$

The mapping here is represented by a 2×3 matrix – that is, 2 rows and 3 columns. Note that 2 gives the dimensions of the output, 3 the dimensions of the input.

When we form a product ST it is understood the output of T becomes the input of S. It will be impossible to form the product ST if the output of T is not acceptable as an input for S; that is to say, if it is of different dimension. We can form this product if S is an $m \times n$ matrix and T an $n \times p$ matrix. For instance, S could be 3×2 and T could be 2×4, as here:

$$\left. \begin{aligned} \xi &= aX + bY, \\ \eta &= cX + dY, \\ \zeta &= eX + fY, \end{aligned} \right\} \tag{3}$$

$$\left. \begin{aligned} X &= gx + hy + iz + jt, \\ Y &= kx + my + nz + pt. \end{aligned} \right\} \tag{4}$$

Here X, Y form a link between x, y, z, t and ξ, η, ζ. If we substitute for X, Y we can obtain equations for the mapping $(x, y, z, t) \rightarrow (\xi, \eta, \zeta)$. If this is done, and all equations are then expressed in matrix form, we arrive at the result:

$$\begin{pmatrix} a & b \\ c & d \\ e & f \end{pmatrix} \begin{pmatrix} g & h & i & j \\ k & m & n & p \end{pmatrix} = \begin{pmatrix} ag + bk & ah + bm & ai + bn & aj + bp \\ cg + dk & ch + dm & ci + dn & cj + dp \\ eg + fk & eh + fm & ei + fn & ej + fp \end{pmatrix}. \tag{5}$$

It will be seen that this resembles very closely the procedure for matrix multiplication that we have been using until now. We go across the rows in the first matrix, down the columns of the second matrix.

If we try to form an impossible product, we soon notice it. Combining a row with a column, we find that we run out of letters in one before the other is finished, and we are at a loss how to proceed. Thus we automatically receive warning that we are writing nonsense.

If S is the matrix corresponding to equations (3), and T to equations (4), the sum $S + T$ is meaningless. We cannot work from the definition $(S + T)v = Sv + Tv$. For S accepts only vectors v from a 2-dimensional space and T accepts only vectors from a 4-dimensional space. Thus we cannot find an input v that makes both Sv and Tv intelligible. Nor could it help us if we could. For Sv, an output from S, is a vector in 3 dimensions, while Tv, an output from T is in 2 dimensions. We cannot add vectors of different dimensionality. For $S + T$ to be meaningful S and T must have the same input space (otherwise we cannot find v to feed in) and the same output space (otherwise we cannot add Sv and Tv).

Thus $S + T$ can be defined if both S and T are 2×3 matrices. If

$$S = \begin{pmatrix} a & b & c \\ d & e & f \end{pmatrix} \qquad T = \begin{pmatrix} g & h & i \\ j & k & l \end{pmatrix}$$

then

$$S + T = \begin{pmatrix} a + g & b + h & c + i \\ d + j & e + k & f + l \end{pmatrix}.$$

Here again the rule closely resembles what we have already been doing. We have only to remember that addition is not always possible, and this is fairly easy since here too the procedure becomes impossible if we apply it to a meaningless situation.

Multiplication by a number k remains exactly as before. For instance, with the matrix S as in the last example

$$kS = \begin{pmatrix} ka & kb & kc \\ kd & ke & kf \end{pmatrix}.$$

This operation can always be done. The mapping kS means that we apply the mapping S, and then multiply the resulting vector by k. There is no obstacle to carrying this out.

The inverse matrix is not defined when the matrix is not square. For instance, with the matrix equation (2), corresponding to the equations (1), we can go from x, y, z to s, t but not in the reverse direction. Given s and t, we cannot find the 3 unknowns x, y, z.

In a mapping of the type in equation (3), such as, for example

$$\xi = X + 2Y,$$
$$\eta = 3X + 5Y,$$
$$\zeta = 4X + 7Y,$$

the difficulty of getting back from ξ, η, ζ to X, Y is of the opposite kind. Now we have too many equations to satisfy rather than too few. For example, we cannot find any X, Y to give $\xi = 3, \eta = 8, \zeta = 2$.

Determinants also are not defined except for square matrices.

Partition of matrices

In §27 dealing with determinants, we sometimes spoke of determinant $\begin{pmatrix} p & q \\ r & s \end{pmatrix}$ and sometimes of $\det(u, v)$ where

$$u = \begin{pmatrix} p \\ r \end{pmatrix}, \qquad v = \begin{pmatrix} q \\ s \end{pmatrix}.$$

Thus as it were, we drew a line down the middle of the matrix, and fused the entries p and r together into the vector u, and the entries q and s into the vector v.

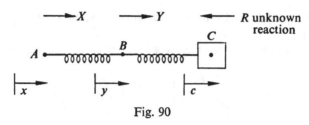

Fig. 90

We can in fact write the matrix $\begin{pmatrix} p & q \\ r & s \end{pmatrix}$ as (u, v) and the procedures of matrix algebra will work just as well. If this matrix represents a transformation mapping x, y to x^*, y^* we can write

$$\begin{pmatrix} x^* \\ y^* \end{pmatrix} = (u\ v)\begin{pmatrix} x \\ y \end{pmatrix} = ux + vy$$

by the usual row and column multiplication procedure. This again can be written

$$\begin{pmatrix} x^* \\ y^* \end{pmatrix} = \begin{pmatrix} p \\ r \end{pmatrix}x + \begin{pmatrix} q \\ s \end{pmatrix}y = \begin{pmatrix} px \\ rx \end{pmatrix} + \begin{pmatrix} qy \\ sy \end{pmatrix} = \begin{pmatrix} px + qy \\ rx + sy \end{pmatrix}$$

by the usual procedures for calculations involving vectors. These are the correct equations, corresponding to the matrix in its original form. Note incidentally how well the expression $ux + vy$ occurring above brings out the meaning of the columns u, v, which we noted in §22.

In fact we have considerable freedom to chop a matrix up and to regard it as a mosaic of smaller matrices. This process is known as *partitioning*.

We may wish to partition a matrix because the quantities on which it acts fall into separate classes. For instance, consider a very much simplified problem in the theory of structures.* We have two springs, which lie in a line, as shown in Fig. 90.

The point C is fastened; it has a prescribed displacement, c. Known forces X and Y act at A and B. They call into existence an unknown reaction R at C. The points A and B experience unknown displacements x and y. Thus both the displacements and the forces fall naturally into two classes, known and unknown. We naturally separate x, y from c; X, Y from R.

The quantities in this problem are linked by the equations

$$\begin{aligned}
X &= k_{11}x + k_{12}y \;\vdots\; + k_{13}c \\
Y &= k_{21}x + k_{22}y \;\vdots\; + k_{23}c \\
\hline
R &= k_{31}x + k_{32}y \;\vdots\; + k_{33}c
\end{aligned}$$

where the coefficients k_{rs} can be calculated from the spring constants.

In view of the separation we have noticed, it would be natural to introduce a vector

* See H. C. Martin – *Introduction to Matrix Methods of Structural Analysis* chapter 2 (McGraw-Hill, 1966).

$v = \begin{pmatrix} x \\ y \end{pmatrix}$ to specify unknown displacements, and a vector $F = \begin{pmatrix} X \\ Y \end{pmatrix}$ to specify known forces. In the equations above, the first two columns on the right-hand side are associated with the vector v, the first two rows with the vector F. The dotted lines indicate this situation. We now look at the various compartments in these equations. The contents of each compartment go very easily into matrix notation:

$$\begin{matrix} k_{11}x + k_{12}y \\ k_{21}x + k_{22}y \end{matrix} = \begin{pmatrix} k_{11} & k_{12} \\ k_{21} & k_{22} \end{pmatrix} \begin{pmatrix} x \\ y \end{pmatrix} = Hv, \quad \text{say.}$$

This deals with the north-west box. In the north-east we see

$$\begin{pmatrix} k_{13}c \\ k_{23}c \end{pmatrix} = \begin{pmatrix} k_{13} \\ k_{23} \end{pmatrix} c = Lc, \quad \text{say.}$$

Similarly we have

$$k_{31}x + k_{32}y = (k_{31} \quad k_{32}) \begin{pmatrix} x \\ y \end{pmatrix} = Gv, \quad \text{say.}$$

We do not need any new notation for $k_{33}c$ which is a single number. Note incidentally that a single number, k_{33}, can be regarded as a 1×1 matrix, mapping 1-dimensional space to 1-dimensional space.

Our equations can now be written

$$F = Hv + Lc, \tag{6}$$

$$R = Gv + k_{33}c. \tag{7}$$

These can be combined into the form of a single matrix

$$\begin{pmatrix} F \\ R \end{pmatrix} = \begin{pmatrix} H & L \\ G & k_{33} \end{pmatrix} \begin{pmatrix} v \\ c \end{pmatrix}. \tag{8}$$

Equations (6) and (7) might be convenient for finding v and R. If we solve equation (6) for v we find $v = H^{-1}(F - Lc)$. All the quantities on the right-hand side are known. Note here that we are making use of the inverse, H^{-1}. Substituting in equation (7) gives us the unknown reaction R as $GH^{-1}(F - Lc) + k_{33}c$.

Of course this problem is so simple that we would in fact never solve it this way. It is intended to provide an illustration and a model for much more complicated problems involving hundreds of variables. The importance of equation (8) is that the whole procedure becomes stereotyped. The computer will be programmed to deal with this single matrix equation, the form of which is such that the machine can recognize which quantities are known and which are unknowns. The computer would be programmed to calculate the matrices H, L, G and the numbers k_{33} from raw data about the structure involved – the number of bars and plates involved, the manner in which they are joined, their elastic constants and so forth.

In this particular example the columns and rows have been split in the same manner. In each case we have separated the first two from the third. It is not in any way necessary that this similarity should exist. For instance, in our first example of partition we had the division

$$\begin{pmatrix} p & \vdots & q \\ r & \vdots & s \end{pmatrix}.$$

We divided the columns; the rows were not divided at all. So long as we are dealing with a single matrix, we can divide the rows in any way we like and the columns in any way we like.

When we are dealing with the product of matrices, or with a matrix acting as a vector, we must make sure that the partition is such that the necessary multiplications can in fact be carried out. For instance in the product

$$\begin{pmatrix} 1 & 2 & 3 & \vdots & 4 & 5 \\ 6 & 7 & 8 & \vdots & 9 & 10 \\ 11 & 12 & 13 & \vdots & 14 & 15 \\ 16 & 17 & 18 & \vdots & 19 & 20 \end{pmatrix} \quad \begin{pmatrix} 1 & 2 & 3 \\ 4 & 5 & 6 \\ 7 & 8 & 9 \\ \hline 10 & 11 & 12 \\ 13 & 14 & 15 \end{pmatrix}$$

if we have decided to split the second matrix into 3 rows and 2 rows, we are forced to split the first matrix into 3 columns and 2 columns. Otherwise we shall not be able to form the necessary products in the scheme

$$(S \,\vdots\, T)\begin{pmatrix} U \\ \hline V \end{pmatrix} = SU + TV.$$

In a product of two matrices, such as the example just given, the restriction does not in any way affect our freedom to chop up the rows of the first matrix and the columns of the second. The partition shown above is perfectly satisfactory: we can, if we wish, leave it as it is. However, if we have some reason for doing so, we can draw further horizontal divisions in the first matrix, and further vertical divisions in the second, in any way we like.

Questions

1 Calculate the products

(a) $\begin{pmatrix} 1 & 1 & 1 \\ 1 & 1 & 0 \end{pmatrix} \begin{pmatrix} 1 & 4 & 6 & 3 \\ 2 & 5 & 5 & 2 \\ 3 & 6 & 4 & 1 \end{pmatrix}$ (b) $\begin{pmatrix} 1 & 1 & 1 \\ 1 & 1 & 0 \end{pmatrix} \begin{pmatrix} 1 & 4 \\ 2 & 5 \\ 3 & 6 \end{pmatrix}$

(c) $\begin{pmatrix} 1 & 1 \\ 1 & 1 \\ 1 & 0 \end{pmatrix} \begin{pmatrix} 1 & 3 & 5 \\ 2 & 4 & 6 \end{pmatrix}$ (d) $\begin{pmatrix} 1 \\ 2 \\ 3 \end{pmatrix} (1 \quad 3 \quad 5)$

(e) $(1 \quad 3 \quad 5) \begin{pmatrix} 1 \\ 2 \\ 3 \end{pmatrix}$
(f) $\begin{pmatrix} a & b & c \\ x & y & z \end{pmatrix} \begin{pmatrix} a & x \\ b & y \\ c & z \end{pmatrix}$

(g) $\begin{pmatrix} a & x \\ b & y \\ c & z \end{pmatrix} \begin{pmatrix} a & b & c \\ x & y & z \end{pmatrix}$.

2 In Question 1, nine different matrices occur as factors of the various products. Discuss which of these matrices are capable of being added together.

3 Calculate

$$\begin{pmatrix} 1 & 1 & 1 \\ a & b & c \end{pmatrix} \begin{pmatrix} 1 & a \\ 1 & b \\ 1 & c \end{pmatrix} \quad \text{and} \quad \begin{pmatrix} 1 & 1 & 1 & 1 \\ a & b & c & d \\ a^2 & b^2 & c^2 & d^2 \end{pmatrix} \begin{pmatrix} 1 & a & a^2 \\ 1 & b & b^2 \\ 1 & c & c^2 \\ 1 & d & d^2 \end{pmatrix};$$

the answers can be written conveniently by using the abbreviation S_n for $a^n + b^n + \cdots$ The products so found belong to a type of matrix that occurs when curves are fitted to observed data by the method of Least Squares.

4 Calculate

$$\begin{pmatrix} 1 & 1 & 2 \\ 1 & -2 & 1 \end{pmatrix} \begin{pmatrix} -1 & -3 \\ 0 & -1 \\ 1 & 2 \end{pmatrix} \quad \text{and} \quad \begin{pmatrix} 1 & 1 & 2 \\ 1 & -2 & 1 \end{pmatrix} \begin{pmatrix} 4 & 2 \\ 1 & 0 \\ -2 & -1 \end{pmatrix}.$$

Does the result throw any light on the question of whether a matrix with 2 rows and 3 columns can have an inverse?

5 P is a matrix with 2 rows and 4 columns. Q is a matrix, and the products PQ and QP are both meaningful. How many rows and columns has Q? What are the numbers of rows and columns in PQ? And in QP?

6 Let

$$\mathbf{O} = \begin{pmatrix} 0 & 0 \\ 0 & 0 \end{pmatrix} \quad \text{and} \quad \mathbf{I} = \begin{pmatrix} 1 & 0 \\ 0 & 1 \end{pmatrix}.$$

Let

$$P = \begin{pmatrix} \mathbf{O} & \mathbf{I} \\ \mathbf{I} & \mathbf{O} \end{pmatrix}, \quad Q = \begin{pmatrix} \mathbf{O} & -\mathbf{I} \\ \mathbf{I} & \mathbf{O} \end{pmatrix}, \quad R = \begin{pmatrix} \mathbf{I} & -\mathbf{I} \\ \mathbf{O} & \mathbf{I} \end{pmatrix}$$

so that P, Q, R are matrices with 4 rows and 4 columns. Calculate P^2, Q^2, R^2, R^3, PQ, PR and QR by means of the partitioned forms given above. Also write P, Q, R out in full and calculate the results directly, without the use of partitions. Compare the labour involved in the two methods.

The solution for the first of the required products is given here.

Solution: to find P^2.
First method.

$$P^2 = \begin{pmatrix} \mathbf{O} & \mathbf{I} \\ \mathbf{I} & \mathbf{O} \end{pmatrix} \begin{pmatrix} \mathbf{O} & \mathbf{I} \\ \mathbf{I} & \mathbf{O} \end{pmatrix} = \begin{pmatrix} \mathbf{I} & \mathbf{O} \\ \mathbf{O} & \mathbf{I} \end{pmatrix} = \begin{pmatrix} 1 & 0 & 0 & 0 \\ 0 & 1 & 0 & 0 \\ 0 & 0 & 0 & 0 \\ 0 & 0 & 0 & 1 \end{pmatrix}.$$

Second method.

$$P^2 = \begin{pmatrix} 0 & 0 & 1 & 0 \\ 0 & 0 & 0 & 1 \\ 1 & 0 & 0 & 0 \\ 0 & 1 & 0 & 0 \end{pmatrix} \begin{pmatrix} 0 & 0 & 1 & 0 \\ 0 & 0 & 0 & 1 \\ 1 & 0 & 0 & 0 \\ 0 & 1 & 0 & 0 \end{pmatrix} = \begin{pmatrix} 1 & 0 & 0 & 0 \\ 0 & 1 & 0 & 0 \\ 0 & 0 & 1 & 0 \\ 0 & 0 & 0 & 1 \end{pmatrix}.$$

7 Let

$$I = \begin{pmatrix} 1 & 0 \\ 0 & 1 \end{pmatrix}, \quad A = \begin{pmatrix} p & q \\ r & s \end{pmatrix} \quad \text{and} \quad O = \begin{pmatrix} 0 & 0 \\ 0 & 0 \end{pmatrix}.$$

Find the product

$$\begin{pmatrix} I & A \\ O & I \end{pmatrix} \begin{pmatrix} I & -A \\ O & I \end{pmatrix}.$$

What is the inverse of the matrix

$$\begin{pmatrix} 1 & 0 & 2 & 3 \\ 0 & 1 & 4 & 5 \\ 0 & 0 & 1 & 0 \\ 0 & 0 & 0 & 1 \end{pmatrix}?$$

8 Calculate the product

$$\begin{pmatrix} a_1 & a_2 & a_3 \\ \hline b_1 & b_2 & b_3 \end{pmatrix} \begin{pmatrix} u_1 & v_1 \\ u_2 & v_2 \\ u_3 & v_3 \end{pmatrix}$$

directly, i.e. without any reference to partitioning. By partitioning in the manner indicated by the broken lines, this product may be written as $\begin{pmatrix} a' \\ b' \end{pmatrix}(u \quad v)$ where a', b' represent the rows of the first matrix and u, v the columns of the second. Use this form to obtain the product. Observe that the product, in this second form obtained by partitioning, gives a useful abbreviation for the product as obtained directly.

9 Let

$$M = \begin{pmatrix} a & d \\ b & e \\ c & f \end{pmatrix} \begin{pmatrix} 1 & 3 & 5 \\ 2 & 4 & 6 \end{pmatrix}.$$

We can write M in the abbreviated form

$$M = (u \, v) \begin{pmatrix} 1 & 3 & 5 \\ 2 & 4 & 6 \end{pmatrix}$$

where u and v represent the columns of the first factor in M. Use this last equation to write the product M. What can be said, in geometrical terms, about the three vectors that occur in the columns of M? What is the value of det M? Can any similar conclusion be drawn about det (PQ) where P is any 3×2 matrix and Q any 2×3 matrix?

29 Subscript and summation notation

In §28, equations (3) and (4) came close to using the entire resources of the alphabet. In problems where large numbers of variables are involved, we are faced not only with the difficulty of finding enough symbols; we also need some systematic way of choosing and employing the symbols. For instance, in an expression such as $ax + by + cz$ we can see that a is the coefficient of the first variable x, b of the second variable y and c of the third variable z. However, we cannot proceed in this way if we have 100 variables and 100 coefficients corresponding to them. If we are communicating with a human being we may write $ax + by + \cdots cz$ where the dots indicate 'and so on, until the 100th variable z is reached'. Such explanations are useless if we are dealing with a computer: explicit instructions are needed then.

The natural numbers provide a convenient means for generating arbitrarily large collections of symbols. If we have 100 unknowns, we can write them as x_1, x_2, x_3 up to x_{100}. If we wish to form a linear expression we can use $a_1x_1 + a_2x_2 + \cdots + a_{100}x_{100}$, and it is immediately clear that the coefficient of, say, x_{57} is a_{57}. We still have dots indicating 'and so on', but it is quite easy to programme instructions that make a computer do exactly what is required.

The summation sign, $\sum$ or S, provides a convenient abbreviation. '$\sum$' is the Greek equivalent of 'S', which is the initial letter of 'Sum'. By the expression

$$\sum x_r, \qquad 1 \leqslant r \leqslant 3$$

we understand $x_1 + x_2 + x_3$. The instruction $1 \leqslant r \leqslant 3$ tells us that in x_r we are to replace r in turn by 1, 2 and 3, so we get x_1, x_2, x_3. The instruction $\sum$ tells us that these are to be summed to give $x_1 + x_2 + x_3$. The linear expression mentioned in the previous paragraph would be shown as $\sum a_rx_r$; $1 \leqslant r \leqslant 100$. Sometimes the numbers, at which r is to begin and end, are put below and above the summation sign. This is convenient enough for handwriting but is very inconvenient for the printer.

When we are dealing with a linear transformation, that is to be expressed by a matrix, a certain scheme of subscripts is almost invariably used. It is seen in the following example:

$$\left.\begin{aligned} x_1^* &= a_{11}x_1 + a_{12}x_2 + a_{13}x_3 + a_{14}x_4 \\ x_2^* &= a_{21}x_1 + a_{22}x_2 + a_{23}x_3 + a_{24}x_4 \\ x_3^* &= a_{31}x_1 + a_{32}x_2 + a_{33}x_3 + a_{34}x_4 \end{aligned}\right\} \tag{1}$$

Here the first subscript tells us which row we are in; for example, a_{31}, a_{32}, a_{33}, a_{34} all occur in the equation for x_3^* and so are found in the third row. The second subscript tells us in which column the coefficient occurs; thus a_{12}, a_{22}, a_{32} all occur as coefficients of x_2 and so are in the second column of the coefficient arrangement.

The whole of equations (1) can be written in the very compact form

$$x_r^* = \sum_s a_{rs}x_s; \qquad 1 \leqslant r \leqslant 3; 1 \leqslant s \leqslant 4.$$

Here we have to write s under the summation sign to indicate that we are summing only for s, not for r. If we plan to make a further transformation

$$\left.\begin{array}{l} x_1^{**} = b_{11}x_1^* + b_{12}x_2^* + b_{13}x_3^* \\ x_2^{**} = b_{21}x_1^* + b_{22}x_2^* + b_{23}x_3^* \end{array}\right\} \qquad (2)$$

this transformation can be written

$$x_q^{**} = \sum_r b_{qr}x_r^*; \qquad 1 \leqslant q \leqslant 2; 1 \leqslant r \leqslant 3.$$

Let A and B represent transformations (1) and (2), and let C stand for the product transformation BA. It is useful for anyone unfamiliar with this symbolism to write the matrices for B and A, and to calculate sample entries in the matrix for C. There are 8 entries in C and each involves the sum of 3 products. Nothing is gained by writing out the entire matrix for C. Enough should be written for the pattern to become apparent. For instance, we can check that c_{23}, the entry in the second row and third column of C, is given by

$$c_{23} = b_{21}a_{13} + b_{22}a_{23} + b_{23}a_{33}.$$

The pattern here may appear more clearly if we suppress the middle subscripts, and write first

$$c_{23} = b_{2.}a_{.3} + b_{2.}a_{.3} + b_{2.}a_{.3}$$

The pattern in the last equation should be evident. To get to the complete equation, we replace the dots in the first term by 1 and 1, in the second term by 2 and 2, and in the third term by 3 and 3. Our results may be shown in compact form as

$$c_{23} = \sum_r b_{2r}a_{r3}.$$

The numbers 2 and 3 enjoy no special privileges. If we replace them by the symbols q and s we obtain the formula for any entry in the matrix for C;

$$c_{qs} = \sum_r b_{qr}a_{rs}. \qquad (3)$$

In our particular example we would have to indicate that q could stand for 1 or 2, s for 1, 2, 3 or 4, and that we had to sum r over the values 1, 2, 3.

For *any* matrix product, $C = BA$, equation (3) is valid. It needs to be supplemented only by details as to the values through which q, r and s are to run.

Obviously this notation is particularly valuable when matrices with many rows and columns are involved. It can also be useful when certain theoretical results are being obtained. For instance, we shall need it when we come to define and to calculate with the transpose of a matrix.

Notice that, with this symbolism, the subscripts over which we sum are always adjacent. In equation (3) above, no other subscript separates the two subscripts r involved in the summation. Similarly, in the compact form of equations (1),

$$x_r^* = \sum_s a_{rs} x_s,$$

we are summing over s, and we find the subscripts s close together. (This does not apply to the letter s written under the summation sign; this is not a subscript.)

Repeated summations may be met. For instance, if in the equations $x_q^{**} = \sum_s c_{qs} x_s$, which represent the transformation C we substitute the values of c_{qs} given by equation (3), we find

$$x_q^{**} = \sum_r \sum_s b_{qr} a_{rs} x_s. \tag{4}$$

These equations give x_1^{**}, x_2^{**} in terms of x_1, x_2, x_3, x_4 and could be obtained by a primitive method, namely by substituting in equations (2) the values of x_1^*, x_2^*, x_3^* given by equations (1). Anyone who is not clear about the effect of the double summation sign might find it helpful to carry through this primitive method far enough to see what the expressions mean.

Quite generally, if some symbolism involving summation signs does not convey a clear message to the student, the best thing to do is to go back from the compact notation to the more cumbersome, but more familiar, elementary notation, such as that in equations (1) and (2).

Notice that equation (4) still shows the subscripts, over which we are to sum, as neighbours. We are summing over r and s; the sequence of subscripts is q; r, r; s, s.

30   Row and column vectors

At some stage of his reading a student is likely to meet references to row vectors and column vectors. For instance in 2 dimensions we have already spoken of the 2×1 matrix $\begin{pmatrix} x \\ y \end{pmatrix}$ as a column vector. The 1×2 matrix (a, b) is often spoken of as a row vector.

In matrix formalism it is possible to add two column vectors: we have already done this on many occasions. It is also possible to add two row vectors,

$$(a_1, b_1) + (a_2, b_2) = (a_1 + a_2, b_1 + b_2).$$

However, it is meaningless to speak of the sum of a row vector and a column vector, for the lay-out of the entries differs in the two cases. With multiplication the situation is rather the reverse. Matrix rules give no meaning to the product of (a_1, b_1) and (a_1, b_2). We cannot carry out row and column multiplication. Similarly no meaning attaches to the product of two column vectors. However, it is possible to form the product of a row vector and a column vector $(a, b)\begin{pmatrix} x \\ y \end{pmatrix} = ax + by$.

It may seem very strange that there should exist two different types of vector, and that there should be such restrictions on operations with them. However, there are situations in which such a distinction arises quite naturally. Suppose for instance that a and b represent forces acting on some structure, while x and y represent displacements. It is quite natural that we should add displacements together, or add forces together, but never add displacement to force. Again force multiplied by displacement represents work done, and in fact the expression $ax + by$ does arise in this connection. However, force multiplied by force does not suggest any significant physical property, nor, in this immediate context, does displacement multiplied by displacement. Matrix formalism therefore would seem to fit quite snugly to the idea of identifying forces with row vectors and displacements with column vectors.

The idea comes out even more strongly in a very elementary application. Suppose $\begin{pmatrix} x \\ y \end{pmatrix}$ represents a purchase of x nuts and y bolts, and that (a, b) represents a price table – a dollars for a nut, b dollars for a bolt. Then $ax + by$, which is $(a, b)\begin{pmatrix} x \\ y \end{pmatrix}$, represents the cost of a purchase $\begin{pmatrix} x \\ y \end{pmatrix}$ when the price table is (a, b). We do not expect to multiply purchases by purchases or dollars by dollars. Equally we may add purchases together, or we may add price schemes, but we do not add purchases to price schemes.

This application is helpful when we come to consider the graphical representation of column and row vectors. A purchase, we represent, as usual, by a point on the graph paper. Thus the point shown in Fig. 91 represents a purchase of 4 nuts and 1 bolt.

How are we to represent a price scheme such as $(0.5, 1)$, \$0.50 for a nut and \$1 for a bolt? We cannot mark the point with $x = 0.5$ and $y = 1$. That would represent a purchase (if such a thing were possible) of half a nut and a bolt. We cannot mark a

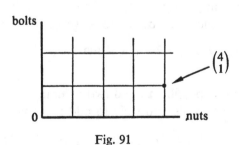

Fig. 91

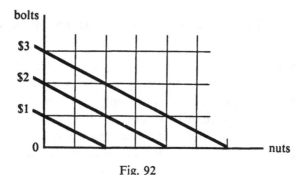

Fig. 92

point on the graph for purchases; knowing the prices at which objects were bought tells us nothing about how many were bought.

We could meet the difficulty by having an entirely separate chart on which prices were marked, and this idea can be developed in a fruitful manner. However, we shall not do this here. We want to bring out the relevance of the price scheme to the purchase diagram, and this is easily stated. A price scheme does not tell us what we have purchased; it does tell us what it would cost us if we decided to make some particular purchase. This we can show on the graph. We could mark on each point the cost corresponding to the purchase that point represents. It might be rather more convenient to follow the procedure of a weather map, in which pressures are indicated by isobars linking points with equal pressures. The diagram would then appear as in Fig. 92.

From this diagram we can read off the cost of each purchase. Thus the point for 4 nuts and 1 bolt, which we saw on the diagram before this, falls on the line marked 3. It would cost $3. Some points of course fall between the lines marked. Thus 3 nuts and 1 bolt cost $2.5. It can be seen that the corresponding point falls halfway between the lines marked 2 and 3. Thus in fact this graph does convey all the information contained in the price scheme.

The equations of the lines are $0.5x + y = 1$, $0.5x + y = 2$, $0.5x + y = 3$. Thus, in reading off the numbers marked on the lines we are simply obtaining the value of $0.5x + y$, the expression with coefficients 0.5 and 1. Thus the row vector $(0.5, 1)$ can be regarded as a coefficient scheme for such an expression.

A price scheme tells us how to work out the cost of any purchase. Thus a price scheme establishes a mapping purchase $\rightarrow$ cost. In our example above this mapping may be specified as

$$\begin{pmatrix} x \\ y \end{pmatrix} \rightarrow 0.5x + y.$$

Quite generally we can regard the row vector (a, b) as something that acts on a column vector and gives a number as the output; this ties in easily with the equation

$$(a, b)\begin{pmatrix} x \\ y \end{pmatrix} = ax + by.$$

We may interpret this as 'the operation (a, b) acting on the vector $\begin{pmatrix} x \\ y \end{pmatrix}$ gives the number $ax + by$'.

The fact that a column vector is represented by a point while a row vector is seen as a system of lines brings out the fact that they are two very different things. What they have in common is that they both require only two numbers to specify them. In 3 dimensions they would each require 3 numbers. In 3 dimensions, the column vector would still be represented by a point; the row vector would be represented by a system of evenly spaced parallel planes corresponding to the equations $ax + by + cz = n$, with n an integer.

Change of axes

In some piece of work, we may begin with a certain system of axes, and points specified by co-ordinates in those axes. We think of these co-ordinates as written in the form of a column. At some stage we may wish to introduce a new system of axes. We still wish to deal with the same points as before, and so we ask what new co-ordinates will now be needed to represent the same points. In this way we arrive at formulas connecting the co-ordinates in the old and the new system. This was done in §9.

In the same way, there may be some physical object which is specified in the original axes by a row vector. In 2 dimensions, this row vector may be represented by a system of lines, that enable us to attach a number to each point of the plane. These numbers have physical significance, so naturally, if we decide to use co-ordinates in some other system of axes, we must make sure that the row vector used gives the same numbers, the same system of lines. Thus the lines provide the link between the two systems.

An example may be helpful. Suppose our two systems are as shown in Fig. 93.

The dotted lines, labelled with the numbers 1, 2, 3, represent a row vector. It can be checked that in the first system of co-ordinates (based on squares) the dotted lines correspond to the expression $(x/2) + (y/2)$. In the second system (oblique co-ordinates) the line marked 1 contains the points $X = 2, Y = 0$ and $X = 0, Y = 1$. It is not hard to show that the equation of this line must be $(X/2) + Y = 1$ and in fact the expression

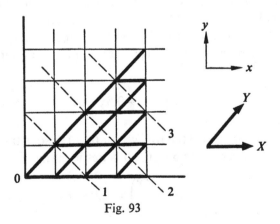

Fig. 93

$(X/2) + Y$ is the one we want for this system of co-ordinates. Reading off the coefficients in the two cases, we see that the row vector $(\frac{1}{2}, \frac{1}{2})$ in the first system represents the same object as the row vector $(\frac{1}{2}, 1)$ in the second system.

Exercise

By the method of §9, find equations giving x, y in terms of X, Y and verify that, for any X, Y, $(x/2) + (y/2) = (X/2) + Y$. This shows that the conclusion reached above could also have been found directly by algebra.

We will now consider the general case. Suppose we have a row vector (a, b) in a system with co-ordinates x and y, and that new co-ordinates, X and Y, are brought in by means of the equations

$$\left.\begin{array}{l} x = pX + qY, \\ y = rX + sY. \end{array}\right\} \tag{1}$$

We wish to find (A, B), the row vector as specified in the new system. This can be found very quickly. Any difficulty that may arise in this connection is not in carrying out the calculations, but in seeing clearly what is being done.

The row vector (a, b) assigns to the point $\begin{pmatrix} x \\ y \end{pmatrix}$ the number $ax + by$. By the equations above, this number is the same as $a(pX + qY) + b(rX + sY)$, which in turn equals $(ap + br)X + (aq + bs)Y$, where $\begin{pmatrix} X \\ Y \end{pmatrix}$ is the specification of the same point in the second system. Now, as seen by users of the second system, this point should be assigned the number $AX + BY$. These two results will agree if $A = ap + br$ and $B = aq + bs$. These equations tell us how to deal with row vectors when we change from one system to the other.

It may be noticed that the equations just obtained can be written in matrix form as

$$(A \quad B) = (a \quad b) \begin{pmatrix} p & q \\ r & s \end{pmatrix}. \tag{2}$$

Now equations (1) can be written as

$$\begin{pmatrix} x \\ y \end{pmatrix} = \begin{pmatrix} p & q \\ r & s \end{pmatrix} \begin{pmatrix} X \\ Y \end{pmatrix}. \tag{3}$$

Thus the same matrix appears in equations (2) and (3); in equations (3) it is written *before* the column vector, in equations (2) *after* the row vector. Note also a difference between equations (2) and (3); in equations (3) we can substitute the *new* co-ordinates (X and Y) to find the old co-ordinates (x and y), but in equations (2) it is the other way round – we can substitute the *old* in order to obtain the new.

All of this can be shown very conveniently by bringing in vector and matrix abbreviations; this incidentally provides a proof of these results for any number of dimensions.

We write

$$v = \begin{pmatrix} x \\ y \end{pmatrix}, \qquad V = \begin{pmatrix} X \\ Y \end{pmatrix}, \qquad M = \begin{pmatrix} p & q \\ r & s \end{pmatrix},$$

$$h = (a, b), \qquad H = (A, B).$$

Then $ax + by$ is written hv; this is the number associated with the vector that is called v in the first system. That vector is called V in the second system and is assigned the number HV. The number marked at any point is independent of the co-ordinates that some observer is using to describe that point, so we must have $hv = HV$.

Equations (3) may be written $v = MV$. If we substitute for v in the equation $hv = HV$, we get $hMV = HV$. Since this is to be so for every V, we must have $hM = H$, and this is equivalent to equations (2).

The equation $hM = H$ may also be written $h = HM^{-1}$. (Since M is a matrix suitable for changing axes, it must have an inverse; it must be possible to go back to our original system if we want to.)

Thus we can write both equations so as to give old in terms of new. The equations are thus, for column vectors $v = MV$, for row vectors $h = HM^{-1}$. It will be seen that our basic requirement, $hv = HV$, is automatically met, for

$$hv = (HM^{-1})(MV) = H(M^{-1}M)V = HV,$$

since $M^{-1}M = I$; the inverse, M^{-1}, wipes out M.

31   Affine and Euclidean geometry

The geometry systematized by Euclid and still taught in schools is essentially mathematical physics; that is to say, it is an attempt to give a logical account of our actual experiences with shapes and sizes. Both in geometry lessons and in everyday life we take it for granted that if there are 2 points there must be a distance between them, and if there are 2 lines that intersect, there must be an angle between them, that we can measure with a protractor. And of course, if we are surveying a region or designing some structure, these are perfectly sound and satisfactory ideas.

However, as we have seen, the idea of vector space is a very general one. The geometry of everyday life gives us an example of a vector space of 3 dimensions but this is far from being the only example. Many of the vector spaces we have met have appeared very remote from geometry as it is usually understood. We have in fact defined a vector space as being any collection of objects, $u, v, \ldots$, with a satisfactory definition of sums, such as $u + v$, and multiplication by a number, leading to expressions such as kv. In such a collection it by no means follows that we are able to define the length of a vector or the angle between two vectors.

For instance we have found it perfectly possible to work with a vector space in which $xA + yB$ signifies 'x copies of an article A and y copies of an article B'. In diagrams related to this space, we may very well use ordinary graph paper simply because it is readily available. In such a diagram the lines QA and QB will be of equal length and perpendicular. It is important to realize that the diagram has brought in aspects which are entirely foreign to the vector space in question, in which such ideas as *length* and *angle* are undefined and are totally without meaning. Suppose, for instance, we mark a dot on the point $(3, 4)$ of ordinary graph paper to represent the vector '3 nuts and 4 bolts'. This dot will be at a distance 5 units from the origin, and we may be tempted to think of 5 as the length or size of the vector. But there is no sensible procedure that would lead us to introduce 5 as a measure of magnitude for the consignment, 3 nuts and 4 bolts. Any meaningful property of the vector space, in which x nuts and y bolts is the typical vector, will appear equally well with graph paper in which the axes are not perpendicular and different units are employed on the x-axis and the y-axis. We shall still find that $u + v$ completes the parallelogram with 3 corners at $0, u, v$, and that kv is reached by taking k steps, each equal to v, from the origin. Observe for instance that in §9, dealing with change of axes, in neither of the systems shown are the axes perpendicular.

A mathematical theory deals with situations in which certain features can be recognized. It may demand a long list of features; in this case very many results will be proved, but it may not be possible to apply the theory very often, since it will be rarely that all the necessary features will be found together in an actual situation. At the other extreme, very few features may be required; then relatively few results will follow but many applications may be expected, since only a few tests have to be passed. Of course, the nature of the features demanded is important. Mathematicians, and particularly of course those mathematicians of most interest to engineers and scientists, study the logical consequences of those features which have frequently occurred in real problems in the past or are likely to occur in the future. Great mathematicians sometimes seem to have an uncanny instinct for studying what will prove important in the future, as Riemann in 1854 laid one of the foundations for Einstein's theory of general relativity, and Hilbert, around 1905, provided mathematical procedures needed for quantum theory after 1925.

Most of the work we have done so far has been based on very modest demands; the only features we have required have been reasonable definitions of $u + v$ and kv. We may refer to this as *pure vector theory*. The geometry that develops from pure vector theory is known as Affine geometry. The word 'affine' is derived from 'affinity'; it deals with figures that resemble each other in a rather loose sense, as contrasted with the relations of congruence and similarity studied in Euclid's geometry. In affine geometry there is no mention of length or perpendicularity. The basic ideas are those

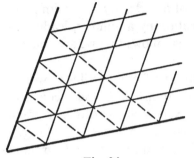

Fig. 94

of *line* and *parallel lines.* A grid, suitable for co-ordinate work, can be specified purely in terms of these two ideas, as Fig. 94 suggests. Notice how the parallel dotted lines here fix the whole grid once the first parallelogram has been drawn; they prevent an uneven spacing of the intervals along the axes.

From time to time we have used concepts that belong to Euclidean rather than affine geometry; we may have used squared graph paper, or spoken of reflections and rotations. This has been done purely in order to draw examples from situations familiar to students. All such references could have been omitted, so far as mathematical development was concerned. A linear mapping was defined by requiring

$$u + v \to u^* + v^*,$$

$ku \to ku^*$; the only operations involved here are the two basic operations of pure vector theory. When dealing with determinants, we seemed to use ideas involving perpendicularity, such as area = base times height. But this could have been avoided. The ratio of areas is a perfectly good affine concept. In Fig. 95, we can see that the parallelogram $OA^*C^*B^*$ can be chopped up into four pieces which can be moved by translation (an affine concept) so as to cover the parallelogram OAA^*B twice. Thus we can conclude, without going outside affine geometry, that the transformation $A \to A^*$, $B \to B^*$, doubles all areas and hence has determinant 2.

Students will no doubt meet the expressions $u \cdot v$, the scalar or dot product, and $u \times v$, the vector or cross product. It should be understood that these do not belong to pure vector theory and affine geometry. The dot product, $u \cdot v$, can be defined only in

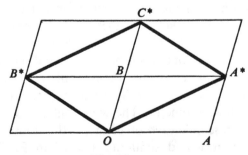

Fig. 95

situations to which Euclidean geometry applies; it requires for its definition the ideas of *length* and *perpendicular*. The cross product $u \times v$ is very much more restricted; it can be defined only in Euclidean space of 3 dimensions.

Until now, apart from occasional references and examples, we have kept within the bounds of affine geometry. From now on, we shall make use of Euclidean concepts whenever they are required.

32 Scalar products

In order to pass from affine geometry to Euclid's geometry a new idea has to be brought in, the *length* of a vector.

If we are using ordinary graph paper we can define the length, L, of a vector by the equation $L^2 = x^2 + y^2$. Similarly in 3 dimensions we would use $L^2 = x^2 + y^2 + z^2$ and in any number, n, of dimensions $L^2 = \sum x_r^2, 1 \leqslant r \leqslant n$.

In affine geometry, we had a great freedom to change axes; we could go from any basis of a space to any other basis without affecting our formulas. In Euclidean geometry this freedom is considerably less. Suppose for instance we decided to introduce new co-ordinates in 2 dimensions by means of the equations $x = X + Y, y = Y$. We would then find $L^2 = x^2 + y^2 = X^2 + 2XY + 2Y^2$. It would therefore be quite incorrect to take $X^2 + Y^2$ as giving L^2; the definition, in the form $L^2 = x^2 + y^2$ holds only in certain privileged co-ordinates systems.

From the definition of length we derive the definition of distance; *the distance between P and Q is the length of the vector Q − P*. If we use s for this distance we find in 2 dimensions $s^2 = (x_2 - x_1)^2 + (y_2 - y_1)^2$ in a well known notation. In 3 dimensions the definition gives $s^2 = (x_2 - x_1)^2 + (y_2 - y_1)^2 + (z_2 - z_1)^2$. In n dimensions there is a similar result; we have to change our notation so as to make use of subscripts; $s^2 = \sum (u_r - v_r)^2, 1 \leqslant r \leqslant n$ for vectors u, v.

The symbol $\|v\|$ indicates the length of the vector v. In the traditional notation of geometry, there is some uncertainty. We may write AB to indicate the line AB, the vector AB that runs from A to B, or the length AB. The length AB is measured by a number, and in our work it is always important to be clear whether a symbol stands for a transformation, a vector or a number. The symbols $\|v\|$ or $\|AB\|$ make it clear that a number is intended; $\|v\|^2$ or $\|AB\|^2$ indicates the square of that number. The symbol v^2 is meaningless; we cannot multiply a vector by a vector. Probably AB^2 would convey something to most readers, owing to their recollections of traditional geometry. However, writing $\|AB\|^2$ we put the matter beyond doubt, if it is understood here that AB means the vector from A to B.

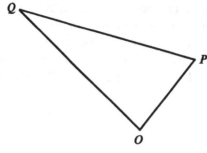

Fig. 96

Our definition of length is obviously suggested by our earlier experiences with Pythagoras' Theorem. We can use Pythagoras' Theorem, so to speak in reverse, to get a definition of perpendicular. We will say that $OP \perp OQ$ if $\|PQ\|^2 = \|OP\|^2 + \|OQ\|^2$, Fig. 96. This defines 'perpendicular' in terms of lengths, and as we have a formula for the length of any vector, we can use this definition to obtain an algebraic condition for OP and OQ to be perpendicular.

Exercises

1 In 3 dimensions, if P has co-ordinates (p_1, p_2, p_3) and Q has (q_1, q_2, q_3) show that $OP \perp OQ$ when $p_1q_1 + p_2q_2 + p_3q_3 = 0$.

2 In n dimensions, find the condition for $u \perp v$, where u has co-ordinates $(u_1, \ldots, u_n)$ and v has $(v_1, \ldots, v_n)$.

Work done by a force

Suppose that, in 3 dimensions a force (F_1, F_2, F_3) moves its point of application through a displacement (x_1, x_2, x_3).

In Fig. 97, suppose the particle on which the force is acting moves from A to D by the

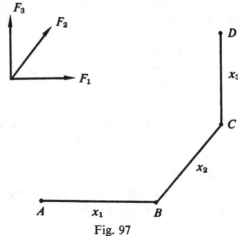

Fig. 97

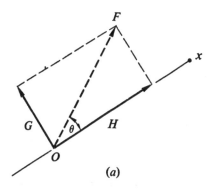

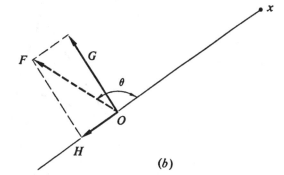

Fig. 98

path $ABCD$, in which AB, BC and CD are parallel to the first, second and third axes respectively. In the displacement AB only the component F_1 does any work; the work done in this part of the displacement is F_1x_1. Similar considerations apply to BC and CD. The total work done, as the particle goes from A to D, is thus $F_1x_1 + F_2x_2 + F_3x_3$.

If the force F were perpendicular to the displacement AD, the work done of course would be zero. This ties in well with the condition proved in Exercise 1 above. It shows further that this kind of expression has a significance even when the vectors are not perpendicular.

The work done by the force F in the displacement x is commonly explained in one of two ways. We may suppose F equivalent to a force G perpendicular to x, and a force H in the same line as x. In the first form of the definition the work done by F in the displacement x is defined as $\|H\| \cdot \|x\|$ in case (a), and $-\|H\| \cdot \|x\|$ in case (b) of Fig. 98. The second definition follows easily from this. If θ denotes the angle between x and F, in both cases we find the work done is given by $\|x\| \cdot \|F\| \cos \theta$, the product of the lengths of the vectors and the cosine of the angle between them. We are thus led to associate the quantity just discussed with the expression found earlier. $F_1x_1 + F_2x_2 + F_3x_3$, and either one of these may be referred to as the 'scalar product' or 'dot product' $F \cdot x$.

The argument given above, for $F_1x_1 + F_2x_2 + F_3x_3$ as being the work done, is useful as a way of seeing why such an expression is to be expected and as a way of remembering it. However, it is not entirely convincing as a proof. How do we know that the work done, if we go from A to D by the route $ABCD$, is the same as if we went from A to D in the most natural way, along the straight line from A to D?

Accordingly, some alternative ways of proving this result will be indicated. These do not make any appeal to mechanics. Apart from leading to the desired result, these proofs are useful exercises in themselves and are presented as such; the general procedure is outlined and the details are to be provided by the student.

Exercises

1 (a) The sides of a triangle are of lengths a, b, c. State, or derive, the formula for $\cos C$ in terms of a, b, c.

(b) Deduce the value of $ab \cos C$ in terms of a, b, c.

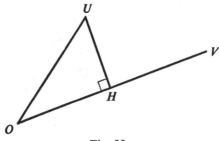

Fig. 99

(c) A triangle is formed by the origin, O, the point $u = (u_1, u_2, u_3)$ and the point $v = (v_1, v_2, v_3)$. If $a = \|u\|$, $b = \|v\|$ and $c = \|u - v\|$, find a^2, b^2 and c^2 in terms of the co-ordinates of u and v.

(d) Use (b) and (c) to calculate $\|u\| \cdot \|v\| \cdot \cos C$ in terms of the co-ordinates of u and v. Here C is the angle between the vectors u and v, in accordance with the usual conventions for the triangle considered in (c); so here we are calculating $u \cdot v$ from the second definition given above.

2 The first definition allows us to find $u \cdot v$ without any appeal to trigonometry. We do, however, have to use the condition for perpendicularity found in the first exercise of this section.

Let U have co-ordinates (u_1, u_2, u_3) and V have co-ordinates (v_1, v_2, v_3). Let H be the foot of the perpendicular from U to OV, Fig. 99. We shall switch freely from geometrical to vector notation. By OV we mean the line joining the points O and V; but we shall also speak of the vector V, the vector H and the vector U.

(a) The fact that H lies on the line OV may be expressed by the vector equation $H = tV$, where t stands for some number that will be found later.

(b) What vector represents the step from H to U? Find its co-ordinates; take account of the equation $H = tV$; that is, we want an answer depending only on t and the co-ordinates of U and V.

(c) Express algebraically $HU \perp OV$, and deduce the value of t in terms of the co-ordinates of U and V.

(d) On the assumption that OH and OV point in the same direction so that $t > 0$, find $\|H\| \cdot \|V\|$, which gives $U \cdot V$ for this case. Fortunately the case $t < 0$, corresponding to case (b) in Fig. 99, leads to the same answer, the algebra automatically compensates for this change.

3 Show that the procedure of Exercise 2 applies equally well in n dimensions, for any natural number n.

Applications of scalar product

When we are dealing with a number of points, specified by their co-ordinates in 3 dimensions, we may wish to visualize the figure they form. This can be done with the aid of the distance formula and the scalar product, as is shown in the following worked examples.

Worked example 1

What figure is formed by $OPQR$ where

$$O = (0, 0, 0), P = (-13, 18, 6), Q = (6, -3, 22), R = (18, 14, -3)?$$

Solution   This question can be answered by using the distance formula alone. We have $\|OP\|^2 = 13^2 + 18^2 + 6^2 = 529$, so $\|OP\| = 23$.

Also $\|PQ\|^2 = (-13 - 6)^2 + (18 + 3)^2 + (6 - 22)^2 = 1058$. Hence $\|PQ\| = 23\sqrt{2}$. In the same way we can find the lengths of the remaining edges of the figure. The result is that PO, OQ and OR are all of length 23, while PQ, QR and PR are all of length $23\sqrt{2}$. It will be seen that PQR is an equilateral triangle; also the lengths of the sides of the triangles OPQ, OPR and OQR indicate that each of these is right-angled with the right angle at O. Thus a cube of side 23 could be placed with one corner at O and adjacent corners at P, Q, R.

We could also have discovered that OP, OQ and OR are mutually perpendicular by using scalar products. For instance

$$P \cdot Q = (-13)(6) + (18)(-3) + (6)(22) = -78 - 54 + 132 = 0$$

which shows $OP \perp OQ$. The scalar products $P \cdot R$ and $Q \cdot R$ will also be found to be zero. Establishing perpendicularity by means of scalar products is here seen to be somewhat quicker and easier than the method first used. Logically the methods are closely related, since the properties of scalar products are proved by appealing to the distance formula.

Worked example 2

Let $A = (10, 5, 5)$, $B = (-11, 2, 5)$, $C = (1, -7, -10)$, $N = (1, -7, 5)$. Compute the scalar products involved and find the lengths of OA, OB, OC and ON. Hence discuss the figure formed by A, B, C, N and O.

Solution

$$A \cdot N = 10 - 35 + 25 = 0. \quad B \cdot N = -11 - 14 + 25 = 0. \quad C \cdot N = 1 + 49 - 50 = 0.$$

Thus ON is perpendicular to OA, to OB and to OC, i.e. A, B and C lie in the plane through the origin perpendicular to ON.

We find also $A \cdot B = -75$, $A \cdot C = -75$, $B \cdot C = -75$ and that OA, OB and OC are all of length $\sqrt{150}$.

Let α denote the angle between OA and OB. We know $A \cdot B = \|OA\| \cdot \|OB\| \cos \alpha$. As $A \cdot B = -75$ and $\|OA\| = \|OB\| = \sqrt{150}$, we find $\cos \alpha = -75/150 = -0.5$. Hence the angle between OA and OB is 120°. Exactly the same calculation shows the angle between OB and OC, and that between OC and OA, to be 120°.

It is now seen that ABC is an equilateral triangle with O as its centre, lying in the plane perpendicular to ON.

The length of ON is found to be $\sqrt{75}$.

Scalar products provide a simple way of dealing with the equations of planes in 3 dimensions. This topic will be taken up again in §39, but it seems useful to introduce it now, by the Worked examples 3 and 4.

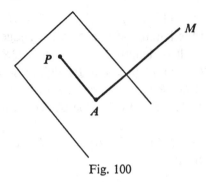

Fig. 100

Worked example 3

Find (*a*) the equation of the plane through the origin perpendicular to ON, where $N = (1, 2, 3)$, (*b*) the equation of the plane through $(4, 5, 6)$ perpendicular to $(1, 2, 3)$.

Solution   (*a*) Let $P = (x, y, z)$. P will lie in the plane through O perpendicular to ON if OP is perpendicular to ON. This will be so if the scalar product $P \cdot N$ is zero, that is, if $x \cdot 1 + y \cdot 2 + z \cdot 3 = 0$. Thus $x + 2y + 3z = 0$ is the required equation.
(*b*) Let $A = (4, 5, 6)$ and let AM be the normal to the required plane, as in Fig. 100. Thus the vector AM is $(1, 2, 3)$.
   Students sometimes find difficulty with vectors that start from points other than the origin. In this example it may help to imagine a mass acted upon by a constant force $(1, 2, 3)$; then AM represents the force acting on the mass when the mass is at A. The force does no work if the mass is displaced from A to P, where P lies in the plane through A perpendicular to AM. As $A = (4, 5, 6)$ and $P = (x, y, z)$, the displacement from A to P is $(x - 4, y - 5, z - 6)$. If the force $(1, 2, 3)$ does no work in such a displacement we have $(x - 4) \cdot 1 + (y - 5) \cdot 2 + (z - 6) \cdot 3 = 0$. This then is the condition for P to lie in the plane specified. It simplifies to $x + 2y + 3z - 32 = 0$.

Worked example 4

What is the orientation of the plane $2x - 5y + 7z = 0$?

Solution   It is evident from the work in the solution to Worked example 3(*a*) that we would arrive at this equation if we sought the equation of the plane through the origin perpendicular to $(2, -5, 7)$. So our answer is immediate; this is the plane with the normal $(2, -5, 7)$.

Worked example 5

A crystal has a cubic lattice; there is an atom at every point (x, y, z) for which x, y and z are whole numbers. Draw a diagram showing how atoms occur in the plane $x = y$.

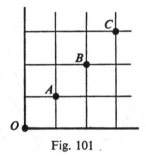

Fig. 101

Solution   Consider first the atoms lying in the plane $z = 0$. If these also lie in the plane $x = y$, they will be at positions such as $(0, 0, 0)$, $(1, 1, 0)$, $(2, 2, 0)$, $(3, 3, 0)$, as shown in Fig. 101. If this book is lying on a table, so that this illustration is in a horizontal plane, the plane $x = y$ will be vertical. It will contain the points O, A, B, C. It will also contain atoms at $(0, 0, 1)$, $(1, 1, 1)$, $(2, 2, 1)$, $(3, 3, 1)$, one unit vertically above O, A, B, C. Rising a further unit we shall find atoms at $(0, 0, 2)$, $(1, 1, 2)$, $(2, 2, 2)$, $(3, 3, 2)$. Thus, as we go up a vertical line, we find atoms spaced at unit intervals.

The points O, A, B, C are spaced at intervals of length $\sqrt{2}$. Thus the section of the crystal by the plane $x = y$ shows atoms arranged as in Fig. 102.

Algebraic properties of scalar products

From the formula for the scalar product it is easy to verify that results such as $u \cdot v = v \cdot u$ and $u \cdot (av + bw) = a(u \cdot v) + b(u \cdot w)$ hold, whatever the vectors u, v, w, and the numbers a, b. Thus scalar products can be handled very much like products of numbers, as is done in the following discussion.

Fig. 103 shows the point P with co-ordinates x, y, z. We usually visualize the numbers x, y, z with the help of a picture such as this, involving the three vectors OM, MN and NP. However, it is possible to visualize the co-ordinate x without bringing in the point N. For we can see that OMP is a right angle. Thus OM is the projection of OP on the first axis; the number x is obtained by taking $\|OM\|$, the length of this projection, and

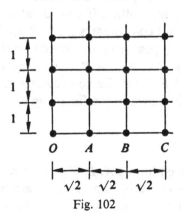

Fig. 102

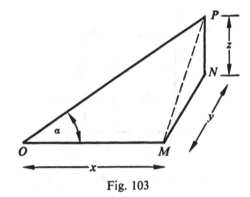

Fig. 103

attaching the appropriate sign, + or −, to it. In fact $x = \|OP\| \cos \alpha$ and this equation gives x with the correct sign. The appearance of projection in our considerations suggests that scalar products may be helpful. In fact, let OA be the unit vector along the first axis, so $\|OA\| = 1$. Then $P \cdot A = \|OA\| \cdot \|OP\| \cos \alpha = \|OP\| \cos \alpha = x$. Similarly if OB and OC are unit vectors along the second and third axes we have $P \cdot B = y$ and $P \cdot C = z$.

This argument can be put in purely algebraic form, without any appeal to a geometrical illustration. As was indicated in §6, we can compress the statement, 'P has co-ordinates (x, y, z) in the system based on A, B, C' into the equation

$$P = xA + yB + zC.$$

Take the scalar product of both sides of the equation with A. It follows that

$$P \cdot A = x(A \cdot A) + y(B \cdot A) + z(C \cdot A).$$

As A, B and C are perpendicular axes, $B \cdot A = 0$ and $C \cdot A = 0$. What is $A \cdot A$? The angle between A and itself is zero and $\cos 0 = 1$. Thus $A \cdot A$ reduces simply to $\|OA\|^2$, and as OA is of unit length, this means $A \cdot A = 1$. So our equation for $P \cdot A$ reduces to $P \cdot A = x$, as expected.

The scalar product approach is particularly useful when we wish to change from one set of perpendicular axes to another. If co-ordinates x, y, z are based on perpendicular unit vectors A, B, C and co-ordinates X, Y, Z are based on perpendicular unit vectors D, E, F, we can quickly pass from one system to the other. The argument used above to prove $x = P \cdot A$, $y = P \cdot B$, $z = P \cdot C$ for any point P shows equally well $X = P \cdot D$, $Y = P \cdot E$, $Z = P \cdot F$. Thus the co-ordinates can be found immediately by forming scalar products, as is shown in the following example.

Worked example 6

Let the vectors D, E, F be given (in the A, B, C co-ordinate system) by $D = (-\frac{1}{3}, \frac{2}{3}, \frac{2}{3})$, $E = (\frac{2}{3}, -\frac{1}{3}, \frac{2}{3})$, $F = (\frac{2}{3}, \frac{2}{3}, -\frac{1}{3})$. Verify that D, E, F are perpendicular vectors of unit length and find the X, Y, Z co-ordinates of the point $(1, 4, 10)$.

Solution Arithmetic only is involved in checking that $D \cdot D = E \cdot E = F \cdot F = 1$ and $D \cdot E = D \cdot F = E \cdot F = 0$, so D, E, F are of unit length and mutually perpendicular. Let $P = (1, 4, 10)$.
We have

$$X = P \cdot D = 1 \cdot (-\tfrac{1}{3}) + 4 \cdot (\tfrac{2}{3}) + 10 \cdot (\tfrac{2}{3}) = 9,$$

$$Y = P \cdot E = 1 \cdot (\tfrac{2}{3}) + 4 \cdot (-\tfrac{1}{3}) + 10 \cdot (\tfrac{2}{3}) = 6,$$

$$Z = P \cdot F = 1 \cdot (\tfrac{2}{3}) + 4 \cdot (\tfrac{2}{3}) + 10 \cdot (-\tfrac{1}{3}) = 0.$$

So P is specified by $X = 9$, $Y = 6$, $Z = 0$. We can check this by calculating

$$9D + 6E + 0 \cdot F$$

and seeing that it does give P.

It would have been possible to answer this question by the method of §9 which applies to *any* change of axes. That method, however, gives the 'old' co-ordinates x, y, z in terms of the 'new' co-ordinates X, Y, Z. To find X, Y, Z given $x = 1$, $y = 4$, $z = 10$ we would have to solve 3 simultaneous equations. The present method is very much quicker but it is important to note that it applies *only when we are changing from one system of perpendicular unit vectors to another system of the same kind.*

Exercises

1 In each of the following cases find the scalar product of the two vectors given. Find the magnitude of each vector. Deduce the cosine of the angle between them.

(a) $(1, 1, 0)$ and $(1, 0, 1)$ (b) $(1, -2, 1)$ and $(3, 4, 5)$
(c) $(1, 1, 1)$ and $(1, 0, 0)$ (d) $(3, 4, 5)$ and $(5, 5, 0)$
(e) $(1, 2, -2)$ and $(1, -1, 4)$ (f) $(-2, 3, 6)$ and $(4, 1, 9)$.

2 According to a student, the diagonal of a cube makes an angle of $45°$ with each edge of the cube, Discuss this remark.

3 Find the lengths and the scalar products of the vectors $(1, 1, 0)$, $(-1, 0, -1)$, $(0, -1, 1)$, $(1, -1, -1)$ and discuss the figure they form.

4 Find the lengths of the vectors $P = (1, 5, 1)$, $Q = (-5, -1, 1)$, $R = (1, -1, -5)$, $S = (3, -3, 3)$ and the values of the 6 scalar products that can be formed from them. Describe in geometrical terms the figure formed by the line segments OP, OQ, OR and OS. What figure is formed by the points $PQRS$? How could the answer to this last question be checked directly?

To what chemical situations is this question related?

5 Check that the vectors $D = (-\tfrac{1}{3}, -\tfrac{2}{3}, \tfrac{2}{3})$, $E = (\tfrac{2}{3}, \tfrac{1}{3}, \tfrac{2}{3})$, $F = (-\tfrac{2}{3}, \tfrac{2}{3}, \tfrac{1}{3})$ are perpendicular unit vectors. Let $V = (1, 1, 1)$. Find X, Y, Z, the co-ordinates of V in the system based on D, E, F. Check that $XD + YE + ZF$ does give V.

6 Proceed as in question 5, but with the values given below for D, E, F and V.

(a) $D = (-\tfrac{7}{9}, \tfrac{4}{9}, -\tfrac{4}{9})$, $E = (-\tfrac{4}{9}, \tfrac{1}{9}, \tfrac{8}{9})$, $F = (\tfrac{4}{9}, \tfrac{8}{9}, \tfrac{1}{9})$, $V = (-2, 2, 1)$.
(b) $D = (-\tfrac{6}{7}, -\tfrac{3}{7}, \tfrac{2}{7})$, $E = (\tfrac{3}{7}, -\tfrac{2}{7}, \tfrac{6}{7})$, $F = (-\tfrac{2}{7}, \tfrac{6}{7}, \tfrac{3}{7})$, $V = (-1, 1, 2)$.
(c) $D = (-\tfrac{4}{9}, \tfrac{1}{9}, \tfrac{8}{9})$, $E = (\tfrac{7}{9}, -\tfrac{4}{9}, \tfrac{4}{9})$, $F = (\tfrac{4}{9}, \tfrac{8}{9}, \tfrac{1}{9})$, $V = (3, 7, 4)$.
(d) $D = (-\tfrac{9}{11}, -\tfrac{6}{11}, \tfrac{2}{11})$, $E = (\tfrac{6}{11}, -\tfrac{7}{11}, \tfrac{6}{11})$, $F = (-\tfrac{2}{11}, \tfrac{6}{11}, \tfrac{9}{11})$, $V = (6, 4, -5)$.
(e) $D = (-\tfrac{2}{3}, -\tfrac{2}{3}, \tfrac{1}{3})$, $E = (\tfrac{2}{3}, -\tfrac{1}{3}, \tfrac{2}{3})$, $F = (-\tfrac{1}{3}, \tfrac{2}{3}, \tfrac{2}{3})$, $V = (1, 6, 5)$.

7   In each section of question 6, and in question 5, replace the particular vector V by the general vector $V = (x, y, z)$, and obtain equations giving X, Y, Z in terms of x, y, z.

Also find the equations giving x, y, z in terms of X, Y, Z. This could be done by solving the first set of equations but is much easier by the method of §9.

Compare the matrices associated with the two sets of equations. What do you notice?

8   In a crystal with a cubic lattice there is an atom at every point (x, y, z) for which x, y, z belong to the whole numbers 0, 1, 2, 3, ... Draw diagrams to show the positions of the atoms in the following planes;

    (a) $x = 0$,   (b) $x + y + z = 3$,   (c) $x + y + z = 4$.

9   In a body-centred cubic lattice atoms occupy all the points (x, y, z) with x, y, z whole numbers, and also points such as $(1\frac{1}{2}, 5\frac{1}{2}, 2\frac{1}{2})$ for which x, y, z are *all* of the form 'whole number plus a half'. Draw diagrams showing the atoms in the following plane sections;

    (a) $z = 0$ (b) $z = \frac{1}{2}$ (c) $x = y$ (d) $x + y = 3$

    (e) $x + y + z = 3$ (f) $x + y + z = 3\frac{1}{2}$ (g) $x + y + 2z = 4$.

10   In a face-centred cubic lattice, atoms are found at all points (x, y, z) with x, y, z whole numbers, and also at points such as $(1\frac{1}{2}, 3, 5\frac{1}{2})$ for which one co-ordinate is a whole number and the other two co-ordinates differ by $\frac{1}{2}$ from whole numbers. Draw diagrams for sections by the planes;

    (a) $z = 0$ (b) $z = \frac{1}{2}$ (c) $x = y$ (d) $y = x + \frac{1}{2}$ (e) $x + y + z = 3$.

33   Transpose; quadratic forms

In §30 we saw that the matrix notation we have had so far gave us no way of multiplying two column vectors together. If we had a column vector $\begin{pmatrix} x \\ y \end{pmatrix}$ we could get an expression such as $ax + by$ for the product with a row vector (a, b) but there was no way of arriving at quadratic terms such as x^2, xy, y^2, for these would imply that we had formed the product of the vector with itself and this, in pure vector theory (affine geometry), is impossible.

Now there are cases in which we want to consider quadratics. In Euclid's geometry we need $x^2 + y^2$ for the definition of length. If we have two springs of stiffness k_1 and k_2, with extensions x_1 and x_2, the potential energy is given by the quadratic expression $\frac{1}{2}(k_1 x_1^2 + k_2 x_2^2)$. Similarly, if we have currents, i_1 and i_2, in resistances r_1 and r_2, the rate at which heat is generated is given by $r_1 i_1^2 + r_2 i_2^2$. If we are to cope with such situations, we must bring some new machinery into matrix algebra.

The situations mentioned in the last paragraph can be conveniently represented within Euclidean geometry. If the displacements are indicated by plotting the point (x_1, x_2) on ordinary graph paper, the forces acting on the springs can be represented by the point (F_1, F_2) where $F_1 = k_1 x_1$, $F_2 = k_2 x_2$. The expression $k_1 x_1^2 + k_2 x_2^2$ that occurs in the formula for the potential energy is then the scalar product $F_1 x_1 + F_2 x_2$. Similarly if we represent the currents by the point (i_1, i_2), the voltage drops in the resistances can be represented by the point (V_1, V_2) where $V_1 = r_1 i_1$, $V_2 = r_2 i_2$; heat generation is then given by the scalar product $V_1 i_1 + V_2 i_2$. In the purely geometrical question, if we write u for the vector with co-ordinates x, y, then $x^2 + y^2$ is $u \cdot u$, the scalar product of u with itself.

Accordingly, we can cope with situations of this kind if we can find some way of writing scalar products in matrix notation.

Suppose then that we have some point $\begin{pmatrix} a \\ b \end{pmatrix}$, plotted on ordinary graph paper. If $\begin{pmatrix} x \\ y \end{pmatrix}$ denotes a variable point, the scalar product of $\begin{pmatrix} a \\ b \end{pmatrix}$ and $\begin{pmatrix} x \\ y \end{pmatrix}$ is $ax + by$. In this expression, the numbers a and b appear as coefficients. §30 has taught us to associate such coefficients with a row vector, in fact with the row vector (a, b). Thus in Euclidean geometry to each column vector $\begin{pmatrix} a \\ b \end{pmatrix}$ there corresponds a row vector (a, b). This situation is entirely different from that in affine geometry where, fortunately, an order for 200 nuts and 100 bolts (a column vector) does not imply a price scheme of $200 per nut and $100 per bolt (a row vector). But in Euclidean geometry, because scalar product is defined, there is a direct correspondence between column vectors and row vectors. Provided a suitable co-ordinate system (based on squared graph paper) is used, this correspondence is very simple; it is given by $\begin{pmatrix} a \\ b \end{pmatrix} \rightarrow (a, b)$.

If $v = \begin{pmatrix} a \\ b \end{pmatrix}$, then (a, b) is known as the *transpose* of v, and is written v' or v^T.

Transposition can be defined for any matrix (compare Exercise 3 on determinants). It consists in interchanging rows and columns. Thus we may also say that v is the transpose of v'.

Note that transposition is something done to the numbers that represent a vector. If we see $\begin{pmatrix} 7 \\ 10 \end{pmatrix}$ we can write down its transpose $(7, 10)$. Whether this transpose will have a significant geometrical or physical meaning will depend on the co-ordinate system in which it occurs. But whether it is meaningful or not, it is still called the transpose.

With the help of the transpose we can express scalar products in matrix notation, as is verified in the following exercises.

Exercises

1 If u and v are two column vectors in 2 dimensions, the co-ordinate system being based on conventional (squared) graph paper, show that $u'v = v'u =$ the scalar product $u \cdot v$.

2 Verify that the same holds in n dimensions for any natural number n.

Quadratic forms

At the beginning of this section we considered quadratic expressions that arose in connection with springs and with electric circuits. The mathematics in the two cases was almost identical. It will be sufficient to study one of these examples, say that of the springs. A student interested in electricity can easily translate the discussion into electrical terms. What processes were involved in forming the expression for the potential energy?

We began with a column vector $\begin{pmatrix} x_1 \\ x_2 \end{pmatrix}$ which gave the extensions of the two springs. We then used Hooke's Law to find F_1, F_2 with $F_1 = k_1 x_1$, $F_2 = k_2 x_2$. We thus arrive at a new vector, F, by means of a linear transformation. As a linear transformation can be specified by a matrix we can write our equations in the form

$$\begin{pmatrix} F_1 \\ F_2 \end{pmatrix} = \begin{pmatrix} k_1 & 0 \\ 0 & k_2 \end{pmatrix} \begin{pmatrix} x_1 \\ x_2 \end{pmatrix}.$$

We may write this more briefly as $F = Kv$, with F the vector specifying forces, v specifying extensions, and K the stiffness matrix. As was pointed out earlier, the expression $k_1 x_1^2 + k_2 x^2$, which equals $2E$, is the scalar product $F \cdot v$. Our last exercise showed that $F \cdot v$ could be written $v'F$. Since $F = Kv$, this expression could be written $v'Kv$. This is the kind of symbol we usually employ for a quadratic form, that is to say an expression in which every term is of the second degree in the variables. Thus $ax^2 + bx + c$ is *not* a quadratic form, since bx is only of the first degree, while c is simply a constant. However, $ax^2 + bxy + cy^2$ is a quadratic form, and so is $x^2 - yz$. With the summation notation, a quadratic form can always be expressed as

$$\sum \sum a_{rs} x_r x_s; \qquad 1 \leqslant r \leqslant n, 1 \leqslant s \leqslant n.$$

Exercises

1 $v'Kv$ means $(x_1 \quad x_2) \begin{pmatrix} k_1 & 0 \\ 0 & k_2 \end{pmatrix} \begin{pmatrix} x_1 \\ x_2 \end{pmatrix}.$
Multiply this out by the usual rule for matrix multiplication and check that it does give the result stated above.

2   Multiply out $(x \quad y) \begin{pmatrix} a & h \\ h & b \end{pmatrix} \begin{pmatrix} x \\ y \end{pmatrix}.$

3   Multiply out $(x \quad y) \begin{pmatrix} 0 & 1 \\ -1 & 0 \end{pmatrix} \begin{pmatrix} x \\ y \end{pmatrix}.$

4   Multiply out $(x \quad y \quad z) \begin{pmatrix} 1 & -1 & -1 \\ -1 & 1 & -1 \\ -1 & -1 & 1 \end{pmatrix} \begin{pmatrix} x \\ y \\ z \end{pmatrix}.$

5   Write $x^2 + 6xy + y^2$ in the form $v'Mv$, with the matrix M of the type used in question 2 above.

Quadratic expressions of the form $v'Mv$ are very important. As we shall see in §34, there are situations in which the whole problem can be specified by giving a certain expression, which is either quadratic or partly quadratic and partly linear. Very often at some stage of the work it becomes necessary to bring in some new co-ordinate system. Instead of v we may now use V to specify some physical situation. We shall have a formula for change of axes, $v = SV$, where S is a matrix. In $v'Mv$ we readily see that Mv is to be replaced by MSV, but what are we to do about v'?

Let us look at a particular example. Suppose

$$v = \begin{pmatrix} x \\ y \\ z \end{pmatrix} \quad \text{and} \quad V = \begin{pmatrix} X \\ Y \end{pmatrix}.$$

Here we have not merely a change of axes but even a change of dimension. Such things do happen in engineering practice, as will be seen in an electrical example in §34. We suppose

$$\begin{aligned} x &= aX + bY \\ y &= cX + dY \\ z &= eX + fY \end{aligned} \qquad v = \begin{pmatrix} x \\ y \\ z \end{pmatrix} = \begin{pmatrix} a & b \\ c & d \\ e & f \end{pmatrix} \begin{pmatrix} X \\ Y \end{pmatrix} = SV.$$

Now v' is the row vector (x, y, z), which, from the equations above, means

$$(aX + bY, cX + dY, eX + fY).$$

This has 1 row and 3 columns, and 2 products appear in each entry. Such an expression can only arise if the first factor has 1 row and 2 columns, while the second factor has 2 rows and 3 columns. It is pretty clear that the first factor must be (X, Y), and in fact we find we can obtain the result above by multiplying out

$$(X, Y) \begin{pmatrix} a & c & e \\ b & d & f \end{pmatrix}.$$

Both of these we can identify. (X, Y) is V', the transpose of V. The 2×3 matrix is S', the transpose of S, for the columns of this matrix are the same as the rows of S, and the rows of this matrix contain the same elements as the columns of S. Thus $v' = V'S'$. As $v = SV$, this means $(SV)' = V'S'$.

The result we have found here is in fact true for any product of matrices; *the transpose of a product is the product of the transposes, in the reverse order.*

This general statement may be proved in two ways.

In the first way, we begin by satisfying ourselves that the result $(SV)' = V'S'$ above does not depend on the fact that S is a 3×2 matrix, but that it would hold for any $m \times n$ matrix S and any vector V in n dimensions. Provided we are able to do this, we then consider a vector acted upon by any succession of mappings, say, for instance $ABCV$. If we put $S = ABC$, we can see that the transpose of $(ABC)V$ is $V'(ABC)'$. But we can also consider the mappings A, B, C applied one at a time. By using the result $(SV)' = V'S'$ three times we see

$$[A(BCV)]' = (BCV)'A' = (CV)'B'A' = V'C'B'A'.$$

Accordingly $V'(ABC)' = V'C'B'A'$ and since this holds for any vector V we must have $(ABC)' = C'B'A'$. Clearly the same argument could be applied to any number of matrices, and so the general result is established.

The second method of proof uses subscript notation and is found in many books.

Let the matrix A have a_{pq} in the pth row and qth column. Since A' is obtained from A by interchanging rows and columns. A' will have a_{qp} in row p and column q. We express this by writing $a'_{pq} = a_{qp}$. Let $D = ABC$. Then $d_{ps} = \sum_q \sum_r a_{pq}b_{qr}c_{rs}$. Accordingly, for the transpose D' we have $d'_{ps} = d_{sp} = \sum_q \sum_r a_{sq}b_{qr}c_{rp}$. Here we have interchanged p and s in the previous formula. We are working towards a result involving A', B' and C', so naturally our next step is to bring these into the picture by using $a_{sq} = a'_{qs}$, $b_{qr} = b'_{rq}$ and $c_{rp} = c'_{pr}$. Making these substitutions we obtain $d'_{ps} = \sum_q \sum_r a'_{qs}b'_{rq}c'_{pr}$. This result is not at present in a form suitable for a matrix product since (see §29) in a matrix product the subscripts over which we sum are always immediate neighbours. We can remedy this by reversing the order of the factors. We then have

$$d'_{ps} = \sum_q \sum_r c'_{pr}b'_{rq}a'_{qs}.$$

Now we have the correct form for a matrix product; the repeated subscripts, r and q, are not separated and p and s, the subscripts of d', occur on the right-hand side in the first and last places respectively, as they should. In fact the last equation expresses $D' = C'B'A'$, the result we wish to prove.

This result covers the case of a matrix and a vector, $(SV)' = V'S'$, for a vector, V, may be regarded as a matrix with one column, and its transpose, V', as a matrix with one row.

Symmetric and antisymmetric matrices

The matrix $\begin{pmatrix} 1 & 3 \\ 3 & 4 \end{pmatrix}$ is the same as its transpose. A matrix with this property, $M' = M$, is said to be *symmetric*.

A matrix for which $M' = -M$ is said to be *antisymmetric*. An example of such a matrix is

$$\begin{pmatrix} 0 & c & -b \\ -c & 0 & a \\ b & -a & 0 \end{pmatrix}.$$

An antisymmetric matrix is bound to have zeros down the main diagonal, as in the example shown. For, if M has m_{pq} as its entry in row p and column q, $m_{pq} = -m_{qp}$, for each p and each q. If $q = p$, this means $m_{pp} = -m_{pp}$, so $m_{pp} = 0$.

Only square matrices can be symmetric or antisymmetric.

In question 3 of the exercises above, on quadratic forms, an expression $v'Mv$ with antisymmetric M occurred. It gave zero. In fact, this always happens. In questions 2 and 5 the matrix was symmetric. *Any quadratic form can be represented by $v'Mv$ with M* *symmetric, and it is the invariable custom (for which there are good reasons) always to use a symmetric matrix for this purpose.*

A single number, a, can be regarded as a matrix of 1 row and 1 column. It is automatically symmetric. We can use this to prove certain results. For instance, the scalar product $u \cdot v$ can be written $u'v$. Its transpose is $v'u$, by our usual rule – form transposes and reverse order. But the scalar product is a single number, hence symmetric, hence equal to its transpose. So $u'v = v'u$, a result that was checked earlier in an exercise.

Again $u'Mv$ is a single number, hence equal to its transpose $v'M'u$. So $u'Mv = v'M'u$. This result holds for any vectors u, v and any matrix M such that the product $u'Mv$ makes sense.

Exercises

1 Prove $(S + T)' = S' + T'$.

2 Prove $(kS)' = kS'$.

3 If M is any square matrix, prove $\frac{1}{2}(M + M')$ is symmetric, and $\frac{1}{2}(M - M')$ is antisymmetric. Hence show that any matrix is the sum of a symmetric and an antisymmetric matrix.

4 If A is an antisymmetric square matrix, prove $v'Av = 0$ whatever the vector v. (*Hint.* Consider the transpose of the number $v'Av$.).

5 If M is any matrix, prove that $M'M$ is symmetric. (Consider its transpose.)

6 If R is a symmetric $m \times m$ matrix, and T is any $m \times n$ matrix, prove that $T'RT$ is symmetric.

7 In the quadratic form $v'Sv$, where S is symmetric, we make the substitution $v = TV$. Find what $v'Sv$ becomes when this substitution is made. Will the matrix that appears in the answer be symmetric or not?

8 Find the result when in the scalar product $u'v$ we substitute $u = TU$, $v = TV$. What condition must T satisfy if for all vectors we are to have $u'v = U'V$?

34 *Maximum and minimum principles*

Early in a calculus course a student learns how to find the places at which some function takes its maximum or minimum values. He works a number of problems which have some interest – the largest parcel permitted by the postal regulations, the most economical shape for fencing a chicken farm, the shape of a soup tin that uses least metal. It is seen that there may be problems of design in which calculus indicates the most efficient or economical way of dealing with some detail. But very few students appreciate the

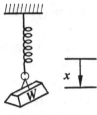

Fig. 104

immense vista that this work in calculus opens up. One of the strange things about this universe is that, in every science that has become sufficiently precise to admit mathematical formulation, there is a principle that some quantity is to be made a maximum or a minimum, and this principle, by itself, sums up the laws of that science. Thus in statics stable equilibrium occurs when potential energy is a minimum; in optics, light travels by the path that takes least time; in dynamics there is the principle of Least Action, which can be adapted so as to apply to electromagnetic phenomena also; in thermodynamics, an isolated system is in equilibrium when the entropy is a maximum, and from this a whole theory applicable to physical and chemical processes can be deduced; in an electrical network the currents distribute themselves in such a way that the generation of heat is a minimum.*

Here we have an outstanding example of the point made in §1, 'that mathematics is concerned not with particular situations but with patterns that occur again and again'. A student who has mastered the basic mathematical tools – differentiation and integration, partial differentiation, calculus of variations – has secured a foothold on an enormous variety of sciences.

We will consider two very simple examples. Suppose a weight W hangs from a spring of stiffness k, and x denotes the extension of the spring (Fig. 104). The potential energy, P, of the system is given by $P = \frac{1}{2}kx^2 - Wx$. This will be a minimum where $0 = dP/dx = kx - W$. Thus the equilibrium condition $W = kx$ follows the principle that potential energy is to be minimized.

Now consider the simple electrical circuit in which a voltage V drives a current i through a resistance r (Fig. 105). Ohm's Law tells us $V = ri$.

If we compare Ohm's Law $V = ri$ and $W = kx$ we see that they have the same

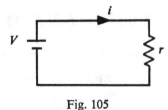

Fig. 105

* Strictly speaking, we should perhaps speak of 'stationary' rather than maximum or minimum principles. For example a bead on a wire can be in equilibrium if it is at a point of horizontal inflexion.

mathematical form. If, starting from Hooke's Law, $W = kx$, we replace W by V, k by r, and x by i, we reach Ohm's Law, $V = ri$.

Now we saw that Hooke's Law was a consequence of the requirement that potential energy should be minimized. If we replace the expression $\frac{1}{2}kx^2 - Wx$ by $\frac{1}{2}ri^2 - Vi$, making this latter expression a minimum will ensure that Ohm's Law is obeyed. (This can easily be verified directly.)

Now of course the principle in statics that potential energy is to be minimized does not apply only to a simple system with one spring. It applies equally well to a structure in which there may be hundreds of elastic members. In the same way, for electrical circuits we can find a function that is to be minimized, even though there may be a complicated network of batteries and resistances. This function may be written $\sum \frac{1}{2}ri^2 - \sum Vi$. By this we mean that for each resistance r in the circuit with current i flowing through it, we form the term $\frac{1}{2}ri^2$ and add all these terms together, to give $\sum \frac{1}{2}ri^2$. Similarly $\sum Vi$ means that for each battery of voltage V with current i flowing through it, we form the product Vi, and add all such products together. This is a very natural generalization of the function that was minimized in the case of the simple circuit.

This kind of result allows us to make a certain type of analogue computer. Suppose we have some complex structure. Its equilibrium is determined by minimizing a certain function. If we now design an electrical circuit that depends on essentially the same function (with displacements replaced by currents, stiffness constants by ohmic resistances and applied forces by battery voltages), the behaviour of this circuit will predict the behaviour of the structure. The same idea allows us to establish analogies between other branches of science, and to replace awkward experiments by convenient ones.

The existence of analogies also allows us to make advances in our theories. If some method of calculation has been found useful in the study of electrical circuits, it should be useful in the theory of structures. And in fact precisely this argument was used in a classical paper, 'Analysis of elastic structures by matrix transformation with special regard to semimonocoque structures' by Borje Langefors of the Saab Aircraft Company, in the *Journal of Aeronautical Sciences* for July 1952 (page 451). His article begins by pointing out that identical principles apply to electrical circuits and to elastic structures. Now Kron in 1939 had published a method for solving circuit problems by using matrices. It should therefore be possible to solve problems about structures by matrices. And indeed Kron in 1944 had published the electrical equivalents of various elastic structures. Langefors then proceeded to develop this idea in relation to the design of aircraft.

It seems worthwhile to consider in detail a particular electrical example, since this shows the use both of the quadratic form and the transpose, the two topics discussed in §33.

We will consider the circuit shown in Fig. 106. The resistances r_1, r_2, r_3 and the voltages of the batteries V_1, V_2, V_3 are known; we wish to determine the currents i_1, i_2, i_3. Our purpose of course is not to solve this simple problem, but to find a procedure that can be applied to much more complicated cases.

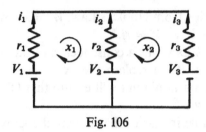

Fig. 106

By considering the potential difference between the wire at the top of the diagram and the wire at the bottom we see

$$V_1 - r_1 i_1 = V_2 - r_2 i_2 = V_3 - r_3 i_3. \tag{1}$$

We have three currents all shown as flowing into the top wire; one of these, at least, must be negative. The currents in fact must satisfy the condition $i_1 + i_2 + i_3 = 0$. This condition can be automatically satisfied by supposing that a current x_1 circulates in the left-hand rectangle, and a current x_2 circulates in the right-hand rectangle. Superposing these we find $i_1 = x_1$, $i_2 = -x_1 + x_2$, $i_3 = -x_2$.

We can express these equations in vector and matrix notation as $i = Tx$, where

$$i = \begin{pmatrix} i_1 \\ i_2 \\ i_3 \end{pmatrix}, \qquad x = \begin{pmatrix} x_1 \\ x_2 \end{pmatrix}, \qquad T = \begin{pmatrix} 1 & 0 \\ -1 & 1 \\ 0 & -1 \end{pmatrix} \tag{2}$$

If we look at the batteries in the left-hand loop, we see V_1 opposed by V_2, so it is natural to write $e_1 = V_1 - V_2$. (The letter e is the initial of 'electromotive force'.) Similarly for the right-hand loop we write $e_2 = V_2 - V_3$. In matrix form these equations are

$$\begin{pmatrix} e_1 \\ e_2 \end{pmatrix} = \begin{pmatrix} 1 & -1 & 0 \\ 0 & 1 & -1 \end{pmatrix} \begin{pmatrix} V_1 \\ V_2 \\ V_3 \end{pmatrix}. \tag{3}$$

The 2×3 matrix that occurs here is T', the transpose of the matrix already introduced. Thus, with a notation that will be understood we may write $e = T'V$.

From equation (1), and the definitions of e_1 and e_2, we find

$$\left. \begin{aligned} e_1 &= V_1 - V_2 = r_1 i_1 - r_2 i_2, \\ e_2 &= V_2 - V_3 = \quad\quad r_2 i_2 - r_3 i_3. \end{aligned} \right\} \tag{4}$$

We introduce the resistance matrix

$$R = \begin{pmatrix} r_1 & 0 & 0 \\ 0 & r_2 & 0 \\ 0 & 0 & r_3 \end{pmatrix}. \tag{5}$$

Observe that Ri gives the voltage drops across each of the three resistances.

Now the coefficients of i_1, i_2 and i_3 in equations (4) are closely related to the matrices R and T', as may be seen from the equation

$$
\begin{pmatrix} r_1 & -r_2 & 0 \\ 0 & r_2 & -r_3 \end{pmatrix} = \begin{pmatrix} 1 & -1 & 0 \\ 0 & 1 & -1 \end{pmatrix} \begin{pmatrix} r_1 & 0 & 0 \\ 0 & r_2 & 0 \\ 0 & 0 & r_3 \end{pmatrix}.
$$

The matrix on the left-hand side of this equation is formed from the coefficients of i_1, i_2 and i_3 in equations (4). The right-hand side is $T'R$.

Thus equations (4) may be put in the compact form $e = T'Ri$.

We intend to work with the loop currents, x, rather than the original currents, i, so we make use of our result $i = Tx$ and write $e = T'RTx$.

Our problem is now expressed purely in matrix form. We introduce new variables for currents by means of the equation $i = Tx$. We introduce new specifications of voltage by the equation $e = T'V$. We then have the equation $e = T'RTx$ in which e is known and the vector x has to be found.

The main advantage of this method is its systematic nature. We can programme a computer to form the necessary matrices, even for a very complicated circuit, and to solve the resulting equations.

Clearly some questions arise. Why is the matrix in the voltage equation, $e = T'v$, the transpose of the matrix in the current equation $i = Tx$? Can we be sure that $e = T'RTx$ will be the correct equation for any circuit however complicated?

A well known and good introduction to our present topic is *Matrix Analysis of Electric Networks* by P. le Corbeiller (Harvard University Press, 1950). Kron's work had been presented in rather an abstruse form. Le Corbeiller set out to explain Kron's ideas in a form that the majority of engineers would be able to understand, and this in general he did very well. However, to answer the two questions above required what he himself called 'a rather elaborate train of reasoning'. The demonstration filled three short sections, 19, 20 and 21, and used a strange argument in which a circuit was dissected into many pieces and then peculiar short-circuit arrangements were introduced.

The value of the minimum principle is shown by the ease with which it allows us to answer the two questions. The minimum principle itself we shall assume as having been established in some standard book on electrical theory.

In our explanation, we shall continue to use the particular circuit considered above, and the symbols introduced in connection with it. However, there is nothing special about this circuit; it should be clear how to apply the same procedure to any circuit whatever.

For our circuit we have to minimize the quantity

$$
\tfrac{1}{2}r_1 i_1^2 + \tfrac{1}{2}r_2 i_2^2 + \tfrac{1}{2}r_3 i_3^2 - V_1 i_1 - V_2 i_2 - V_3 i_3,
$$

which may be abbreviated to $\tfrac{1}{2}i'Ri - i'V$. It is understood that we only consider as possible those currents in which the amount of electricity flowing into any point is balanced by the amount flowing out. In our example, this means $i_1 + i_2 + i_3 = 0$.

Thus we can only choose two of the quantities i_1, i_2 and i_3; the value of the third is then fixed. This condition is automatically met by bringing in the loop currents, x_1 and x_2. They can vary in any way they like; there are no hidden conditions, and this of course simplifies the work considerably.

Accordingly, we get rid of i_1, i_2, i_3 in favour of x_1, x_2 by means of the equation $i = Tx$. Then, by the reversal rule, $i' = x'T'$. Substituting these in $\frac{1}{2}i'Ri - i'V$, we obtain the quantity to be minimized in the form $\frac{1}{2}x'T'RTx - x'T'V$. If we bring in e as an abbreviation for $T'V$, we obtain the expression $\frac{1}{2}x'T'RTx - x'e$.

Our task then is to minimize $\frac{1}{2}x'Mx - x'e$ where $M = T'RT$.

Notice that in this expression we have no terms of more than second degree. We are dealing with a quadratic expression; the problem is a generalization of finding the minimum value of $\frac{1}{2}ax^2 - bx$. It should not be too difficult.

Minimum of a quadratic expression

In calculus we have learnt to find maximum and minimum values in cases where x represented a number; we have not met examples where x was a vector. It is therefore necessary to leave our electrical problem for a moment, and consider how calculus can be adapted to vector problems.

The usual calculus argument amounts more or less to the following. If, in $\frac{1}{2}ax^2 - bx$, we replace x by $x + h$ we get $\frac{1}{2}a(x + h)^2 - b(x + h)$, which can be written as $\frac{1}{2}ax^2 - bx + h(ax - b) + \frac{1}{2}ah^2$. From this we can see the effect of a small change h. When h is small, h^2 is even smaller. For example, if $h = 0.001$, then $h^2 = 0.000001$. If $ax - b$, the coefficient of h, is not zero, for sufficiently small h the term $h(ax - b)$ will outweigh the term $\frac{1}{2}ah^2$, and so will decide whether the change from x to $x + h$ makes the quantity increase or decrease. Now $h(ax - b)$, in these circumstances, changes sign if h changes sign. Accordingly, by choosing the sign of h appropriately, we can make the quantity increase or decrease, whichever we want. In this case, we have neither a maximum nor a minimum. Such things can happen only if $ax - b = 0$. Then our quantity is $(\frac{1}{2}ax^2 - bx) + \frac{1}{2}ah^2$. If $a > 0$, this will be larger than $\frac{1}{2}ax^2 - bx$, whether h is positive or negative. That is to say, we have a minimum corresponding to the value x; any small change from x to $x + h$ produces an increase in the quantity given by our quadratic expression. Of course, if $y = \frac{1}{2}ax^2 - bx$, then $dy/dx = ax - b$. Our discussion, purely in terms of algebra, is establishing that, in this particular case, minimum can occur only if $dy/dx = 0$.

Let us try to apply this procedure to the expression $\frac{1}{2}x'Mx - x'e$, where x now represents a vector. We suppose x replaced by $x + h$, where h is a vector too; all the numbers appearing in h are to be small. We thus get $\frac{1}{2}(x' + h')M(x + h) - (x' + h')e$, which multiplies out to give

$$\tfrac{1}{2}x'Mx - x'e + (\tfrac{1}{2}h'Mx + \tfrac{1}{2}x'Mh - h'e) + \tfrac{1}{2}h'Mh.$$

Here we see the original expression at the beginning; the bracket in the middle contains the terms which, apart from the exceptional case, determine whether we have an increase; finally we have $\frac{1}{2}h'Mh$, which involves the squares and products of small quantities. We need to do a little work on the middle term. Now $x'Mh$ represents simply a number; it therefore is equal to its transpose $(x'Mh)'$. This is $h'M'x$ and, as we have agreed always to use a symmetric matrix when we are dealing with quadratic forms, we know $M' = M$. Accordingly $\frac{1}{2}x'Mh = \frac{1}{2}h'Mx$. Thus the first two terms in the bracket are equal. Their sum is $h'Mx$. Accordingly the bracket contains the quantity $h'Mx - h'e$ or $h'(Mx - e)$.

We now argue, just as we did in the case where x was a number, that this term will control the size of the change, and can be made positive or negative at will, except when $Mx - e = 0$. (Check for yourself that these assertions remain true in the vector situation.)

We need not go on to investigate whether we have a maximum, minimum or other situation. It will be sufficient for us to have the result corresponding to the calculus result that stationary points occur only when $dy/dx = 0$, namely –

The expression $\frac{1}{2}x'Mx - x'e$, where M is a symmetric matrix, is stationary only when $Mx - e = 0$.

The electrical problem, resumed

In the electrical question, the expression $\frac{1}{2}x'Mx - x'e$ does in fact have a minimum. But we need not go into that. If it is to be a minimum, it certainly must be stationary and so, by our result above, x must satisfy $Mx = e$.

We must tidy up one point. The theorem above requires M to be symmetric. Now, in our electrical question, M is an abbreviation for $T'RT$. Is this in fact symmetrical? It certainly is. In question 6 of the exercises at the end of §33 we saw that, if R is symmetric, then $T'RT$ is automatically symmetric. And our matrix R (see equation (5)) obviously is symmetric.

We have now proved everything we had to do. We have shown that if we put $i = Tx$, where the matrix T is fixed when we dissect our circuit into loops, then it is natural to introduce a vector e defined by $e = T'V$ and a matrix $M = T'RT$, and that this will lead to the equation $Mx = e$.

The power of matrix notation is shown not only by its application to electrical, structural and other problems, but also by the general result we obtained, on the way, about the stationary points for $\frac{1}{2}x'Mx - x'e$. Here, by working on the analogy of a proof in calculus, concerned with a single number x, we obtained a proof that would apply, for instance, to a problem involving a thousand variables, for the notation $Mx = e$ remains the same whether x denotes a vector in one, two or a thousand dimensions.

Exercises

1 For the circuit shown in Fig. 107, write equations expressing the currents i_1, i_2, i_3, i_4, i_5, i_6 in

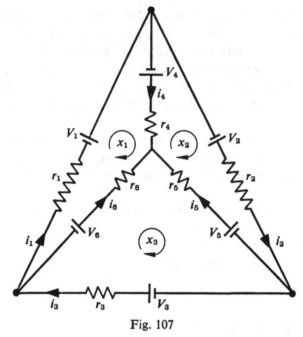

Fig. 107

terms of the loop currents x_1, x_2, x_3. Also write equations giving the loop voltages e_1, e_2, e_3 in terms of V_1, V_2, V_3, V_4, V_5, V_6. Check that the matrices corresponding to these systems of equations are transposes, each of the other.

2 Fig. 108 shows three springs joined together and lying in a straight line. The end points are

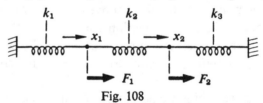

Fig. 108

fixed and the system is not self-stressed. Constant forces F_1 and F_2 act at the junctions of the springs. Write the expression for the potential energy of the system when these junctions have displacements x_1 and x_2, the stiffness constants of the springs being k_1, k_2 and k_3.

3 Fig. 109 shows an electrical circuit. Is it possible to choose the voltages V_1, V_2 and the resistances

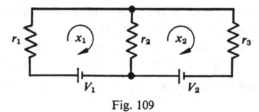

Fig. 109

r_1, r_2, r_3 in such a way that function to be minimized in the electrical situation will correspond exactly to the potential energy which is minimized for equilibrium of the system of springs? (If this can be done, the behaviour of the electrical circuit will predict the behaviour of the structure.)

Check your answer by writing the equations for the two systems, without any appeal to minimum principles, and comparing the equations obtained.

35 Formal laws of matrix algebra

So far our work with matrix algebra has been informal. In §20, §21 and §22 exercises were provided by which the student could see that transformations, and the matrices that represented them, when expressed in algebraic symbolism sometimes behaved very much like numbers in elementary algebra, and sometimes behaved very differently. Thus for example in the exercises at the end of §22, question 4 led to

$$(A - I)^2 = A^2 - 2A + I,$$

which we would expect by analogy with elementary algebra, while question 5 showed that of $(A + B)(A - B)$, $(A - B)(A + B)$ and $A^2 - B^2$, no two were equal – an unexpected result. In Exercise 20, we had various examples, intended to give the student a feeling for the way things worked, and on the basis of this experience judgements were invited – does it seem likely that $ST = TS$, that $S + T = T + S$? We have worked with matrix algebra on the basis of these judgements, but we have never set out its laws explicitly and considered their justification. It is now time to turn to this.

If we consider the properties of numbers assumed in elementary algebra, and ask whether similar properties hold for matrices, we find the properties fall into three classes: (1) those that do not apply to matrices, and that we must remember *not* to assume, (2) those that are most easily verified by calculation, (3) those that are most easily verified by a more or less geometrical argument.

Properties that do not apply are easily disposed of. We simply have to produce an example that shows matrices behaving in a way incompatible with those properties. Thus, in the symbolism of §20 and §21, the fact that $MJ_1 \neq J_1M$ immediately indicates that the commutative property of multiplication, $ST = TS$, is *not* universally true for matrices. In fact it is exceptional for matrices to commute when multiplied.

Again, we cannot conclude from $ST = 0$ that $S = 0$ or $T = 0$. Questions 12 and 15 at the end of §21 clearly rule this out.

The commutative and associative properties for the addition of matrices can be checked by routine algebra. These two properties together amount to the 'in-any-order-rule' – if matrices are added in any order the final result is the same. If matrices of specific shape and size are given, the entries being either particular numbers or algebraic symbols such as a, b, c, d, the student will have no difficulty in checking that the order does not affect the sum. Difficulty may be found in establishing the result quite generally, for $m \times n$ matrices. The idea is understood, but it may be hard to clothe it in language. The subscript symbolism may be helpful. If $S = A + B + C$, where A, B, C are matrices with entries a_{rt}, b_{rt}, c_{rt}; $1 \leqslant r \leqslant m$, $1 \leqslant t \leqslant n$, the typical entry in S can be calculated, and it will be seen that no change in the order of A, B, C can lead to a different sum.*

The proof will not be set out in this book, since it uses only routine algebra. A student who cannot work it out for himself will not be able to follow the argument as set out by someone else.

Students who find great difficulty in the linguistic aspects of mathematics – in expressing ideas in words and symbols – may find that, after checking the in-any-order property of addition for say 2×2, 2×3 and 4×2 matrices, with arbitrary entries, $a, b, c, \ldots$, they achieve an insight which enables them to be confident that this property holds for any $m \times n$ matrix. Such students may find it possible to work on, without bothering about the details of the formal general proof. However, it is advantageous if students can master the language and symbolism of formal proofs, since on occasion they may need to consult works using such words and symbols.

Very similar remarks apply to the distributive law, $A(B + C) = AB + AC$. A student can easily verify that this law does in fact hold for, say, all 2×2 matrices. He can soon convince himself that this verification does not depend on the particular number of rows or columns considered, provided of course that the matrices are of types such that the additions and multiplications required are meaningful. Finally, the student should, if possible, write a proof for the general case, where A is an $m \times n$ matrix and B and C are $n \times p$ matrices.

In arithmetic, from $a(b + c) = ab + ac$ we can deduce $(b + c)a = ba + ca$ since multiplication is commutative. With matrices we cannot use this argument. It is therefore necessary to state and prove separately $(B + C)A = BA + CA$. This second distributive law holds for all matrices for which the products involved are meaningful.

In arithmetic, the 'in-any-order' law holds for multiplication. If we have to calculate $2 \times 3 \times 4$ there are six different ways of doing it and they all lead to the same answer 24. For a child learning arithmetic there is no obvious reason why the 'in-any-order' rule should be split into two properties, commutative and associative. But when we come to matrices, this distinction does arise in a very natural way. As we have seen, the 'in-any-order' property ceases to hold; as a rule it is not even true that $ST = TS$ for matrices. So, for a product involving three matrices A, B and C, we might guess that no law at all would hold, that every way of combining them would lead to a distinct result. Actually, this is not so; there is something that can be said. The point in question arises in practice. In an earlier section we were concerned with the quadratic expression $v'Mv$, which in 2 dimensions has the form

$$(x, y) \begin{pmatrix} a & h \\ h & b \end{pmatrix} \begin{pmatrix} x \\ y \end{pmatrix}.$$

---

* It will be necessary to assume the commutative and associative laws for the addition of *numbers*. Students sometimes object that we are assuming what we want to prove and that this is unfair. But we are not 'begging the question'. At this stage, we *know* that the *real numbers* have the properties in question; we are trying to prove that *matrices*, which are built from numbers but are not themselves numbers, have some of these properties.

When confronted with such an expression students frequently ask in what way they are supposed to multiply this out. Should they first work out the product of the row vector and the square matrix? This would give them

$$(ax + hy, hx + by) \begin{pmatrix} x \\ y \end{pmatrix}.$$

Or should they first work out the product of the square matrix and the column vector? This would give

$$(x, y) \begin{pmatrix} ax + hy \\ hx + by \end{pmatrix}.$$

It turns out not to matter. By either route, the final result is $ax^2 + 2hxy + by^2$.

It should be noted that, in the calculations above, we have nowhere changed the *order* of the matrices. Throughout, the row vector came in front of the square matrix, and the square matrix in front of the column vector; that is, we always kept the order of the product $v'Mv$. The difference in the two calculations above lay, not in changing the order of the matrices, but in the procedure for combining them. In the first calculation, we find $v'M$ and then combine this with v. Symbolically this is recorded as $(v'M)v$. In the second calculation we find Mv and then bring in v'; that is, we calculate $v'(Mv)$. The fact that both ways lead to the same final result is expressed by the equation $(v'M)v = v'(Mv)$.

This is an example of the associative law, and in fact for any three matrices A, B, C it is true that $(AB)C = A(BC)$; here of course we are assuming that the numbers of rows and columns in A, B and C are such as to make these products meaningful.

It is possible to check the assertion just made by direct calculations, but even for moderate-sized matrices the work is very tedious. For instance, if A, B and C are all 2×2 matrices, with arbitrary entries $a, b, c, \ldots$, to express the product $(AB)C$ we have to write 48 letters and 12 plus signs. The general result can be proved by using subscripts and summation signs, and is so proved in many textbooks.

It seems easier and more instructive to note that matrices represent linear mappings, and that all mappings, whether linear or not, automatically obey the associative law. In Fig. 110, p represents any kind of object. Some mapping C sends p to q, then B sends q to r and finally A sends r to s. It seems reasonable that we should be able to write

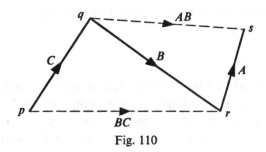

Fig. 110

$s = ABCp$, and that the mapping ABC should have a clear meaning and not require any brackets in its symbol. We can in fact show that $(AB)C$ and $A(BC)$ represent the same mapping. By BC we understand the effect of C followed by B. Either from the diagram, or from the equations $q = Cp$, $r = Bq$, we can see that BC sends p to r. As A sends r to s, it follows that $A(BC)$ sends p to s. Again we can see that AB sends q to s. Thus $(AB)C$, which indicates C followed by AB, sends p to s. Either way, p goes to s. Now p could be any object for which the mappings in question are defined. Thus $(AB)C$ and $A(BC)$ have the same effect on each object; that is, $(AB)C$ and $A(BC)$ represent the same mapping, which was what we wanted to show.

Accordingly, as far as addition and multiplication are concerned, the main thing to remember is that $ST = TS$ is *as a rule, not true* for matrices. We must be careful to preserve the order in which matrix products are written. Apart from this, we can carry out calculations very much as in the elementary algebra of numbers.

In matrix algebra we may meet expressions such as $2A + 3B$. The presence of the numbers, such as 2 and 3, does not lead to any surprises. We have, for example, $4A + 5A = 9A$ and $(2A)(3B) = 6AB$. We can form general laws to cover such results, and these can be checked or given formal proof.

We think of $2A$ as the product of the *number* 2 and the *matrix* A. There is a matrix closely associated with the number 2. For instance, if $A = \begin{pmatrix} a & b \\ c & d \end{pmatrix}$, it is easily verified that

$$\begin{pmatrix} 2 & 0 \\ 0 & 2 \end{pmatrix} \begin{pmatrix} a & b \\ c & d \end{pmatrix} = \begin{pmatrix} a & b \\ c & d \end{pmatrix} \begin{pmatrix} 2 & 0 \\ 0 & 2 \end{pmatrix} = \begin{pmatrix} 2a & 2b \\ 2c & 2d \end{pmatrix} = 2A.$$

Quite generally, if we let $K = kI$, we can verify that, for any square matrix A, $KA = AK = kA$. (If A is $n \times n$, we understand that I also is $n \times n$.)

A matrix of the form kI is known as a *scalar matrix*. Such a matrix commutes with every matrix A.

Subtraction also behaves as expected. We can regard $A - B$ as $A + (-1)B$, so that it is covered by the remarks made above about numerical coefficients.

Accordingly in matrix algebra we can make calculations such as the following:

(1) $(A + B)^2 = (A + B)(A + B) = A(A + B) + B(A + B) = A^2 + AB + BA + B^2$,

(2) $(A + B)(A - B) = A(A - B) + B(A - B) = A^2 - AB + BA - B^2$.

The absence of a commutative law prevents us from simplifying these any further.

Polynomials in one matrix

There is an important case in which matrix algebra is particularly simple; this occurs when we are dealing with expressions involving only one matrix. For example, the matrices A^2 and A^3 *do commute*. Here A^2 indicates that a certain transformation has to be applied twice, A^3 that it has to be applied three times. Both $A^2 \cdot A^3$ and $A^3 \cdot A^2$

indicate that the transformation has to be applied five times. Thus $A^2 \cdot A^3 = A^3 \cdot A^2 = A^5$. The same argument shows that for any natural* numbers m and n we have

$$A^m \cdot A^n = A^n \cdot A^m = A^{m+n}$$

exactly as in elementary algebra. *Thus the powers of a single matrix commute among themselves.* The identity operation I may be regarded as A^0 – applying the transformation 'no times', to put it ungrammatically. (I commutes with every matrix; $IM = MI$; it does not matter whether you 'leave things alone' before applying M or after.)

Thus for polynomials in a single matrix A we have *all the properties on which elementary algebra is based*; the computations are identical in form with those in the algebra of numbers.

Division

Division is essentially different for matrices from what it is for numbers. In arithmetic $12 \div 3$ corresponds to the question '3 times what is 12?' In slightly more sophisticated terminology this question invites us to solve the equation $3x = 12$. In the algebra of real numbers, the equation $ax = b$ always has a solution, except when $a = 0$. Thus b/a, the solution of this equation, is defined unless $a = 0$.

With matrices the situation is far otherwise. If we wish to be sure that the equation $AX = B$ has a solution X, (A, B, X being matrices), it is not by any means sufficient to check $A \neq 0$. We have already seen a particular case of this in §26, when we were considering the inverses of square matrices. The inverse of A, when it exists, satisfies the equation $AX = I$. No inverse exists when $\det A = 0$, that is, when A maps the space into a space of lower dimension. Fig. 111 indicates what happens in such a case. Several vectors u, v, w will be sent by A to the same point p, while no vector is sent to some other point q. Notice in this situation we have $p = Au = Av = Aw$. However, it would be incorrect to 'cancel A' in the equation $Au = Av$ and conclude $u = v$.

Exercise

Construct an example to illustrate the possibility $Au = Av$ with $u \neq v$ and $A \neq 0$.

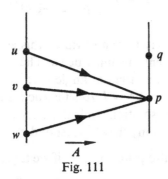

Fig. 111

* A natural number is a positive whole number 1, 2, 3, . . .

A vector can be regarded as an $m \times 1$ matrix. If such complications can arise in the equation $AX = B$ even when X and B are vectors, we would naturally expect complications in the general case, and in fact such complications do arise: the equation may have no solution at all, or it may have an infinity. Either way, we are unable to define 'B divided by A'.

The only straightforward case is when A has the inverse A^{-1}. Then $AX = B$ can have only one solution. For if $AX = B$, then $A^{-1}(AX) = A^{-1}B$, and this simplifies to $X = A^{-1}B$. So $AX = B \Rightarrow X = A^{-1}B$. By substitution we verify that

$$X = A^{-1}B \Rightarrow AX = B.$$

With ordinary numbers we could do without fraction symbolism b/a and write ba^{-1} or $a^{-1}b$ for this fraction. With numbers ba^{-1} and $a^{-1}b$ mean the same thing. With matrices, as multiplication is not commutative, BA^{-1} and $A^{-1}B$ are liable to mean different things; $A^{-1}B$ gives the solution of $AX = B$ and BA^{-1} the solution of $XA = B$.

It will be realized that the inverse A^{-1} can exist only for a square matrix A. The inverse exists if $\det A \neq 0$. However, a determinant is usually a very inconvenient thing to calculate; later we will give more suitable ways for seeing whether an inverse exists, and for calculating it when it does exist.

If A^{-1} exists, it is legitimate to 'cancel A' in an equation such as $AX = AY$. For we have $A^{-1}(AX) = A^{-1}(AY)$, which simplifies to $X = Y$. However, it is wisest to use the brief argument just given rather than to think of any 'rule for cancelling'. If we proceed from $AX = AY$ to $A^{-1}(AX) = A^{-1}(AY)$ we are reminded that this step is allowable only when A^{-1} exists, and we are less likely to commit the error of supposing X must equal Y in other circumstances.

In elementary algebra a quadratic equation never has more than two solutions. For example, if $x^2 = 1$, we can deduce $(x - 1)(x + 1) = 0$. The product is zero only when one factor is zero, so $x - 1 = 0$ or $x + 1 = 0$ and the only solutions are 1 and -1. *For matrices this argument fails.* It is easily verified that, for

$$X = \begin{pmatrix} \cos \theta & \sin \theta \\ \sin \theta & -\cos \theta \end{pmatrix},$$

we have $X^2 = I$. Thus, as θ may have any value, this equation has an infinity of solutions. Geometrically, it is evident this must be so. The matrix X given above represents reflection in the line through the origin at angle $\theta/2$; any reflection, done twice, brings us back to our starting point, so $X^2 = I$. It is instructive to see at which point the proof that $x^2 - 1 = 0$ has only two solutions fails if we try to apply it to $X^2 - I = 0$. Now I and $-I$ do satisfy the equation; they correspond to the solutions 1 and -1 in the algebra of numbers. But we also get a solution if we take, say, $X = \begin{pmatrix} 0 & 1 \\ 1 & 0 \end{pmatrix}$, and we can use this to find out where the argument breaks down. The early parts are quite in order.

With this value for X, we do in fact have $(X - I)(X + I) = 0$. For

$$X - I = \begin{pmatrix} -1 & 1 \\ 1 & -1 \end{pmatrix} \quad \text{and} \quad X + I = \begin{pmatrix} 1 & 1 \\ 1 & 1 \end{pmatrix}.$$

It is easily checked that the product of these two matrices is 0. The argument breaks down at the next step; the product is zero but neither factor is zero. (It may be noted that neither $X - I$ nor $X + I$ has an inverse.)

Exercise

Show that if $AB = 0$ for two matrices A, B, and one of the matrices has an inverse, then the other matrix must be 0.

The simplest example in which $PQ = 0$ without either factor being zero is obtained by taking

$$P = \begin{pmatrix} 1 & 0 \\ 0 & 0 \end{pmatrix}, \quad Q = \begin{pmatrix} 0 & 0 \\ 0 & 1 \end{pmatrix}.$$

P represents projection on the x-axis; it sends the point (x, y) to the point $(x, 0)$; that is, it wipes out the y co-ordinate. Similarly Q wipes out the x co-ordinate. Thus the operation PQ wipes out both co-ordinates, and sends every point to the origin; thus $PQ = 0$. But this effect is achieved in two stages; first Q replaces y by 0, then P replaces x by 0. But neither P nor Q is the matrix 0.

Exercise

Interpret geometrically the transformations $(X - I)/2$ and $(X + I)/2$, where $X = \begin{pmatrix} 0 & 1 \\ 1 & 0 \end{pmatrix}$, in terms of standard graph paper. What is their product, in either order?

Functions involving inverses

If two matrices commute, so do their powers. If $AB = BA$, we have

$$A^2 B = A(AB) = A(BA) = (AB)A = (BA)A = BA^2$$

and it is not difficult to extend this argument to show $A^m B^n = B^n A^m$ for any natural numbers m, n.

It follows from this, that, if we are working with polynomials in two commuting matrices, A and B, we can carry out additions, subtractions and multiplications exactly as in elementary algebra.

Now we saw in §26 that, if A has an inverse A^{-1}, then $AA^{-1} = A^{-1}A = I$. So $AA^{-1} = A^{-1}A$, that is, A commutes with its inverse. Accordingly, we may take $B = A^{-1}$

in the result of the previous paragraph. A polynomial in A and A^{-1} is an expression involving positive and negative powers of A such as, for instance,

$$5A^2 + 3A - 4I + 7A^{-1} - A^{-3}.$$

Any two such expressions commute, and we can work with them by the familiar rules of elementary algebra, as regards addition, subtraction and multiplication. As we saw earlier, caution is required in any process involving division of matrices.

For reference purposes it may be helpful to list the properties of real numbers that, consciously or unconsciously, we assume when doing algebra, and the situation in matrix algebra in relation to each.

Real numbers	*Matrices*
$a + b$ defined always	$A + B$ defined always
$a - b$ defined always	$A - B$ defined always
'In-any-order rule' holds for $a + b + c$	The same for $A + B + C$
ab defined always	AB defined always
$ab = ba$ always	$AB = BA$ only in special cases
$a(bc) = (ab)c$ always	$A(BC) = (AB)C$ always
$a(b + c) = ab + ac$ always	$A(B + C) = AB + AC$ always
$(b + c)a = ba + ca$ always	$(B + C)A = BA + CA$ always
a/b defined unless $b = 0$	AB^{-1} and $B^{-1}A$ defined when B^{-1} exists
$bx = by$ implies $x = y$, if $b \neq 0$	$BX = BY$ implies $X = Y$ only when B^{-1} exists
$xy = 0$ implies $x = 0$ or $y = 0$	$XY = 0$ can happen with $X \neq 0$ and $Y \neq 0$
—	If $K = kI$, $KA = AK$ for any A
—	Powers, positive or negative, of a single matrix commute $A^m \cdot A^n = A^n \cdot A^m = A^{n+m}$ for any integers m, n

36   Orthogonal transformations

In §5 we considered the geometrical effect of a linear transformation of a plane. This was shown by a diagram in which a network of squares was transformed into a network of parallelograms. If we wanted to demonstrate such a transformation by applying it to an actual object, we would have to choose something, like an elastic sheet, that was capable of considerable deformations. Some linear transformations, such as rotations and reflections, can be demonstrated with a rigid body, such as a piece of cardboard. (Cardboard of course is not absolutely rigid – nothing is! – but its behaviour is sufficiently different from that of an elastic sheet to make our distinction meaningful.) Such transformations, and their generalizations to space of any number of dimensions, form a special class, known as orthogonal transformations. These transformations naturally occur when we are dealing with rigid bodies, or indeed with any objects in actual physical space, which we think of as being Euclidean. Orthogonal transformations also have certain mathematical properties that cause them to occur in situations that have no direct connection with rigid bodies or physical space. Statistics is one example of this. Vibration problems are another; it may be objected that vibrations are concerned with actual rigid bodies (though this hardly applies to electrical vibrations), but the relevance of orthogonal transformations does not arise from this aspect of the situation. It is due rather to analogies that exist between vibration problems and the purely mathematical idea of rotations in space of n dimensions (where n may be bigger than 3).

Let us consider the simplest and most familiar case, a rotation in a plane, specified with the help of conventional graph paper.

Suppose we rotate the plane until the point P, which originally had co-ordinates (x, y) lands at the position (X, Y) as shown in Fig. 112. If the angle of rotation is θ, it can be shown (by projections, or by considering the fate of the unit basis vectors) that

$$\begin{pmatrix} X \\ Y \end{pmatrix} = \begin{pmatrix} \cos\theta & -\sin\theta \\ \sin\theta & \cos\theta \end{pmatrix} \begin{pmatrix} x \\ y \end{pmatrix}. \tag{1}$$

If we wished to obtain (x, y) in terms of (X, Y) we might consider that $(X, Y) \to (x, y)$ is a rotation of $-\theta$, or we might use projection, or we might solve the equations (1). Whichever we did, we would arrive at the result

$$\begin{pmatrix} x \\ y \end{pmatrix} = \begin{pmatrix} \cos\theta & \sin\theta \\ -\sin\theta & \cos\theta \end{pmatrix} \begin{pmatrix} X \\ Y \end{pmatrix}. \tag{2}$$

The matrix occurring in equations (2) is the inverse of the matrix occurring in equations (1), for it represents the transformation that undoes the effect of rotation through θ. Thus if L denotes the matrix in equations (1), L^{-1} denotes the matrix in equations (2). But if we examine these matrices as they appear in the two equations, we notice that the second matrix is simply the transpose of the first. Accordingly we have $L' = L^{-1}$. *This is the characteristic property of an orthogonal matrix; its inverse is the same as its transpose.* This incidentally is an extremely convenient property. In 2 dimensions it is easy to calculate the inverse of a matrix, but with increasing number of dimensions the

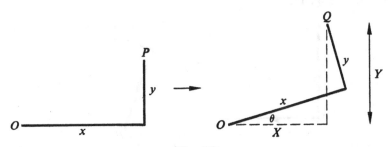

Fig. 112

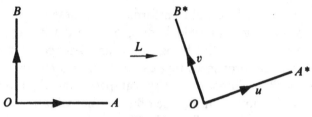

Fig. 113

calculation of an inverse becomes extremely messy. On the other hand, a transpose can be written down immediately. Accordingly, the appearance of an orthogonal matrix in a problem is always welcomed.

Work in 2 dimensions uses very special methods. In 2 dimensions we can specify a rotation by giving a single number, θ; we can specify a direction by giving a single number, the value of m in $y = mx$. In 3 or more dimensions such simple procedures cease to be possible.

It is therefore useful to re-examine the situation in 2 dimensions, and to try to explain $L' = L^{-1}$ by methods that work equally well in 3 or more dimensions.

Accordingly, let us consider a 2×2 matrix

$$L = \begin{pmatrix} u_1 & v_1 \\ u_2 & v_2 \end{pmatrix}$$

and try to interpret what it means if for this matrix $L^{-1} = L'$. This means $L'L = I$, so we have

$$\begin{pmatrix} u_1 & u_2 \\ v_1 & v_2 \end{pmatrix} \begin{pmatrix} u_1 & v_1 \\ u_2 & v_2 \end{pmatrix} = \begin{pmatrix} u_1^2 + u_2^2 & u_1v_1 + u_2v_2 \\ u_1v_1 + u_2v_2 & v_1^2 + v_2^2 \end{pmatrix} = \begin{pmatrix} 1 & 0 \\ 0 & 1 \end{pmatrix}.$$

This means $u_1^2 + u_2^2 = 1$, $u_1v_1 + u_2v_2 = 0$, $v_1^2 + v_2^2 = 1$. These three equations are readily interpreted geometrically. The first and the last indicate that the vectors u and v are of unit length; the middle equation involves a scalar product and indicates that u and v are perpendicular.

Now the vectors u and v occur in the columns of the matrix L. In §22, under the heading 'The columns in a matrix' it was pointed out that these columns represented the vectors to which the unit vectors along the axes were mapped. Thus we may represent the effect of L as in Fig. 113.

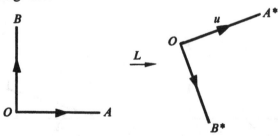

Fig. 114

The condition $L'L = I$ assures us that the lines OA^* and OB^* will be of unit length and at right angles, which of course is so for a rotation. The condition, however, does *not* tell us that we *must* be dealing with a rotation. It is perfectly consistent with a diagram such as Fig. 114. Here a reflection is involved.

The argument we have used does not in any way depend on the special features of 2 dimensions. It is easily adapted to show that, in any number of dimensions, the condition $L'L = I$ means that L sends the unit vectors along the axes to vectors that are of unit length and perpendicular to each other.

Exercise

Write the 3 × 3 matrix L, whose columns are the vectors u, v, w. Check that $L'L = I$ does mean that u, v, w are all of unit length and perpendicular to each other.

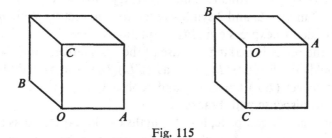

Fig. 115

For each of the two cubes shown in Fig. 115, the lines OA, OB, OC are of equal length and are perpendicular to each other. Yet it is impossible to turn the first cube in such a way that the letters will appear as in the second cube. The second cube is a reflection of the first.

Thus, as was already noted in connection with 2 dimensions, the condition $L'L = I$ does not tell us that L represents a rotation; it may represent a rotation followed by a reflection. But either way we can assert that the transformation represented by L *preserves lengths unchanged*; to use the technical term, it is an *isometry* (iso-, the same; -metry, measure).

It seems reasonable that L should preserve all distances unchanged. If OA, OB, OC are unit vectors along the axes, they can be visualized as the edges of a unit cube. If $L'L = I$, L sends these vectors to OA^*, OB^*, OC^* which are of unit length and perpendicular – that is to say, are edges of some unit cube. We can bring the original unit cube to the new position by a rotation, perhaps combined with a reflection. We saw in §10 that the fate of the unit vectors along the axes determined the fate of every vector, in a linear transformation. So, if we think of all the vectors as being embedded in some rigid body, we can show the effect of L by rotating this rigid body, and, if necessary, following this by a reflection. In such a process, all lengths stay unchanged.

It is sometimes necessary to consider orthogonal transformations in, for example, 6 dimensions, and here we may not feel the same confidence in our physical picture of what is meant by a rotation or a reflection. So it is desirable to provide a proof by algebra that, when $L'L = I$, the transformation L preserves the distances between points. The proof is very simple, and illustrates a type of matrix calculation we may often meet.

We first show that, if L sends $p \to p*$, $q \to q*$, for any vectors p, q, then the scalar product of $p*$ and $q*$ equals the scalar product of p and q. More briefly, L *preserves all scalar products.*

The proof is immediate. $p* = Lp$, $q* = Lq$. The scalar product is, by §33, $(Lp)'Lq$. This is, by the reverse-order rule for transposes, $p'L'Lq$. But $L'L = I$. So we have $p'Iq$, which is $p'q$, the scalar product of the original vectors.

Note the convenience of matrix notation. Without any heavy work or special trick, we have proved the result for any number of dimensions. Even in 3 dimensions the notation effects a considerable economy. This proof, written out in full as was done in the older textbooks on 'co-ordinate solid geometry' would need 9 symbols for the entries in the 3×3 matrix L, and 3 symbols each for the vectors p and q. Several equations would be needed to express our single equation $L'L = I$.

Note also that the proof is making free use of the associative property. If we retained brackets, we would obtain $(Lp)'(Lq)$ first as $(p'L')(Lq)$, and would then have to re-arrange brackets to get $(L'L)$ appearing and replaced by I. The associative property allows us simply to dispense with brackets.

Our proof is now nearly complete, for all lengths can be expressed as scalar products. For any vector p, we have $\|p\|^2 = p'p$. If p represents the vector OP, $\|p\|$ equals the length $\|OP\|$. As this length is expressible by means of the scalar product $p'p$, and as L preserves scalar products, it follows that, when L acts, the distance of any point from the origin is unchanged. We wish also to show that the length $\|PQ\|$ is preserved for any points P, Q. Let q denote the vector OQ. Then $q - p$ denotes the vector PQ. The scalar product of this vector with itself gives $\|PQ\|^2$. So

$$\|PQ\|^2 = (q - p)'(q - p) = (q' - p')(q - p) = q'q - p'q - q'p + p'p.$$

It is sufficient to observe that each term here is a scalar product, hence unchanged when p is replaced by $p*$ and q by $q*$. The expression can in fact be simplified, since $q'p = p'q$. (See §33, near end.) It may be written

$$\|PQ\|^2 = q'q - 2p'q + p'p.$$

This last result in fact is essentially the same as a standard formula in trigonometry, $c^2 = a^2 - 2ab \cos C + b^2$.

Determinants

The determinant of L, $\det L$, gives the ratio in which volumes (or the corresponding concept in n dimensions) are changed. A rotation keeps volumes unchanged, so for a rotation $\det L = +1$. A reflection, with the convention established in §27, multiplies volumes by -1. So if L represents the effect of a rotation followed by a reflection, det $L = -1$. Thus the determinant enables us to tell to which of these two classes any given orthogonal transformation, L, belongs.

Exercise

Apply this test to the matrices

$$\begin{pmatrix} \cos \theta & -\sin \theta \\ \sin \theta & \cos \theta \end{pmatrix} \quad \text{and} \quad \begin{pmatrix} \cos \theta & \sin \theta \\ \sin \theta & -\cos \theta \end{pmatrix}.$$

By a reflection in n dimensions we understand a situation in which one vector is reversed, and all vectors perpendicular to it remain unchanged. Thus, for example, in 4 dimensions the equations $x^* = -x, y^* = y, z^* = z, t^* = t$ would define a reflection. Any reflection can be expressed in this form by choosing axes suitably related to its eigenvectors.

Change of axes

At the beginning of this section we had equations (1), $X = x \cos \theta - y \sin \theta$, $Y = x \sin \theta + y \cos \theta$. These gave us the co-ordinates (X, Y) to which a point (x, y) was sent by a rotation through θ. Part of the diagram used to illustrate this rotation appeared as in Fig. 116.

If we look at this part of the diagram by itself we see the point Q; we do not see the point P from which it came by rotation. Someone seeing only this part of the diagram might easily imagine that equations (1) were intended to specify, not a rotation, but a change of axes. For the point Q has co-ordinates (X, Y) in horizontal and vertical axes, and co-ordinates (x, y) in axes inclined at an angle θ. In fact equations (1) could be used for this purpose. Thus the same equations arise in describing a rotation and in specifying a change of axes. In practice, orthogonal transformations are probably used more often for introducing more convenient axes than for describing a rotation of some object.

In lectures on matrices it is customary to mention a mild joke – that a matrix can either be an alibi or an alias. An alibi means 'somewhere else', and so might refer to a transformation that sends each point to a new position. An alias means a new name, and

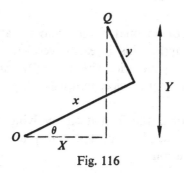

Fig. 116

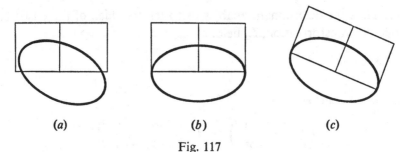

(a)   (b)   (c)

Fig. 117

is appropriate to the situation just discussed where the point Q has two names; Q is called (x, y) in one system of axes and (X, Y) in another.

We can see why the same equations arise in the two cases.

In Fig. 117 (a) we see an ellipse and some squares which we may suppose drawn on a transparent sheet placed over the ellipse. In (b) the ellipse has been rotated until its principal axes coincide with lines on the graph paper. In (c) the same effect has been achieved by bringing the graph paper to the ellipse. Thus from (a) to (b) represents a rotation of the ellipse, with fixed graph paper. From (a) to (c) represents a rotation of the graph paper (a change of axes) with a fixed ellipse. The relative motion of the ellipse and the graph paper is the same in both cases. Note that if in going from (a) to (b) the ellipse rotates through an angle θ, in going to (a) to (c) the graph paper rotates through an angle $-\theta$, equal and opposite. In work of this kind it is always necessary to check carefully that the matrix you have written represents the transformation you want, and not its inverse.

Worked example

Test for orthogonality the matrix

$$M = \begin{pmatrix} -\tfrac{2}{3} & \tfrac{1}{3} & \tfrac{2}{3} \\ \tfrac{1}{3} & -\tfrac{2}{3} & \tfrac{2}{3} \\ \tfrac{2}{3} & \tfrac{2}{3} & \tfrac{1}{3} \end{pmatrix}.$$

Investigate fully the geometrical meaning of the transformation of 3-dimensional space it represents.

Solution The matrix will be orthogonal if the vectors that occur in its columns are perpendicular and of unit length. This is easily verified. So M is orthogonal and $M'M = I$. But we notice M is symmetric, with $M' = M$. Accordingly $M^2 = I$. So M represents some operation that, done twice, returns each point of space to its original position.

Is the operation a rotation or a transformation involving reflection? The sign of the determinant will tell us which; $\det M = (4 + 4 + 4 + 8 - 1 + 8)/27 = 1$. As $\det M$ is positive, M represents a rotation.

Any rotation in 3 dimensions must be about some axis. Every point on the axis stays still, so any point v on the axis must satisfy $Mv = v$. If $v = (x, y, z)$ this means

$$x = (-\tfrac{2}{3})x + (\tfrac{1}{3})y + (\tfrac{2}{3})z,$$
$$y = (\tfrac{1}{3})x + (-\tfrac{2}{3})y + (\tfrac{2}{3})z,$$
$$z = (\tfrac{2}{3})x + (\tfrac{2}{3})y + (\tfrac{1}{3})z.$$

These equations are equivalent to $x = y$, $z = 2x$. So, for example, if we choose $x = 1$, we see that $(1, 1, 2)$ is on the axis of rotation.

Earlier we had $M^2 = I$. A rotation which, done twice, brings us back to the starting situation, must be through $0°$ or $180°$. (We need not consider $360°$, $540°$ and so on, since these have the same final effect.) Now $0°$ would indicate $M = I$, which is clearly not so, so M must represent a rotation through $180°$.

We can confirm this conclusion. A rotation of $180°$ reverses every vector in the plane perpendicular to the axis of rotation. Thus every vector in this plane should satisfy the equation $Mv = -v$. If $v = (x, y, z)$ this means

$$-x = (-\tfrac{2}{3})x + (\tfrac{1}{3})y + (\tfrac{2}{3})z,$$
$$-y = (\tfrac{1}{3})x + (-\tfrac{2}{3})y + (\tfrac{2}{3})z,$$
$$-z = (\tfrac{2}{3})x + (\tfrac{2}{3})y + (\tfrac{1}{3})z.$$

Each of these equations reduces to $x + y + 2z = 0$. This is the equation of the plane perpendicular to $(1, 1, 2)$, since it expresses that the scalar product of $(1, 1, 2)$ and (x, y, z) is zero.

It will be noticed that the non-zero vectors lying in the axis of rotation are eigenvectors with $\lambda = 1$, while those perpendicular to the axis are eigenvectors with $\lambda = -1$. This illustrates one of the most important maxims of linear algebra – if you want to know what a matrix does, look for its eigenvectors and eigenvalues.

In this particular case, by observing special features of the problem, we were able to avoid the computation involved in the standard routine of solving $\det(M - \lambda I) = 0$, and then finding the eigenvectors. If this routine had been followed, we would have found $\det(M - \lambda I) = -(\lambda - 1)(\lambda + 1)^2$, and been led to the same results, though with a little more computation.

Questions

1 Let P be the point $(0.8, 0.6)$ and Q the point $(-0.6, 0.8)$ on conventional graph paper. Find the lengths OP and OQ and the angle between OP and OQ. Is the matrix $\begin{pmatrix} 0.8 & -0.6 \\ 0.6 & 0.8 \end{pmatrix}$ orthogonal or not?

What is its inverse?

2 Which of the following matrices are orthogonal?

$$\begin{pmatrix} 0 & 1 \\ 1 & 0 \end{pmatrix}, \quad \begin{pmatrix} 0 & -1 \\ 1 & 0 \end{pmatrix}, \quad \begin{pmatrix} 0.8 & 0.6 \\ 0.6 & -0.8 \end{pmatrix}, \quad \begin{pmatrix} 1 & 1 \\ 0 & 1 \end{pmatrix}.$$

Which of them represent rotations? Which of them represent reflections?

3 Test for orthogonality the matrix

$$\begin{pmatrix} -\frac{1}{3} & \frac{2}{3} & \frac{2}{3} \\ \frac{2}{3} & -\frac{1}{3} & \frac{2}{3} \\ \frac{2}{3} & \frac{2}{3} & -\frac{1}{3} \end{pmatrix}.$$

What is its inverse? What does this matrix do to the vectors

$$\begin{pmatrix} 1 \\ 1 \\ 1 \end{pmatrix}, \quad \begin{pmatrix} 1 \\ -1 \\ 0 \end{pmatrix}, \quad \begin{pmatrix} 1 \\ 0 \\ -1 \end{pmatrix}.$$

How can its effect be described in simple geometrical terms?

4 A transformation T sends (x, y, z) to (x^*, y^*, z^*) where $x^* = y$, $y^* = x$, $z^* = -z$. Write the matrix for T. Is it orthogonal? What is the geometrical meaning of T? What are its eigenvectors and eigenvalues?

5 The matrices L and M are both orthogonal. Which of the following matrices can we be certain will be orthogonal?
    (a) LM,   (b) L^{-1},   (c) $L + M$.
(Much harder.) Can we be certain that any of the above matrices will *not* be orthogonal?

6 Let

$$S = \begin{pmatrix} 0 & k \\ -k & 0 \end{pmatrix}.$$

Calculate the matrix $L = (I + S)(I - S)^{-1}$ and check that L is orthogonal. (It can in fact be proved that for any antisymmetric matrix S, acting in n dimensions, the matrix L given by this formula will be orthogonal.)

7 Is the matrix

$$\begin{pmatrix} -\frac{3}{7} & \frac{6}{7} & \frac{2}{7} \\ -\frac{2}{7} & -\frac{3}{7} & \frac{6}{7} \\ \frac{6}{7} & \frac{2}{7} & \frac{3}{7} \end{pmatrix}$$

orthogonal or not? Apply the transformation specified by this matrix to the points $(-5, 3, 8)$, $(-7, 0, 14)$, $(-8, 9, 10)$, $(-10, 6, 16)$. What geometrical figure results? What can be said about the figure formed by the original four points?

8 Show that the four points $(0, 0, 0)$, $(0, 1, 1)$, $(1, 0, 1)$, $(1, 1, 0)$ are the corners of a regular tetrahedron. Find where these points are sent by the transformation

$$\begin{pmatrix} -\frac{2}{3} & \frac{2}{3} & \frac{1}{3} \\ -\frac{1}{3} & -\frac{2}{3} & \frac{2}{3} \\ \frac{2}{3} & \frac{1}{3} & \frac{2}{3} \end{pmatrix}.$$

Would you expect the transformed points to be at the corners of a regular tetrahedron? If so, why? Are they in fact so situated?

9 Find the length of, and the angles between, the vectors given by the columns of the matrix

$$\begin{pmatrix} -7 & 4 & -4 \\ -4 & 1 & 8 \\ 4 & 8 & 1 \end{pmatrix}.$$

This matrix is not orthogonal. Is there any combination of simple geometrical operations that would describe its effect? To what would it transform the four points mentioned in Question 8?

10 Do either of the matrices mentioned in Questions 7 and 8 represent a reflection or a rotation, and if so about what axis?

11 What vectors are unaltered when the transformation

$$\begin{pmatrix} \tfrac{1}{3} & -\tfrac{2}{3} & -\tfrac{2}{3} \\ -\tfrac{2}{3} & \tfrac{1}{3} & -\tfrac{2}{3} \\ -\tfrac{2}{3} & -\tfrac{2}{3} & \tfrac{1}{3} \end{pmatrix}$$

acts on them? What vectors are reversed by it? What geometrical operation does this matrix represent? Is it an orthogonal matrix? Would it be possible to say, without the labour of carrying out the calculation, whether its determinant is positive or negative?

12 In question 6 at the end of §32, each part specified 3 vectors D, E, F. In each case, if a matrix were formed having D, E, F as its columns, would this matrix be orthogonal? If a matrix were formed having D, E and F as its rows, would this matrix be orthogonal?

13 A rotation is represented by the matrix

$$M = \begin{pmatrix} -\tfrac{6}{7} & \tfrac{3}{7} & -\tfrac{2}{7} \\ -\tfrac{3}{7} & -\tfrac{2}{7} & \tfrac{6}{7} \\ \tfrac{2}{7} & \tfrac{6}{7} & \tfrac{3}{7} \end{pmatrix}.$$

Find the axis about which this rotation takes place. Let P be the point $(1, 3, -2)$. Find P^*, the point to which P is mapped by M. What is the cosine of the angle between OP and OP^*? If M represents a rotation through the angle α, does α equal the size of the angle between OP and OP^*? Justify what you assert.

14 A rotation is represented by the matrix

$$M = \begin{pmatrix} 0 & 0 & 1 \\ 1 & 0 & 0 \\ 0 & 1 & 0 \end{pmatrix}.$$

Find the axis of rotation. Find the angle of rotation produced by M by considering the powers of M. Let $P = (1, -1, 0)$ and $Q = (0, 1, -1)$. Find P^* and Q^*, the points to which P and Q map under the action of M. Find the angle between OP and OP^* and the angle between OQ and OQ^*. Explain the result.

Show that M has only one real eigenvalue and consequently only one real eigenvector. Explain why this result is to be expected on geometrical grounds.

15 The 8 corners of a cube have the co-ordinates $(\pm 1, \pm 1, \pm 1)$, in which it is understood that every possible combination of plus and minus signs is to be taken. Apply to these 8 points the transformation represented by the matrix M in Question 13. Delete the third co-ordinate in each point so obtained, i.e. change (x, y, z) into (x, y), and plot the 8 resulting points (x, y) on ordinary squared paper. Do these points, when suitably joined, look like a picture of a cube? Explain what you observe.

Note. Further exercises of this kind can be obtained, if desired, by taking for M the matrix mentioned in some other question of this section.

37   Finding the simplest expressions for quadratic forms

In most secondary school courses, students meet problems about ellipses and hyperbolas. The ancient Greeks studied these curves and named them conic sections, since they could be obtained by taking a plane section of a cone. Chopping cones by planes is not an activity that occupies a large part of an engineer's time, and one is inclined to regard conic sections as an entertainment for pure mathematicians. This view, though natural, is entirely mistaken. Conics have a way of turning up in all kinds of practical situations.

Before dealing with their applications today, we may note that the whole development of modern science would probably have been considerably delayed if the Greeks had not established conics as a traditional part of mathematical education. In the seventeenth century, when Newton put the science of mechanics into systematic form, he was very much helped by Kepler's work on the orbits of the planets. Now Kepler had been faced by a difficult problem. In astronomy the motion of, say, Mars is observed from the earth, itself travelling in a curved path with varying speed. The problem is to find assumptions about the orbits of Mars and the earth that will explain the observations of Mars made from the earth. At first Kepler tried to fit the data by assuming the orbits to be circles, perhaps with centres not at the sun, and perhaps described with variable speeds. Long and patient calculations showed that circles would not do. But if not circles, what then? There are other possible curves in great variety. How to decide which kind to try next? Very fortunately, the ellipse, which was already known to Kepler as a curve resembling the circle, turned out to be the correct guess. Without this fortunate coincidence, Kepler might have spent his entire life trying different curves and never landing on the right one.

For an understanding of the modern importance of conics the Greek definition 'section of a cone' is not particularly helpful. For us the important thing is that any equation of the second degree gives a conic. The general equation of the second degree is $ax^2 + 2hxy + by^2 + 2gx + 2fy + c = 0$. Here a, b, c, f, g, h are constants. By 'the second degree' we understand that terms may be present, such as x^2 or xy, in which two variables are multiplied together, but that we never have terms, such as x^4, x^2y or y^3, with three or more variables as factors. Such an equation may represent an ellipse, parabola or hyperbola; there are also certain special cases such as, for example, a pair of straight lines. For engineering work the ellipse is probably the case most often met.

The definition just given is easily generalized. In 3 dimensions we can consider equations of the second degree; they may contain linear terms, squares such as x^2, y^2, z^2, and products such as xy, xz and yz, but no higher terms. Such an equation defines a *quadric surface*; in particular it may define an *ellipsoid*. An ellipsoid may be visualized as a sphere that has been subjected to a particular kind of distortion. Suppose we start with the unit sphere, $X^2 + Y^2 + Z^2 = 1$. Then we enlarge the scale on the first axis a times, on the second axis b times, and on the third axis c times. That is, we send (X, Y, Z) to (x, y, z) where $x = aX$, $y = bY$, $z = cZ$. We obtain the ellipsoid with equation $(x^2/a^2) + (y^2/b^2) + (z^2/c^2) = 1$. This surface in 3 dimensions obviously has analogies with the ellipse in 2 dimensions.

We saw in §2 that many scientific laws can be well approximated by linear functions and that this is why we are likely to meet simple linear expressions for such things as momentum, mv, force in a stretched spring, kx, and voltage drop in a resistance, ri. Now the integrals of these expressions also have physical significance; the momentum, mv, is the rate of change of $\frac{1}{2}mv^2$, the kinetic energy; the force, kx, is the rate of change of potential energy, $\frac{1}{2}kx^2$; the voltage drop, ri, is the derivative of $\frac{1}{2}ri^2$, which is proportional to the rate of generation of heat; the angular momentum, $I\omega$, is the derivative of $\frac{1}{2}I\omega^2$, the kinetic energy of a rotating flywheel. It will be noticed that the expressions related to energy are all of the second degree. This happens also when we are considering systems involving several variables. Thus we are not surprised to find that in much scientific work we are confronted by some formula involving squares and products, a formula that can be shown graphically by means of an ellipsoid or other quadric surface. Thus in the theory of elasticity we meet the stress-ellipsoid; both in connection with the balancing of machines and with objects hurled through space we meet the momental ellipsoid; in the theory of vibrations we have to deal with a pair of ellipses or ellipsoids representing kinetic and potential energy; in statistics the dots of scatter diagrams frequently form egg-shaped clusters; in quantum theory we meet something resembling an ellipsoid, adapted to the needs of that strange subject (the Hermitian form).

In §24 we considered a vibration problem with two masses and three springs. It was shown that the differential equations for the motion of these masses could be brought to their simplest form by means of the substitution $x = X + Y, y = X - Y$. We could have been led to this substitution by a different argument, which does not even use calculus, if we had posed the question in another way: find the co-ordinate system in which *the expressions for the kinetic and potential energies take the simplest form.*

Let u and v stand for the velocities of the two unit masses in the problem of §24. (Note that u and v here stand for single numbers, not for vectors in 2 or 3 dimensions as they often have.) The kinetic energy, T, is given by $T = \frac{1}{2}(u^2 + v^2)$. The potential energy is $P = \frac{1}{2}kx^2 + \frac{1}{2}k(y - x)^2 + \frac{1}{2}ky^2$, which simplifies to $P = k(x^2 - xy + y^2)$.

We can show the positions of the two masses by plotting (x, y) on graph paper. The vector (u, v) will then show the velocity with which the plotted point moves (Fig. 118). The quantities u and v are taken parallel to the same axes as those used for x and y. Accordingly, if we introduce new co-ordinates X, Y for position by means of the equations $x = aX + bY, y = cX + dY$ we must also introduce new co-ordinates for velocity, U, V, with $u = aU + bV, v = cU + dV$. (The same conclusion could be reached by the calculus argument that $u = \dot{x}$ and $v = \dot{y}$, where the dot indicates differentiation with respect to time; if $U = \dot{X}$ and $V = \dot{Y}$, we can obtain the equations

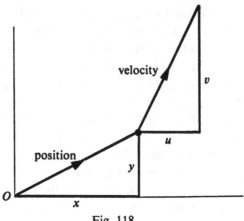

Fig. 118

connecting u, v with U, V by differentiating the equations $x = aX + bY$, $y = cX + dY$.)

Accordingly, we make diagrams to illustrate kinetic and potential energy using the same graph paper to plot (x, y) and (u, v). We will show (u, v) as a vector springing from the origin $(0, 0)$, not as in the last diagram, springing from the point (x, y) and indicating the direction of motion of that point.

The kinetic energy, T, is already in a very simple form. The curves T = constant have equations $u^2 + v^2$ = constant, that is to say they are circles. The curves P = constant have equations $x^2 - xy + y^2$ = constant. These curves are in fact ellipses, tilted as shown in Fig. 119. The diagram shows just one ellipse, and the two circles that touch it. The points of contact lie on the lines $y = x$ and $y = -x$ respectively. Thus if we had some machine that automatically drew the curves T = constant and P = constant for us, we would obtain a diagram containing many circles and many ellipses which were enlarged or reduced copies of those shown. At most points the circles would cross the ellipses, but along the lines $y = x$ and $y = -x$ the curves would touch. Thus the machine would draw our attention to the fact that the north-east and south-east direc-

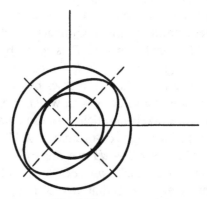

Fig. 119

tions had a special physical significance. The vectors $\begin{pmatrix}1\\1\end{pmatrix}$ and $\begin{pmatrix}1\\-1\end{pmatrix}$ lie in these direc-

tions. If we take them as a basis for new co-ordinates (X, Y) we obtain the equations

$$\begin{pmatrix}x\\y\end{pmatrix} = X\begin{pmatrix}1\\1\end{pmatrix} + Y\begin{pmatrix}1\\-1\end{pmatrix}.$$

That is, we have $x = X + Y$, $y = X - Y$ as in §24. If we substitute these expressions for x and y in our equation for P, and the corresponding expressions $u = U + V$, $v = U - V$ in the equation for T, we find $T = U^2 + V^2$, $P = k(X^2 + 3Y^2)$.

Thus this substitution achieves two goals: (1) it brings P to a simple form, with no product XY, (2) it does this without spoiling the simple form that T had from the outset.

Now of course in practice we do not usually have a machine that will sketch graphs for us; we may have computers that will calculate such transformations for us, but they do it by a rather different method. The purpose of the discussion so far has been, not to provide a method of computation, but to show the kind of situation that exists when we have two physically significant quantities (such as kinetic and potential energy) represented by quadratic forms. In such a situation it is frequently possible to choose axes in such a way that one quadratic form is of the type we associate with a circle or sphere, while the other is of the type we associate with a conic referred to its principal axes. Thus in 3 dimensions we can often find axes that bring two quadratic expressions to the forms $X^2 + Y^2 + Z^2$ and $aX^2 + bY^2 + cZ^2$ when both involve the same variables, or to $U^2 + V^2 + W^2$ and $aX^2 + bY^2 + cZ^2$ in cases, such as that considered above, where one quadratic form involves velocity and the other position.

We cannot prove this here, but in fact by choosing the axes that put kinetic and potential energy into the simplest possible form, we are automatically choosing the axes that put the differential equations for the motion into the simplest possible form.

We have used vibration problems as an illustration because in §24 we worked such a problem out in detail. The mathematical ideas apply equally well to any problem in which two quadratic forms are involved. Sometimes one of the quadratic forms may be present but camouflaged. For instance, in any question involving the motion of a rigid body (balancing of machinery, projectiles, flight of aircraft) the values of the moments and products of inertia can be shown graphically by a certain ellipsoid embedded in the body. We are interested in finding the principal axes of this ellipsoid, that is to say, the axes in which its equation takes the form

$$(X^2/a^2) + (Y^2/b^2) + (Z^2/c^2) = 1.$$

Here we seem to be dealing with only one quadratic form, that appearing in the equation of the ellipsoid. But in fact a second quadratic is involved. The body is *rigid*; the distances between any two particles in it is fixed. Now the distance of a particle at (x, y, z) from the origin is given by $\sqrt{(x^2 + y^2 + z^2)}$. When we bring in new co-ordinates X, Y, Z we want this distance to be given by $\sqrt{(X^2 + Y^2 + Z^2)}$; otherwise we are liable to slide into all kinds of mistakes, because our usual formulas for distances, scalar products, perpendicularity and so forth will cease to apply. So what in fact we are trying to do when we search for the principal axes of the ellipsoid is to find co-ordinates that simplify the equation of the ellipsoid *without spoiling the simplicity of the distance formula* $s^2 = x^2 + y^2 + z^2$. Compare this with our earlier work on the vibration problem, where we managed to bring the potential energy P to a simple form without disturbing the simple form that T already had. It will be seen that the two problems, so different in physical terms, involve essentially the same mathematical problem.

Students sometimes think that all they need learn in engineering mathematics is procedures for carrying out calculations. But this is in fact not enough. We need first of all an assurance that the problem does have a solution. If we apply some standard procedure to an impossible problem difficulties of one kind or another are bound to arise. Again, even when a problem is possible, it may be that there are certain exceptional cases where the solution is of an unexpected kind, or where the computing procedure has to be modified. It is clearly desirable to be aware of these possibilities.

The first task, then, is not to learn an algorithm, but to survey the scene and classify problems so that we can recognize the general case, the exceptional case and perhaps the impossible case.

When dealing with quadratic forms, we can get quite a good idea of how things will work out in any number of dimensions by considering fairly familiar examples in 2 or 3 dimensions. In 2 dimensions we very readily believe that any ellipse has a major and a minor axis, and that these are perpendicular. Accordingly, if we choose points C and D on these axes, at unit distance from the origin O, OC and OD will be unit vectors at right angles (Fig. 120). If OA and OB are the basis vectors of the original co-ordinate system, we need only rotate axes to arrive at OC and OD. Rotation of axes does not alter the equation of a circle. The circle $x^2 + y^2 = 1$ in the system based on OA and OB will have the equation $X^2 + Y^2 = 1$ in the system based on OC and OD.

Thus our impression is that there will be no impossible problems. Given a quadratic form $ax^2 + 2hxy + by^2$, we expect to be able to rotate axes (thus preserving the simplicity of $x^2 + y^2$) so as to obtain an expression of the form $AX^2 + BY^2$. Of course, this expression may not correspond to ellipses. For example, we might obtain $X^2 - Y^2$ and the curves $X^2 - Y^2 = $ constant would be hyperbolas.

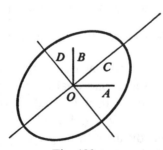

Fig. 120

Exercise

Prove the statement that the problem is never impossible; that is, show that with the substitution $x = X \cos \theta - Y \sin \theta$, $y = X \sin \theta + Y \cos \theta$ we can always choose θ so that $ax^2 + 2hxy + by^2$ becomes an expression in which the XY term is absent. Check also that $x^2 + y^2 = X^2 + Y^2$.

While there is no impossible case, there do exist exceptional cases. A very simple example would be the expression x^2. This is already in its simplest possible form. The equation $x^2 = 1$ represents a pair of vertical lines (Fig. 121). This may be thought of as a limiting case, in which the major axis of the ellipse has become of infinite length. Such a situation arises if in $AX^2 + BY^2$ either $A = 0$ or $B = 0$. Of course it is not necessary that the lines should be vertical as in our example using x^2. By rotating the diagram we can obtain an example with parallel lines pointing in any direction. For example $x^2 - 2xy + y^2 = 1$ is the same as $(x - y)^2 = 1$, which breaks up into the two lines $x - y = 1$ and $x - y = -1$.

In 3 dimensions we have an element of variety in the special cases. We can always rotate axes so as to get the form $AX^2 + BY^2 + CZ^2$. In the general case, A, B, C are all different from zero; if they all happen to be positive, $AX^2 + BY^2 + CZ^2 = 1$ will represent an ellipsoid. If one of them, say C, is zero, the equation reduces to

$$AX^2 + BY^2 = 1.$$

Now Z does not appear in this, so Z is completely arbitrary; by this we mean that if some particular point, $(X_0, Y_0, 0)$ satisfies the equation, then the point (X_0, Y_0, t) will also satisfy it, for every value of t. Thus the surface consists of complete lines parallel to the Z-axis. It is a cylinder – not necessarily a cylinder with circular cross-section (Fig. 122). If A and B are both positive, the section will be an ellipse, as in the illustration here. If A and B have opposite signs, the section will be a hyperbola.

If two of the quantities A, B, C are zero, we have an even more special case. For example, if $A = 1$, $B = 0$, $C = 0$ we have $X^2 = 1$. Now Y and Z can vary at will; the surface breaks up into the two planes $X = 1$ and $X = -1$.

It will be noticed that we have sometimes spoken simply of an *expression* such as

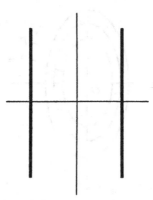

Fig. 121   Graph of $x^2 = 1$.

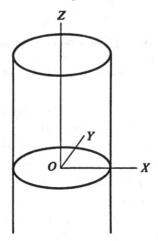

Fig. 122

$ax^2 + 2hxy + by^2$ and sometimes of the *equation* $ax^2 + 2hxy + by^2 = 1$. Let us consider a particularly simple vibration problem, that of a particle sliding around near the bottom of a smooth bowl. This situation is not particularly significant in itself: it is important, however, as giving us a way both of *visualizing and simulating what happens in a large class of vibration problems*. The potential energy of the particle is given by its vertical height above the lowest point of the bowl; suppose that, when it is at the position (x, y), its vertical height is given by $P = 9x^2 + 4y^2$. The shape of the bowl may be represented by drawing the contours $P = 1$, $P = 2$, $P = 3$ and so on. As shown in Fig. 123 these are all ellipses. They have the same axes, and they differ only in the scale on which they are drawn. Accordingly, if we can find the principal axes for any ellipse, say for $P = 1$, we have the axes in which the whole system appears most simply.

It will be noticed that the axes correspond to the points on any ellipse that are

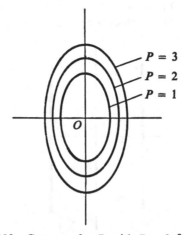

Fig. 123   Contours for P, with $P = 9x^2 + 4y^2$.

closest to and farthest from the origin O. This provides an idea that can be used both to prove that, in any number of dimensions, perpendicular principal axes always exist, and also, on occasion, as a way of calculating the positions of these axes.

Before we develop this idea we had better look at a possible complication. Suppose we are confronted with a situation involving not ellipses but hyperbolas, for example $P = x^2 - y^2$. The contours would now run somewhat as in Fig. 124. We now no longer have a bowl, but something like a saddle or mountain pass. The origin lies in the middle of the mountain pass. If you go east or west from O, you are climbing the sides of the pass. If you go north or south from O you are descending; this is where the road through the mountains would be. It is evident that such a situation is not likely to require detailed analysis in engineering work. It represents an unstable situation; a particle dislodged from O would be most likely to roll down the hillside and never be seen again. As a rule our concern would be simply to avoid a situation in which such a collapse could occur; we would not be interested in knowing exactly how fast the collapse developed.

Even in this case the x-axis and the y-axis can be identified by a minimum principle. The x-axis is given by the points on the contour $P = 1$ nearest to O; the y-axis is given by the points on the contour $P = -1$ nearest to O. Accordingly our guiding idea can be adapted to this case, by considering the two contours $P = 1$ and $P = -1$. This complication would certainly have to be considered in any exhaustive mathematical treatment. As this complication does not occur in the most important practical situations, it seems justified to simplify and shorten our explanation by excluding it. From now on, we will consider only situations in which hyperbolas and their generalizations are excluded; this is equivalent to saying that *we shall consider only quadratic forms that never take negative values*. These are technically known as *non-negative definite quadratic forms*.

We have to establish a connection between the geometrical idea of 'the point nearest to the origin' and the algebraic idea of getting rid of products such as xy in the equation of an ellipse.

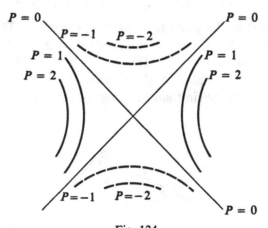

Fig. 124

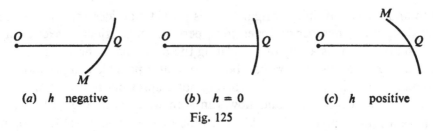

(a) h negative   (b) $h = 0$ (c) h positive

Fig. 125

This connection depends upon the following fact in plane geometry. The behaviour of the curve $ax^2 + 2hxy + by^2 = 1$ near the point where it crosses the x-axis depends on the coefficient h as shown in the three diagrams of Fig. 125 (we suppose $a > 0$). Only when $h = 0$ can Q be at a minimum distance from O. To prove this we change to polar co-ordinates (Fig. 126), by putting $x = r \cos \theta$, $y = r \sin \theta$ in the equation $ax^2 + 2hxy + by^2 = 1$. From this we find

$$\frac{1}{r^2} = a \cos^2 \theta + 2h \cos \theta \sin \theta + b \sin^2 \theta.$$

So

$$\frac{d}{d\theta} \left(\frac{1}{r^2} \right) = 2(b - a) \sin \theta \cos \theta + 2h(\cos^2 \theta - \sin^2 \theta).$$

When $\theta = 0$ this derivative has the value $2h$. When r is a minimum, $1/r^2$ is a maximum, and its derivative must be zero. So only when $h = 0$ can this happen for $\theta = 0$, corresponding to the point Q. (The same argument would show that only when $h = 0$ can the distance be a maximum at Q.)

Thus we have the result: *if $(q, 0)$ is the point of the curve $ax^2 + 2hxy + by^2 = 1$ nearest to the origin, then $h = 0$.*

In 2 dimensions this is enough to show that the equation must be in its simplest form $ax^2 + by^2 = 1$; the result also has consequences in spaces of higher dimension.

Suppose we are told that no point of the surface

$$ax^2 + by^2 + cz^2 + 2fyz + 2gxz + 2hxy = 1$$

is nearer to the origin than the point $(q, 0, 0)$ of this surface. (The point being in the surface implies that $aq^2 = 1$, but we shall not need to use this fact.) Consider the points of the surface that lie in the plane $z = 0$. They satisfy $ax^2 + by^2 + 2hxy = 1$. If $h \neq 0$, by our result there will be some point of the ellipse $ax^2 + by^2 + 2hxy = 1$ in the plane $z = 0$ that lies nearer to the origin than the point of this ellipse on the x-axis. Such a

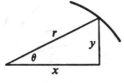

Fig. 126

point would lie on the surface of the ellipsoid, and this would contradict our assumption. Accordingly we must have $h = 0$.

Again consider the intersection of the ellipsoid with the plane $y = 0$. This intersection is the ellipse $ax^2 + cz^2 + 2gxz = 1$. The same argument shows $g = 0$.

Accordingly the ellipsoid must have an equation of the form

$$ax^2 + (by^2 + cz^2 + 2fyz) = 1,$$

if the point $(q, 0, 0)$ is the point closest to the origin. Thus, if the nearest point lies on the x-axis the equation must break into two parts; in one part we have an x^2 term, in the other part we have only the other variables y and z; nowhere do we find the product of x with another variable, such as xy or xz.

This argument generalizes to any number of dimensions. For instance, if we had four variables, x, y, z, t we would consider points with $z = 0$, $t = 0$; then points with $y = 0$, $t = 0$; finally points with $y = 0$, $z = 0$. From these we would deduce in turn that the coefficients of xy, of xz, and of xt were all zero.

Note that all this argument is based on the assumption that we are using a 'conventional' system of co-ordinates, with perpendicular axes and the same units on all axes; otherwise expressions such as $x^2 + y^2$ or $x^2 + y^2 + z^2$ for the square of a distance, and equations such as $x = r \cos \theta$, $y = r \sin \theta$ cease to be true. We assume that we start with such conventional axes and that we try to bring our quadratic expression to its simplest form by rotating the axes, not by distorting them. By a suitable rotation we can always get rid of the product terms xy, xz, yz, and our result above allows us to prove this very easily.

Suppose then we have an ellipsoid $ax^2 + by^2 + cz^2 + 2fyz + 2gxz + 2hxy = 1$. Somewhere on this ellipsoid there will be a point Q at the minimum distance from O. We get a new system of axes by rotating the old ones until the first axis lies along OQ. The other two axes will automatically be in the plane perpendicular to OQ, and for the moment we do not care how they lie in that plane. Thus in this first step we choose any set of perpendicular unit vectors, demanding only that the first vector lies along OQ. In these axes the point Q will be $(q, 0, 0)$; as it is at minimum distance from O, the equation (by our earlier result) must reduce to the form $ax^2 + by^2 + cz^2 + 2fyz = 1$. The quantities a, b, c, f here will of course be different from those in the original equation. We are concerned only with the *type* of equation, not with the actual values of the coefficients.

We are now satisfied with our choice of OQ as the first axis and turn our attention to what is happening in the plane perpendicular to OQ. The equation of that plane is now $x = 0$; by substituting $x = 0$ in the (new) equation for the ellipsoid we see that points in that plane lie on the ellipse $by^2 + cz^2 + 2fyz = 1$. Our next step is to rotate the axes about OQ in such a way as to give this ellipse the simplest possible form. So we are confronted with a problem of the same type, but in one dimension less. Accordingly we proceed in the same way. On the ellipse there will be some point R that is nearest to the origin; that is, it is the point of the *ellipse* nearest to O; it will probably be farther from O than Q, but we are now concerned only with points in the plane $x = 0$ (Fig. 127).

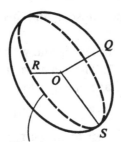

The plane $x = 0$

Fig. 127

We choose OR as our second axis. By our earlier result, this gets rid of the yz term, and the ellipse comes to the form $by^2 + cz^2 = 1$. Rotating the axes about OQ does not change the x co-ordinate in any way, so the equation of the ellipsoid now has the form $ax^2 + by^2 + cz^2 = 1$ that we were hoping for.

There may be some difficulty in seeing that by bringing the *ellipse* in the plane $x = 0$ to its simplest form we are automatically achieving the desired result for the ellipsoid. It may help to look at the work in terms of algebra. The first step brings the equation to the form $ax^2 + by^2 + cz^2 + 2fyz = 1$. When we rotate axes about OQ we are bringing in new co-ordinates, say (X, Y, Z), related to the old by the equations

$$x = X, \qquad y = Y \cos \theta - Z \cos \theta, \qquad z = Y \sin \theta + Z \cos \theta.$$

The first term ax^2 is thus replaced simply by aX^2. The terms $by^2 + cz^2 + 2fyz$ are subjected to exactly the same substitution that we would use in a problem in plane geometry. We know that, by suitable choice of the angle θ, these terms can be made to give a result of the form $BY^2 + CZ^2$. Thus the whole expression on the left-hand side of our equation gives us $aX^2 + BY^2 + CZ^2$, as predicted.

In the most general case, where the principal axes are all of different lengths, the procedure will go through exactly as above. The first step will give us OQ, the shortest semi-axis. The next step will give us OR, the semi-axis of medium length. The longest semi-axis, OS, will remain and will be given automatically by the direction perpendicular both to OQ and OR.

Certain special cases may arise. For instance, it may happen that the ellipsoid has the shape of the football used in Canadian or rugby football. In that case 'a point at minimum distance from O' does not fix a definite point Q (Fig. 128). This does not matter. Anywhere on the circle shown in the illustration will do for Q. The terms xy and xz disappear provided there is no point of the ellipsoid *nearer to O than Q*. It does not matter if there are other points *as near*. Our earlier proof depended on noticing that $2h = d/d\theta(1/r^2)$ at $\theta = 0$. If $1/r^2$ is constant we still have $h = 0$.

The extreme case is that of a soccer football, a sphere. In that case *any* three perpendicular directions will serve as principal axes, OQ, OR and OS.

The general ellipsoid, the rugby football (ellipsoid of revolution) and the soccer football (the sphere) are easily visualized. It is important to learn this easy way of remembering the special cases that may arise. In §38 we shall be concerned with the

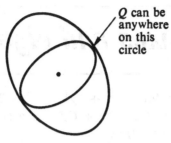

Q can be anywhere on this circle

Fig. 128

actual calculation of the principal axes; naturally these calculations turn out rather differently in the special cases and it is necessary to be prepared for these. Thus, for example, a student who has become accustomed to the general case, in which OQ is a definite line, is likely to be disturbed if he meets the 'rugby football' case in which the equations no longer lead to any one solution for OQ.

Another type of special case has already been mentioned in which the standard form, $AX^2 + BY^2 + CZ^2$, contains one or more zero coefficients. In this case the process of finding OQ, OR, OS unexpectedly terminates. For instance suppose we have the equation $x^2 + y^2 + z^2 + 2xy + 2xz + 2yz = 1$. This in fact is $(x + y + z)^2 = 1$ and represents the planes $x + y + z = 1$ and $x + y + z = -1$. The points in these planes nearest to the origin lie in the direction of the vector $(1, 1, 1)$. If we take the X-axis in this direction we find the equation becomes simply $3X^2 = 1$. The coefficients of Y^2, YZ and Z^2 are all zero. Our work is complete. In the same way, it may happen that, after the second stage of the work, we arrive at an equation of the form $AX^2 + BY^2 = 1$.

We have considered in detail how things work out in 3 dimensions, but the procedure is perfectly general. Given an equation, in any number of dimensions, there will be a point on the corresponding 'surface' nearest to the origin. Take the first axis in the direction of this point, and the remaining axes in any convenient manner (the basis vectors must be perpendicular and of unit length). The equation, in these new axes, will not contain any products x_1x_2, x_1x_3, ..., x_1x_n. We then put $x_1 = 0$ and find the point $(0, x_2, x_3, ..., x_n)$ that is on the surface and nearest to the origin. This point gives the direction for the second axis. We continue in this way. At each stage we are concerned with finding the point nearest to the origin on a certain surface; at each stage, we are working in a dimension one less than in the previous stage.

As mentioned earlier, the object of this section has been to survey the possibilities that may arise when we are seeking the simplest form of a quadratic expression. The next section will be concerned with the actual process of finding the best axes by calculation.

38 Principal axes and eigenvectors

As we saw in §37, a quadratic form $ax^2 + by^2 + cz^2 + 2fyz + 2gxz + 2hyz$ can be brought to the simpler form $AX^2 + BY^2 + CZ^2$ by a suitable change of axes. The quadratic form may be written as $v'Mv$ where

$$v = \begin{pmatrix} x \\ y \\ z \end{pmatrix} \qquad M = \begin{pmatrix} a & h & g \\ h & b & f \\ g & f & c \end{pmatrix},$$

and the simplified form as $V'DV$ where

$$V = \begin{pmatrix} X \\ Y \\ Z \end{pmatrix} \qquad D = \begin{pmatrix} A & 0 & 0 \\ 0 & B & 0 \\ 0 & 0 & C \end{pmatrix}.$$

We have chosen the letter D for this last matrix because this matrix is in *diagonal form*; that is, all its elements are zero except those on the main diagonal.

We met matrices in the first place as ways of specifying mappings, and square matrices in particular were associated with transformations. Thus the matrix M can be used to specify the transformation $v \to v^*$ where $v^* = Mv$, and it can also be used, as above, to specify the bilinear form $v'Mv$ which we visualize by means of some surface, such as an ellipsoid. We naturally wonder whether there is some relationship between the transformation $v \to Mv$ and the surface $v'Mv =$ constant.

There is in fact such a connection, and the transformation $v \to Mv$ in fact gives us a very useful way of finding the simplest expression for the quadratic form $v'Mv$.

The connection between the quadratic form and the transformation was well known long before mathematicians began to use matrix notation,* for a question that arises naturally in many sciences leads to this connection being noticed. It will be easiest to discuss this in 2 dimensions; the argument applies equally well in any number.

In §37, we had a number of ellipses which represented the contours of a bowl. Let us consider the general question of a point on the side of a fairly smooth hill. Through this point there is a line that goes directly up the hill. This is known as the line of greatest slope. If you are very strong and vigorous you may decide to climb in this direction. If you wish for an easier climb, or if you are choosing a route for a road or railway, you may prefer to climb obliquely. The question arises: how does the steepness of the climb vary as you increase the angle between your direction of navigation and the direction of steepest slope?

* Matrix notation was first explained by the English mathematician A. Cayley in 1855, but in looking at mathematical and scientific work during the century before this date one can see that much of it would have gone very nicely in matrix symbolism.

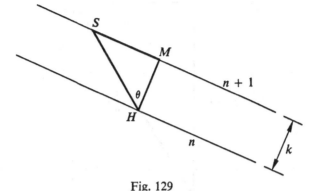

Fig. 129

It is usually sound policy to attack oroblem by considering the simplest possible case. Let us simplify drastically by supposing the hillside to be an inclined plane. The contours will then be parallel straight lines (Fig. 129). We suppose we are at H on the contour for height n and that contour $n + 1$ is a parallel line at distance k from contour n. If we take the shortest and steepest path HM, we shall travel a horizontal distance k in order to rise through unit height: $1/k$ thus measures the steepness of this path. If instead we decide to use path HS, this is of length $k/\cos \theta$. The steepness is obtained by dividing 1 by this quantity; thus the steepness for HS is $(1/k) \cos \theta$. We observe that the steepness falls off with the cosine of the angle, exactly as when we are finding the resolved part of a force. Thus, if we draw a vector of length $1/k$ in the direction of HM, the steepness in any direction is given by the component of the vector in the direction in question (Fig. 130). If $z = f(x, y)$ gives the height z at the point (x, y) of the contour map, the vector just described is called 'the gradient of z' or 'grad z' for short. Grad z is in the direction of greatest slope, and its magnitude equals the steepness (the gradient) in that direction.

In applications of gradient we need not be concerned only with heights of hillsides. In an expression grad V, the symbol V may stand for temperature, gravitational, electric or magnetic potential, or a potential function used in hydrodynamics and aerodynamics.

With all these applications existing, the question has naturally been studied – how do we calculate grad z when we need it? Suppose, in our example of the inclined plane, we have $z = px + qy$ as the equation of the plane. Let H be the point (x_0, y_0). We have

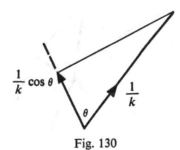

Fig. 130

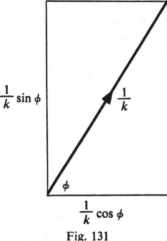

$\frac{1}{k}\sin\phi$

$\frac{1}{k}$

ϕ

$\frac{1}{k}\cos\phi$

Fig. 131

already used $1/k$ to indicate the magnitude of grad z; we suppose that grad z makes the angle ϕ with the x-axis (Fig. 131). Accordingly the components of grad z are $(1/k)\cos\phi$ and $(1/k)\sin\phi$. We know something about these components; they represent the steepness of the hillside in the directions parallel to the x-axis and the y-axis. Now it is not hard to calculate these gradients. If we start at (x_0, y_0) and move parallel to the x-axis, the x co-ordinate will change but the y co-ordinate will stay fixed. Thus we shall reach points given by (x, y_0) with varying x. So, for these points, $z = px + qy_0$. Differentiating, we find $dz/dx = p$, the gradient for motion parallel to the x-axis. In the same way, we find for motion parallel to the y-axis the gradient q. Accordingly the components of grad z are p, q. We may, if we wish, write $p = (1/k)\cos\phi$, $q = (1/k)\sin\phi$ or we may prefer to write simply grad $z = (p, q)$. We notice that grad z is constant; it does not depend on (x_0, y_0); this is reasonable, for in respect of steepness any part of an inclined plane is like any other part.

It was mentioned in §2 that one of the guiding ideas of calculus was that, for a wide class of situations, a small part of a curve could be efficiently approximated by a straight line, and a small part of a surface by a plane. For a curved surface the contours will be as shown in Fig. 132. The arc MS is no longer straight, the angle SMH is no longer

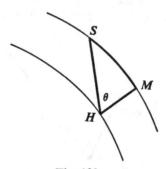

Fig. 132

exactly a right angle. However, it will often happen that, in the limit, as M is taken closer and closer to H, the diagram will become more and more like the one we had earlier. To find out and prove in exactly what conditions that will happen will not be attempted here. We will state, however, that for $z = f(x, y)$, where $f(x, y)$ is a polynomial in x and y, our main conclusions about grad z still hold. In particular, (1) the vector grad z is perpendicular to the contour $z = $ constant, (2) the components of grad z can be found by calculating the rates of change of z, for motion parallel to each axis in turn; that is, by the procedure used above. Of course for a curved surface, these components will not be constants, like p and q, but will depend on the position of the point (x_0, y_0), as will be illustrated in a moment.

We now come to our main question, the behaviour of the contours for

$$z = ax^2 + 2hxy + by^2.$$

If we consider points (x, y_0), with fixed y_0 and varying x, we have

$$z = ax^2 + 2hxy_0 + by_0^2.$$

In this expression a, h, b, y_0 all represent constants, x is the only variable – the problem is only that of differentiating a quadratic (the profusion of symbols sometimes obscures this simple fact). Accordingly $dz/dx = 2ax + 2hy_0$. We are interested in what happens at the point (x_0, y_0), so we put $x = x_0$, and obtain $2ax_0 + 2hy_0$. This gives the x-component of grad z at (x_0, y_0). In the same way, by considering points (x_0, y) and eventually putting $y = y_0$ we find the y-component of grad z to be $2hx_0 + 2by_0$.

Thus at (x_0, y_0) we have grad $z = (2ax_0 + 2hy_0, 2hx_0 + 2by_0)$. (See Fig. 133.) We know that grad z is perpendicular to the contour through (x_0, y_0). Multiplying a vector by a number does not change its direction, we may remove the factor 2, and thus reach the result; *the vector $(ax_0 + hy_0, hx_0 + by_0)$ gives the direction of the perpendicular at (x_0, y_0) to the contour, $ax^2 + 2hxy + by^2 = $ constant, passing through (x_0, y_0).* In the diagram, the vector HJ is perpendicular to the curve. The line HJ is often called 'the normal at H'.

In much of our work we think of a vector as a point, and, if we wish to draw an arrow, we draw it from the origin, O, to that point. In the present case, it is obviously more helpful to draw the vector $(ax_0 + hy_0, hx_0 + by_0)$ with its beginning at the point H. We could, if we liked, draw it from the origin: it would then represent a direction parallel to the normal at H. Pictorially this is less effective.

The following purely graphical exercise may help to fix this result in the memory.

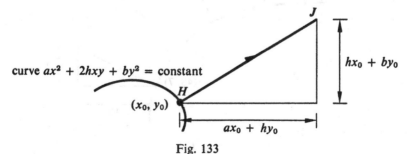

Fig. 133

Exercise

On graph paper plot the following points $(50, 0)$, $(48, 7)$, $(40, 15)$, $(30, 20)$, $(14, 24)$, $(0, 25)$. These lie on the ellipse $(x^2/4) + y^2 = 625$ and thus correspond to $a = \frac{1}{4}$, $h = 0$, $b = 1$. The vectors along the normals are thus given by the formula $(x_0/4, y_0)$. So, for example, from the point $(48, 7)$ we draw the vector with components $(12, 7)$. The resulting diagram appears as in Fig. 134. Small pieces of the tangents are shown, perpendicular to these vectors. These facilitate the freehand drawing of the ellipse.

We will now express this result in matrix notation. If we let

$$v = \begin{pmatrix} x \\ y \end{pmatrix} \qquad v_0 = \begin{pmatrix} x_0 \\ y_0 \end{pmatrix} \qquad M = \begin{pmatrix} a & h \\ h & b \end{pmatrix}$$

then $ax^2 + 2hxy + by^2 =$ constant may be written $v'Mv =$ constant; the constant is to be chosen so that the curve passes through the particular point v_0. The vector $\begin{pmatrix} ax_0 + hy_0 \\ hx_0 + by_0 \end{pmatrix}$ that gives the normal at this point is Mv_0. The transformation $v_0 \to Mv_0$ thus associates with any point v_0 a vector that gives the direction of the normal at that point (Fig. 135).

The matrix M that appears here must be symmetrical. This diagram gives us a useful way of visualizing the effect of a transformation, represented by a symmetric matrix.

What happens if v_0 should chance to lie along a principal axis of the conic, $v'Mv =$ constant? Evidently the normal, Mv_0, will be in the same direction as v_0; that is $Mv_0 = \lambda v_0$, and so v_0 *is an eigenvector of the matrix M* (Fig. 136).

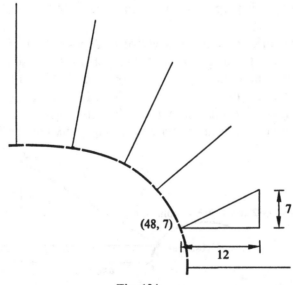

Fig. 134

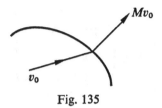

Fig. 135

Accordingly no new method is required for finding the principal axes, which give the simplest form of $v'Mv$. We simply have to look for the eigenvectors of M.

In this section we have considered an expression $ax^2 + 2hxy + by^2$ involving only two variables x, y. The arguments however are perfectly general and apply to $v'Mv$, where v can be a column vector in any number of dimensions.

The previous section showed us that we have to be prepared for various special cases. With only two variables, these special cases can be spotted immediately. By turning the axes, we can bring our expression to the form $AX^2 + BY^2$. If $A \neq B$, $A \neq 0$, $B \neq 0$, we have the general case. If $A = B \neq 0$, we have a system of circles $A(X^2 + Y^2) =$ constant and our original equations must have been $Ax^2 + Ay^2 =$ constant, since turning the axes has no effect on the equation of a circle. We get a pair of lines if $B = 0$, $A \neq 0$. This would happen, for instance, if the original expression were $x^2 + 6xy + 9y^2$; this is the same as $(x + 3y)^2$, and such a situation can always be spotted by the fact that the quadratic form is a perfect square. Finally, if $A = 0$, $B = 0$ in the simplified form, the original expression must have been $0x^2 + 0xy + 0y^2$, which we would immediately recognize (and rarely if ever meet in practice).

Accordingly it is only when we come to quadratic forms in three variables that worthwhile illustrations can be found of the special cases that may arise. These we can visualize as corresponding to the rugby football, the sphere, the cylinder, and the pair of planes. The sphere we need not discuss; it can only arise from a quadratic form of the type $kx^2 + ky^2 + kz^2$, which we recognize at once.

The general case, and the other special cases, we will show by means of worked examples.

Worked example 1: the general case

Find the axes in which $v'Mv = 10x^2 - 12xy + 23y^2 + 12yz + 16z^2$ takes its simplest form

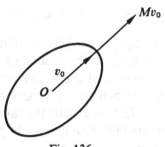

Fig 136

Solution   Here

$$M = \begin{pmatrix} 10 & -6 & 0 \\ -6 & 23 & 6 \\ 0 & 6 & 16 \end{pmatrix}.$$

We are looking for eigenvectors of M, that is, non-zero vectors that satisfy some equation of the form $Mv = \lambda v$. As explained near the end of §26, this equation may be written $(M - \lambda I)v = 0$ and can only have a non-zero solution v if λ is chosen to satisfy the characteristic equation $\det (M - \lambda I) = 0$.

If we write this determinant and work it out as described in §27 (no neat method is available) we get

$$\det (M - \lambda I) = \begin{vmatrix} 10 - \lambda & -6 & 0 \\ -6 & 23 - \lambda & 6 \\ 0 & 6 & 16 - \lambda \end{vmatrix}$$

$$= (10 - \lambda)(23 - \lambda)(16 - \lambda) - 36(10 - \lambda) - 36(16 - \lambda)$$

$$= -\lambda^3 + 49\lambda^2 - 686\lambda + 2744.$$

This expression is zero if λ equals 7, 14 or 28. These then are the eigenvalues.

It will be noticed that the eigenvalues are not very small numbers. In fact arithmetical simplicity cannot be expected here. The characteristic equation is a cubic; in a realistic situation, the solutions of it would probably be irrational numbers, and we would use some form of automatic computation to arrive at them. We have avoided irrational numbers artificially by selecting M carefully, and of course the same has been done in the later examples.

Having the eigenvalues, we now proceed to find the eigenvectors. By solving $Mv = 28v$, we find the column vector with components $-2k, 6k, 3k$, where k may be any non-zero number, corresponds to $\lambda = 28$; similarly any non-zero column vector with components in the ratios $3:-2:6$ is an eigenvector for $\lambda = 14$, and one with ratios $6:3:-2$ is an eigenvector for $\lambda = 7$.

It can be checked that these three vectors are mutually perpendicular, as we expect for the principal axes of an ellipsoid. We could arrive at the simple form

$$AX^2 + BY^2 + CZ^2$$

by changing axes to these principal axes. This is mildly laborious but by no means impossible. When carried out it leads to the result $28X^2 + 14Y^2 + 7Z^2$.

This result suggests very strongly that the rather lengthy calculation for change of axes was quite unnecessary, except perhaps as a check on accuracy, for the coefficients 28, 14 and 7 in $28X^2 + 14Y^2 + 7Z^2$ are exactly the same as the eigenvalues λ that have already been found. This suggests that we might try to prove this always happens. This is done in the following theorem.

Theorem

Let the perpendicular vectors u_1, u_2, u_3, each of unit length, be eigenvectors of the 3×3 matrix M, so that $Mu_1 = \lambda_1 u_1$, $Mu_2 = \lambda_2 u_2$, $Mu_3 = \lambda_3 u_3$. If (X, Y, Z) are co-ordinates based on u_1, u_2, u_3, the quadratic form $v'Mv = \lambda_1 X^2 + \lambda_2 Y^2 + \lambda_3 Z^2$.

Proof If v has co-ordinates X, Y, Z in the system based on u_1, u_2, u_3, we have $v = Xu_1 + Yu_2 + Zu_3$.

Accordingly

$$v'Mv = (Xu_1' + Yu_2' + Zu_3')M(Xu_1 + Yu_2 + Zu_3).$$

As

$$Mu_1 = \lambda_1 u_1, \qquad Mu_2 = \lambda_2 u_2 \quad \text{and} \quad Mu_3 = \lambda_3 u_3,$$

we have

$$v'Mv = (Xu_1' + Yu_2' + Zu_3')(X\lambda_1 u_1 + Y\lambda_2 u_2 + Z\lambda_3 u_3).$$

When this expression is multiplied out, 9 terms result, but 6 of these are zero; for example, since $u_1 \perp u_2$, the scalar product $u_1'u_2 = 0$. By this and similar considerations we see that the expression reduces to $X^2\lambda_1 u_1'u_1 + Y^2\lambda_2 u_2'u_2 + Z^2\lambda_3 u_3'u_3$ Now $u_1'u_1$ is the scalar product of u_1 with itself. As u_1 is of unit length, this scalar product must be 1. Similarly we find $u_2'u_2 = 1$ and $u_3'u_3 = 1$. Accordingly $v'Mv = \lambda_1 X^2 + \lambda_2 Y^2 + \lambda_3 Z^2$.

This was what we wanted to prove – the coefficients are given by the eigenvalues.

Note that the above proof does not employ any ingenuity. We are asked to show that something happens when a certain system of axes is used. The equation

$$v = Xu_1 + Yu_2 + Zu_3$$

is our regular starting point when we want to work with co-ordinates based on u_1, u_2, u_3. (Compare the equation $S = XP + YQ$ in §9, 'Change of axes', to express that the point S has co-ordinates X, Y in axes based on P and Q.) In the rest of the proof, we simply substitute this expression for v in $v'Mv$ and use the information we have; the equations such as $Mu_1 = \lambda_1 u_1$ tell us where M sends u_1, u_2 and u_3. After using this information, we are left simply with numbers and scalar products. The values of the scalar products follow immediately from our information that u_1, u_2 and u_3 are perpendicular and of unit length, i.e. that they are axes of the type conventional in elementary work.

The best way to remember the above theorem, and its proof, is to observe how simple and straightforward the strategy is, and to think the proof out from time to time.

It should be noted that nothing whatever is assumed about the eigenvalues $\lambda_1, \lambda_2, \lambda_3$; it does not matter whether they are equal or unequal, zero or non-zero, positive or negative. Thus this theorem applies equally well to general and special cases, to ellipsoids and to hyperboloids.

Worked example 2: the rugby football

Find the axes for which $v'Mv = 2x^2 + 2y^2 + 2z^2 + 2xy + 2yz + 2zx$ takes the simplest form.

Solution Here

$$M = \begin{pmatrix} 2 & 1 & 1 \\ 1 & 2 & 1 \\ 1 & 1 & 2 \end{pmatrix}.$$

As in the first example, we calculate the characteristic equation, det $(M - \lambda I) = 0$. We find

$$0 = -\lambda^3 + 6\lambda^2 - 9\lambda + 4 = -(\lambda - 4)(\lambda - 1)^2.$$

Taking $\lambda = 4$, we solve $Mv = 4v$ and find that any non-zero column vector with components k, k, k is an eigenvector for $\lambda = 4$. This part of the procedure works out exactly as in the general case; the equation $Mv = 4v$ is equivalent to 2 equations in the 3 unknowns x, y, z. The new feature appears when we come to the other solution, $\lambda = 1$, for $Mv = v$ gives the same equation $x + y + z = 0$ three times. Thus we shall get an eigenvector by choosing any numbers, x, y, z with sum zero. We have an unusual variety of eigenvectors with $\lambda = 1$; they do not all lie in a line, as in the general case. In fact, we can interpret the equation $x + y + z = 0$, for $x + y + z$ is the scalar product of (x, y, z) and $(1, 1, 1)$; the equation tells us that these vectors are to be perpendicular. *Thus any vector perpendicular to $(1, 1, 1)$ is an eigenvector with $\lambda = 1$.* These vectors fill a plane.

What is happening in this plane, that causes *every* line through the origin to be a principal axis? Usually, if we have an ellipse in a plane, there are two quite definite principal axes of this ellipse. The only exception is the circle; for it, any diameter is a principal axis; no direction is favoured above any other. Thus we are led to guess that the intersections of the surfaces $v'Mv = $ constant with the plane $x + y + z = 0$ are circles.

We can establish this quite definitely by using the theorem given above. To apply this theorem, we must consider three perpendicular unit vectors, u_1, u_2, u_3, that are eigenvectors of M. We take u_1 in the direction $(1, 1, 1)$; u_2 and u_3 can be any two unit vectors, that are perpendicular to each other, and lie in the plane perpendicular to u_1. We know, by the work above, that $Mu_1 = 4u_1$, $Mu_2 = u_2$, $Mu_3 = u_3$. So, by our theorem, if we take u_1, u_2, u_3 as a basis for co-ordinates, $v'Mv = 4X^2 + Y^2 + Z^2$. Where a surface $v'Mv = $ constant meets the plane $X = 0$, we have $Y^2 + Z^2 = $ constant, the equation of a circle.

A surface $v'Mv = $ constant could be obtained by spinning an ellipse about the line $x = y = z$; it is an ellipsoid of revolution. This particular ellipsoid is more like a flying saucer than a football.

Worked example 3: the sphere

As already mentioned, there is nothing to say about this. A sphere corresponds to $k(x^2 + y^2 + z^2)$. It is already in its simplest form. If we rotated the axes, it would remain in this form; we would obtain $k(X^2 + Y^2 + Z^2)$. Any three perpendicular directions will serve as principal axes.

Worked example 4: the cylinder

Find the axes for which

$$v'Mv = 2x^2 + 2xy + 2y^2 - 6xz - 6yz + 6z^2$$

takes the simplest form.

Solution Here we find

$$
\det (M - \lambda I) = \begin{vmatrix} 2 - \lambda & 1 & -3 \\ 1 & 2 - \lambda & -3 \\ -3 & -3 & 6 - \lambda \end{vmatrix}
$$

$$= (2 - \lambda)^2(6 - \lambda) + 9 + 9 - 9(2 - \lambda) - 9(2 - \lambda) - (6 - \lambda)$$
$$= -9\lambda + 10\lambda^2 - \lambda^3$$
$$= -\lambda(\lambda - 1)(\lambda - 9).$$

The eigenvalues are 9, 1, 0. $Mv = 9v$ leads us to an eigenvector with components in the ratios $1:1:-2$. $Mv = v$ gives the ratios $1:-1:0$, while $Mv = 0$ gives $1:1:1$. If we take unit vectors with these directions as our basis, we find $v'Mv = 9X^2 + Y^2 + 0 \cdot Z^2$, that is, simply $9X^2 + Y^2$.

It can be checked that the eigenvectors just found are perpendicular to each other.

The only special feature in this situation is that we obtain $\lambda = 0$ as one of the solutions of the characteristic equation.

There is a particular case of the cylinder which also has the peculiarity discussed in Worked example 2, the rugby football. The rugby football had circular sections perpendicular to one of its principal axes; we can quite easily have a cylinder with circular sections perpendicular to its axis – in fact, in everyday life this is what we understand by a cylinder, as when we speak of the cylinders in a car. In such a case we would have $X^2 + Y^2 = $ constant, and any vector in the plane $Z = 0$ will be an eigenvector of M. In such a case, the calculations will run very much as they did in Worked example 2. Instead of a characteristic equation such as $0 = -(\lambda - 4)(\lambda - 1)^2$ we shall have perhaps $0 = -\lambda(\lambda - 1)^2$. The fact that $\lambda = 0$ is a solution will not affect our general strategy at all.

Worked example 5: a pair of planes

Find the axes in which $x^2 + y^2 + z^2 + 2xy + 2xz + 2yz = 1$ takes its simplest form.

Solution Here we have the characteristic equation

$$
0 = \det (M - \lambda I) = \begin{vmatrix} 1 - \lambda & 1 & 1 \\ 1 & 1 - \lambda & 1 \\ 1 & 1 & 1 - \lambda \end{vmatrix}
$$

$$= (1 - \lambda)^3 + 2 - 3(1 - \lambda)$$
$$= -\lambda^3 + 3\lambda^2 = -\lambda^2(\lambda - 3).$$

For $Mv = 3v$ we find the eigenvector with components in the ratios $1:1:1$. If $\lambda = 0$, we get, just as we did in Worked example 2 for $\lambda = 1$, the same equation $x + y + z = 0$ three times. Thus any vector in this plane is an eigenvector with $\lambda = 0$. Here again, the fact that $\lambda = 0$ does not call for any change in our strategy. We can argue exactly as we did in Worked example 2, and conclude that the simplified form must be

$$3X^2 + 0 \cdot Y^2 + 0 \cdot Z^2.$$

Naturally, we write this as $3X^2$.

Summing up

If we look at these examples, we can see that two considerations are involved.

(1) Do we have a repeated root of the characteristic equation? That is, are we dealing with an equation such as $0 = -(\lambda - a)(\lambda - b)^2$ or even $0 = -(\lambda - a)^3$?

(2) Does zero occur as a root of the characteristic equation?

When all the roots are distinct, we get eigenvectors lying in definite directions, which are fixed by the equations $Mv = \lambda v$. When a repeated root occurs, we no longer have definite eigenvectors: we may be free to choose any pair of perpendicular axes in some plane, or, in the extreme case of the sphere, any set of perpendicular axes whatever. Thus repeated roots affect the kind of calculations we have to make.

Quite the opposite is true for the second consideration, whether $\lambda = 0$ is a solution or not. When zero eigenvalues occur the surfaces $v'Mv = $ constant certainly *look* different from the other cases. An infinite cylinder, for example, would hardly be convenient for use as a rugby football. But the procedure for calculating the axes would be exactly the same. The procedure of Worked example 1 can be applied unaltered to the problem stated at the beginning of Worked example 4; all we have to remember is not to be upset by the appearance of the solution $\lambda = 0$. In the same way, the procedure of Worked example 2 (rugby football) can be applied unaltered to the particular case mentioned at the end of Worked example 4 (circular cylinder). In fact, we can usefully think of the elliptical cylinder as a special or limiting case of the general ellipsoid, and the circular cylinder as a limiting case of the rugby football.

How this comes about can be seen by considering a sequence of ellipses in 2 dimensions. The equation $(x^2/a^2) + (y^2/b^2) = 1$ represents an ellipse with axes of length $2a$ and $2b$. We will fix b at the value 1, but consider in turn $a = 10$, $a = 100$, $a = 1000$, and so on. In this sequence of ellipses, the major axis is becoming longer and longer. The equations of the ellipses are $0.01x^2 + y^2 = 1$, $0.0001x^2 + y^2 = 1$, $0.000001x^2 + y^2 = 1$ and so on. Clearly the coefficient of x^2 is approaching 0. If we consider the limiting case, in which the coefficient of x^2 is made actually equal to zero, we have $y^2 = 1$, which represents the pair of lines $y = 1$ and $y = -1$.

The business of finding eigenvectors in the limiting case is no different from what it was for any ellipse in the sequence. For example, the first ellipse corresponds to the matrix

$$\begin{pmatrix} 0.01 & 0 \\ 0 & 1 \end{pmatrix}.$$

We have an eigenvector in the y-axis with $\lambda = 1$, and one in the x-axis with $\lambda = 0.01$. For the pair of lines $y^2 = 1$ we have the matrix

$$\begin{pmatrix} 0 & 0 \\ 0 & 1 \end{pmatrix},$$

which has an eigenvector in the y-axis with $\lambda = 1$ and one in the x-axis with $\lambda = 0$.

Many paradoxes and incorrect results can come from proceeding rashly to a limit or from talking loosely about infinity. In the example just given we have checked carefully that there is no sudden jump made by the eigenvectors or eigenvalues as we go to the limit. It is legitimate to think of the pair of lines, $y^2 = 1$, as an ellipse for which the major axis has become infinite, or as a limiting case of an ellipse.

Thus in 2 dimensions we could classify the possibilities as follows:

Type 1 Roots for λ unequal; ellipse or hyperbola (limiting case, when $\lambda = 0$ is one solution, pair of lines).

Type 2 Roots for λ equal; circle.

For type 1 we have two definite principal axes; for type 2, any pair of perpendicular lines are principal axes.

In 3 dimensions, we can similarly justify the inclusion of limiting cases. Our classification then runs:

Type 1 Roots for λ distinct: the general case (limiting case, elliptic or hyperbolic cylinder).

Type 2 Equation of form $(\lambda - a)(\lambda - b)^2 = 0$; rugby football (limiting cases, circular cylinder if $a = 0$; pair of planes, if $b = 0$.)

Type 3 Equation of form $(\lambda - a)^3 = 0$; sphere.

For type 1, there are definite principal axes; for type 2, one axis is in a definite direction, about which the other axes are free to rotate; for type 3, any three perpendicular directions will do for principal axes.

The cylinders, type 1 or type 2, occur when one axis has become of infinite length; the pair of planes occurs when two axes have become infinite in length.

Note that the rugby football and the flying saucer both belong to the same type, type 2; which case we get depends on whether $a < b$ or $a > b$.

Questions

1   Find the eigenvalues and eigenvectors of the matrix

$$\begin{pmatrix} 5 & 4 \\ 4 & 5 \end{pmatrix}.$$

Hence find the principal axes of the conic $5x^2 + 8xy + 5y^2 = 9$. Find the equation of the conic referred to its principal axes and sketch the curve. Give the (x, y) co-ordinates of the ends of the principal axes, and check that they do satisfy the original equation.

2   Do the same for the matrix

$$\begin{pmatrix} 5 & -4 \\ -4 & 5 \end{pmatrix}$$

and the equation $5x^2 - 8xy + 5y^2 = 9$. How does this curve differ from the curve in Question 1?

3   Do as in Question 1 for the matrix

$$\begin{pmatrix} 4 & 5 \\ 5 & 4 \end{pmatrix}$$

and the equation $4x^2 + 10xy + 4y^2 = 9$.

4   Find the equation of the conic $3x^2 - 4xy = 4$ referred to its principal axes and sketch the curve.

5   As in Question 4 investigate the conic

$$8x^2 - 4xy + 5y^2 = 36.$$

6   Write down a vector that has the same direction as the normal to the conic

$$10x^2 - 12xy + 5y^2 = 13$$

at the point (x_0, y_0) on the curve. Find this normal vector for the point $(2, 3)$ on the curve. What conclusion can be drawn about the directions of the principal axes of this conic?

7   Do as in Question 6, but with the conic $22x^2 - 6xy + 14y^2 = 130$ and the point $(1, 3)$ on it.

8   Find vectors that give the directions of the normals to the conic $16x^2 + 40xy + 25y^2 = 196$ at the points $(1, 2)$ and $(6, -2)$ on this conic. What are the slopes of these normals? Explain your result.

9   As was mentioned earlier there is difficulty in finding 3-dimensional examples in which attention is not distracted from the principles involved by purely arithmetical complications. In the selection of the following examples an attempt has been made to keep the numbers involved as small as possible. The characteristic equation being a cubic, it will often be necessary to search for some simple solution, such as ± 1, ± 2 or ± 3, and then to remove the corresponding factor; a quadratic equation then remains to be solved.

Each of the equations listed below is of the form $v'Mv = $ constant. In each case, write the matrix M and find its eigenvalues and eigenvectors. Write the equation of the quadric referred to its principal axes, and indicate clearly which vectors (as specified in the x, y, z system) are the new X, Y and Z axes. Say to which type the quadric belongs.

(a) $5x^2 + 4xy + 4y^2 - 4yz + 3z^2 = 28$
(b) $3x^2 + 8xy + 5y^2 - 8xz + z^2 = 1$
(c) $2x^2 + 2y^2 + 2z^2 - 2xy - 2xz - 2yz = 1$
(d) $2xy + 2xz + 2yz = 1$
(e) $2x^2 + 4xy + y^2 - 4yz = 4$
(f) $5x^2 + 8xy + 5y^2 - 4xz - 4yz + 2z^2 = 1$
(g) $4x^2 + 4xy + 5y^2 + 4xz + 3z^2 = 28$
(h) $x^2 + 4y^2 + 9z^2 + 4xy + 6xz + 12yz = 25$
(i) $4xy + y^2 + 4xz - z^2 = 3$

(*j*) $x^2 + 4xy - 4yz - z^2 = 3$

(*k*) $7x^2 + 4xy + 6y^2 - 4yz + 5z^2 = 3$, it being given that $(2, 2, -1)$ lies along a principal axis.

10 Stable equilibrium occurs only when the contours for potential energy are ellipsoids. If the expressions in Question 9 represent potential energy, in which cases is equilibrium stable?

39   Lines, planes and subspaces; vector product

When finding the eigenvectors of a matrix, as §38 showed, we may meet various situations. In the general case we find that the eigenvectors, corresponding to some particular value of λ, lie in a line through the origin *O*. But, for the rugby football and the flying saucer, they may lie in a plane. And for the sphere, they are scattered through the whole space of 3 dimensions. Now in all the cases considered in §38, we were solving 3 equations in 3 unknowns; thus all the cases looked very much alike, yet they led to several different kinds of result. In engineering applications, we may meet *m* equations in *n* unknowns, where *m* and *n* are any natural numbers whatever. It is evident, from our experience in §38 with 3 equations in 3 unknowns, that we must expect to meet a variety of situations and be prepared to cope with them. Accordingly the problem arises quite naturally – *to classify the possibilities that may arise in solving systems of linear equations.* This problem we shall take up in §40. Before coming to that it will be wise to clear the ground by considering another question. In §38, the solutions of the equations were associated with a line, a plane, or the whole space. It seems desirable to discuss how we specify a line or a plane in 3 dimensions, and the corresponding ideas for space of *n* dimensions. That will be done in the present section.

Lines and planes can be specified by the methods of affine geometry – that is, by means of the basic vector operations, giving *u* + *v* and *ku*, and without any reference to such Euclidean ideas as length or angle. We shall use this approach first. Its advantage is its flexibility. In 3 dimensions we can use *any* 3 vectors as a basis for co-ordinates, so long as they do not lie in a plane; we need not worry whether they are perpendicular or not. After finding out as much as we can by affine methods, we will turn to Euclidean ideas and look at such questions as, for example, the equation of a plane perpendicular to a given vector. Such results are useful, for example, in problems concerned with actual objects in physical space, which has approximately the properties required by Euclid.

Our main interest will be in lines and planes through the origin.

Suppose then we are dealing with some object that we cannot handle directly, but which we can control by dials on some piece of equipment. To begin with, suppose there is only one dial. The object is mounted on rails. It starts at the origin, and the rails carry it in the direction of some fixed vector u. The rails are very long; they run right through the origin, and by turning our dial suitably, we can get the object anywhere we like on the line containing the vector u. Any point on this line is specified by a vector su, where s is some real number, positive, negative or zero. By turning the dial an amount corresponding to s, we can bring the object to the position su.

Thus s is at our disposal. We can vary it as we like by turning the dial. A quantity that we can vary at will is often called a *parameter*. Thus s is a parameter, and when we say that the line consists of the points su, where s runs through all real values, we are specifying the line *parametrically*.

Since there is *one* number s at our disposal, this situation is described as having *one degree of freedom*.

There is another way of specifying a line. Instead of describing the freedom it has, we may describe the restrictions put on it. Suppose we take any two vectors, v and w, such that u, v, w constitute a basis, and let x, y, z be co-ordinates in this system; that is, (x, y, z) denotes the point $xu + yv + zw$. The points su, that can be reached by the object when we turn the dial, have co-ordinates of the form $(s, 0, 0)$. Thus we have restrictions imposed on us; we can only reach points for which $y = 0$, $z = 0$. Thus the line can be specified by *a pair of equations*. Such specification can be done in many ways; for example, the equations $y - z = 0$, $y + z = 0$ are equivalent to $y = 0$, $z = 0$ and specify the same line.

A convenient way of visualizing and remembering the parametric form su is to imagine an object travelling with constant velocity u; s seconds after passing through the origin it would reach the position su. Its path would be a straight line. Negative values of s would give the places on the line reached before it came to the origin.

Now of course we cannot always arrange to have one axis along the line we are studying; there may be other aspects of the problem that compel us to use some other system of axes. Suppose then that we have to use co-ordinates X, Y, Z based on axes U, V, W, where $U = au + bv + cw$, $V = du + ev + fw$, $W = gu + hv + iw$ connect the two systems. We can change axes by the method of §9. We have

$$xu + yv + zw = XU + YV + ZW$$
$$= X(au + bv + cw) + Y(du + ev + fw) + Z(gu + hv + iw).$$

By considering the amount of V on each side we find $y = bX + eY + hZ$, and from the amount of w we get $z = cX + fY + iZ$. Thus the equations $y = 0$, $z = 0$ are replaced by $bX + eY + hZ = 0$, $cX + fY + iZ = 0$.

So, in the X, Y, Z system of co-ordinates we still have 2 equations for the line through the origin.

It is reasonable that there should be 2 equations. If a point is free to roam anywhere in space of 3 dimensions, it has 3 degrees of freedom; we are free to choose each of the 3 co-ordinates. Each restriction, that is, each equation, destroys one degree of freedom, so to cut down the freedoms of the point from 3 to 1, we require 2 equations. And, incidentally, these two equations must be independent. It will not do to have equations like $X + Y + Z = 0, 2X + 2Y + 2Z = 0$ where the second equation is a consequence of the first. In the work above, one could show that there would be a contradiction if anything like this occurred, for $y = 0$, $z = 0$, the equations from which we started are independent.

The change-of-axes argument used above is a convenient way of showing that a line must have two equations, since it does not involve us in any very detailed algebraic calculations. In any particular example, where we wanted to find 2 equations for a line, we would not go through the trouble of changing axes.

Worked example

Find the equation of the line joining the origin to the point $(2, 3, 4)$.

Solution   If a point travels with velocity $(2, 3, 4)$ for s seconds, it will reach the point with $x = 2s$, $y = 3s$, $z = 4s$. This point satisfies the equations $3x = 2y$, $4y = 3z$.

Conversely it can be seen that any point satisfying these equations does lie on the line in question. For these equations imply that $x/2$, $y/3$ and $z/4$ are all equal. Take this common value for s. Then $x = 2s$, $y = 3s$, $z = 4s$, as required.

It may seem that the algebra used in this example is so simple that we might dispense with the change-of-axes argument altogether and simply do the algebra. However, it should be borne in mind that our study of the straight line, given by 2 equations in 3 unknowns, is only the curtain raiser for the general case of m equations in n unknowns. In such a vague and general situation, it is difficult, perhaps impossible, to perform algebraic calculations in the way we do for simple, particular cases. It is a great advantage if we can suppose axes chosen in such a way that the equations become something like $x_k = 0$, $x_{k+1} = 0, \ldots, x_{n-1} = 0$, $x_n = 0$.

Planes in 3 dimensions

We now return to our apparatus for moving an object, and suppose two dials are provided. The rails now form a kind of grid. When the first dial is turned, the object is displaced in a line parallel to a fixed vector u, when the second dial is turned, the object is displaced parallel to a vector v. Thus, if the object starts at the origin, by turning the dials suitably we can bring it to any position $su + tv$, where s and t are real numbers.

In some cities, such as Toronto, the public transport system permits a passenger to travel a desired distance east or west, and then transfer to another vehicle which carries him or her north or south. This has much in common with the object moving apparatus with two dials; setting the first dial corresponds to deciding how far to travel east or west: setting the second dial corresponds to choosing a distance north or south. Of course the correspondence is not exact; a passenger cannot choose to travel through π stops, not even through $\frac{1}{4}$ of a stop. Again, the apparatus would allow us to simulate conditions in a city that lay on an inclined plane, with roads that ran in two directions not necessarily perpendicular.

For our purposes, it will be convenient to *define* a plane through the origin as consisting of all the points $su + tv$, where u and v are two fixed, independent vectors, and s, t are arbitrary real numbers.

Thus a plane appears first of all in terms of the 2 parameters s, t. It has 2 degrees of freedom; we have 2 numbers at our disposal when we are choosing a point of the plane.

What about specifying it by an equation or equations? We already have 2 fixed vectors u and v, pointing in different directions; if we take any vector w not lying in the plane of u and v, we obtain a basis u, v, w for co-ordinates x, y, z. (Engineers will regard this as geometrically obvious; in a treatise more formal than this, the possibility of obtaining a basis in this way would appear as a theorem, stated in a very general way as applying to any number of vectors in any number of dimensions.)

In the co-ordinate system just introduced the point $su + tv$ will appear as $(s, t, 0)$. Thus the x and y co-ordinates are completely at our disposal. The only restriction is $z = 0$, which is necessary and sufficient; that is to say, the statements '(x, y, z) lies in the plane containing the origin and the vectors u and v' and '$z = 0$' are completely equivalent. If (x, y, z) lies in the plane, then z must be zero; if $z = 0$, then (x, y, z) lies in the plane.

Thus a plane has *one* equation only.

If we are working with axes that are not so conveniently related to the position of the plane, we will do as we did earlier with the line and call our co-ordinates X, Y, Z. With the same symbolism as before, the equation $z = 0$ will appear in our system as $cX + fY + iZ = 0$. Thus we shall still have just one equation for the plane.

This interpretation of a single equation helps us to see the meaning of our earlier equations for a line. We had $y = 0$, $z = 0$. Each of these equations represents a plane.

$y = 0$ is obviously satisfied by the points $(1, 0, 0)$ and $(0, 0, 1)$, thus $y = 0$ is the equation of the plane containing the first and third axes. Similarly $z = 0$ is the equation of the plane containing the first and second axes. When we require $y = 0$ and $z = 0$ simultaneously we are insisting that the point (x, y, z) should lie in both these planes, that is, in their intersection which is in fact the first axis.

Fig. 137 uses a picture of a cube with the origin at one corner to illustrate this, because

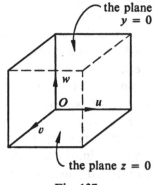

Fig. 137

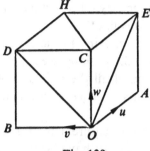

Fig. 138

perpendicular axes are familiar to us and easily visualized. It should be emphasized that perpendicularity is in no way necessary and is indeed totally irrelevant to the result.

The line, with only 1 degree of freedom, in 3 dimensions requires 2 equations to remove 2 freedoms. The plane, having 2 freedoms, is sacrificing only 1 degree of freedom and obeys only 1 equation. As an extreme case, we could consider the whole space of 3 dimensions; then x, y, z are all arbitrary; all 3 freedoms are preserved and no equation restricts our movements. At the other end, a point has sacrificed all its freedom; the origin is specified by $(0, 0, 0)$ in which no parameter occurs, and satisfies the 3 equations $x = 0, y = 0, z = 0$.

When we are counting equations, it is essential to check that they are independent. For instance, if we have $x - y = 0, y - z = 0, z - x = 0$ we seem to be dealing with 3 equations, and we might, at first sight, expect these to fix a single point, the origin. But in fact (s, s, s) satisfies all three equations. Any point on the line joining the origin to $(1, 1, 1)$ lies in the planes $x - y = 0, y - z = 0, z - x = 0$. In fact the first two equations tell us $x = y$ and $y = z$, from which it follows that $x = z$. The third equation does not bring us any new information.

Fig. 138, which uses a cube with the same reservations as before, the equation $x = y$ corresponds to the plane OCH, $y = z$ to ODH and $x = z$ to OEH. These three planes all contain the line OH.

Thus three planes with dependent equations may, as a rule, be pictured as having a line in common, rather like the pages of a pamphlet that have been somewhat spread out. We have to say 'as a rule' because there is an extreme case of dependence in which all 3 equations represent the same plane.

Subspaces

A line is a vector space of 1 dimension, and a plane is a vector space of 2 dimensions.

In this section we have considered both lines and planes as objects in space of 3 dimensions. This is not the only way to study lines and planes. In much of school geometry – for example, in Pythagoras' Theorem – we are simply concerned with the plane itself; we do not think of the diagrams as lying in a horizontal, or a vertical, or an inclined plane. They are just in 'the' plane; we are in a universe of 2 dimensions.

We can work with a line in the same way – for instance, the 'real number line'; no one discusses the compass bearing of this line!

When we consider a vector space as lying inside a vector space of higher dimension (like a line or a plane in a space of 3 dimensions) we call it a *subspace*. Thus, for instance, if we consider the plane, consisting of all the points $xu + yv$, as part of the 3-dimensional space, containing all the points $xu + yv + zw$, then this plane is a subspace of the 3-dimensional space.

Towards the end of §7 we gave the requirements for a vector space; if p and q were two objects in a vector space, we had to have 'sensible definitions' of $p + q$ and kp, for any number k. It was understood that $p + q$ and kp would also belong to the vector space; they would not be something outside it. Those requirements imply that, for any numbers a and b, the 'mixture' $ap + bq$ belongs to the space. For by multiplication of p and q we can obtain ap and bq, and by addition of these we reach $ap + bq$.

A subspace is a collection of some of the points of a vector space, which itself meets the qualifications for a vector space. If we restrict ourselves to vectors in the subspace, we can add, and multiply by a number, without ever needing to go outside the subspace. (Consider, for example, the horizontal plane through the origin in physical space of 3 dimensions.) So, if p and q are in a subspace, $ap + bq$ must also be in that subspace. In particular, we can consider $a = 0$, $b = 0$. This gives us the origin, O. Accordingly, *a subspace is bound to contain the origin O.*

In §12, when we were dealing with eigenvectors and eigenvalues, we made the convention that invariant line was to mean only an invariant line *through the origin*. The reason for this is now beginning to appear. Invariant lines are a particular case of a more general idea, invariant subspace. Since a subspace automatically contains the origin, it was useful to make this stipulation for the invariant lines considered in §12.

We can now generalize our earlier results about the degrees of freedom and equations of lines and planes in 3-dimensional space.

A line appeared earlier as consisting of all the points su, with u a fixed vector and s an arbitrary number; a plane appeared similarly as a collection of points $su + tv$. A line is a subspace of 1 dimension, a plane a subspace of 2 dimensions.

Suppose now we want to define a subspace of r dimensions in a vector space of n dimensions. We suppose r to be smaller than n.

We begin with r independent vectors, $u_1, u_2, \ldots, u_r$, and r arbitrary numbers $s_1, s_2, \ldots, s_r$, and form the mixture $s_1u_1 + s_1u_1 + s_2u_2 + \cdots + s_ru_r$.

All these points form a vector space: it is easy to see that you can add two of these vectors, or multiply one of them by a number, without reaching anything of a different type. So this collection does qualify as a subspace; it is in fact a subspace of r dimensions. Now $r < n$, so, on the analogy with a line or plane in 3-dimensional space, we assume that this subspace, of dimension r, does not in fact fill the whole space of dimension n. We will further assume, on the analogy of what we did with lines and planes, that we can find a basis for *the whole space*, the first r vectors of which are $u_1, \ldots, u_r$, which are the basis of the subspace. The vectors $u_{r+1}, \ldots, u_r$ will lie outside the subspace and can be chosen in many different ways. Thus, in the whole space we have co-ordinates $(x_1, x_2, \ldots, x_n)$ for the point $x_1 u_1 + x_2 u_2 + \cdots x_n u_n$. The points of the subspace appear as $(s_1, s_2, \ldots, s_r, 0, 0, \ldots, 0)$. The first r co-ordinates are arbitrary numbers, the rest are all zero. Thus the subspace gives us r degrees of freedom, corresponding to the r parameters $s_1, s_2, \ldots, s_r$. There are $n - r$ equations, expressing the fact that for points in the subspace, the remaining co-ordinates are zero. These equations are $x_{r+1} = 0$, $x_{r+2} = 0, \ldots, x_{n-1} = 0$, $x_n = 0$. If we change axes, we shall obtain equations of a less simple type, but there will still be $n - r$ independent equations.

The converse can be proved; if we start with $n - r$ independent linear equations (with no constant terms), each equation kills one degree of freedom, and we are left with r degrees of freedom in a subspace of r dimensions.

To prove, or even to justify in part, the assumptions made above requires machinery that will be developed in §40. For the moment, we are not attempting to prove the majority of our statements; we are surveying the scene and reporting on it. Most of our remarks deal with results which a mathematician, investigating this topic for the first time, would guess to be true on the basis of analogy and a few experiments in calculation.

All these considerations are relevant to our work in §38. When we are looking for eigenvectors, we first find a suitable value for λ and then solve the equations $(M - \lambda I)v = 0$. Here we have (in the general case) n equations in n unknowns. Some of these equations will be redundant, that is, they will be consequences of other equations in this set. Weeding these out, we are left with a certain number of independent equations, say m equations. Accordingly the solutions will form a subspace of $n - m$ dimensions.

Lines and planes not through the origin, in 3 dimensions

To deal with lines not through the origin we need only very small modifications of our earlier work. If a point moving with constant velocity, given by the vector u, starts not at the origin but at a position, given by the vector p, then, after s seconds it will be at the position $p + su$. Thus the parametric specification of the line it moves along can be written down immediately.

The equations of the line can be obtained from the parametric form. These equations express the condition that the line will pass through some given point (x, y, z). Suppose for example, a point starts at $(1, 2, 3)$ and moves with constant velocity $(4, 5, 6)$. After s seconds it will be at $(1 + 4s, 2 + 5s, 3 + 6s)$. Now suppose the question arises; does the line pass through the point $(13, 12, 9)$? This is the same as asking, will there be a time, s seconds, when the point is at $(13, 12, 9)$? This question can be answered in more than one way. We might, for instance, find s by equating the x co-ordinates. This gives us $1 + 4s = 13$, so the moving point has $x = 13$ when $s = 3$. Putting $s = 3$, we find the point is then at $(13, 17, 21)$ which is not the same as $(13, 12, 9)$; choosing the time that makes the x value right makes y and z wrong, so the point is not on the line.

Another way would be to write down the 3 equations that hold if $(1 + 4s, 2 + 5s, 3 + 6s)$ coincides with $(13, 12, 9)$. These equations are $1 + 4s = 13$, $2 + 5s = 12$, $3 + 6s = 9$. The first equation requires $s = 3$, the second requires $s = 2$, the third

$s = 1$. Thus the co-ordinates x, y, z take the required values *at different times*; there is no time when they all have the required values; this means the line does *not* go through the point.

We could use either method to find conditions for the line to go through *any* point (x, y, z). The second method leads to a result that is simpler and more easily remembered, and most textbooks accordingly give the equations of a line in the form obtained by this method.

Suppose then we are given a certain point (x, y, z) and asked to find whether the line passes through it. We write the 3 equations $x = 1 + 4s, y = 2 + 5s, z = 3 + 6s$ and solve each of them. The first gives $s = (x - 1)/4$, the second $(y - 2)/5$, the third $(z - 3)/6$. The point will pass through (x, y, z) if these 3 equations are satisfied *at one and the same time*. So the 3 values of s found above must be the same, for this to happen. That means

$$\frac{x - 1}{4} = \frac{y - 2}{5} = \frac{z - 3}{6}.$$

Note the significance and position of the numbers here; $(1, 2, 3)$ is the place where the point starts, (x, y, z) is a point that it reaches (if the conditions are satisfied); $(4, 5, 6)$ specify the direction of the line, by giving the velocity of the moving point. What is the significance of the vector $(x - 1, y - 2, z - 3)$ whose components appear in the numerators? If the equations are satisfied, in what relation does this vector stand to the velocity? Is the result a reasonable one?

Note also that these equations were obtained by finding a common value for s. If we want to get back from the equations to the parametric form, all we need do is to write s for this common value: let

$$\frac{x - 1}{4} = \frac{y - 2}{5} = \frac{z - 3}{6} = s.$$

Then $x = 1 + 4s, y = 2 + 5s, z = 3 + 6s$ follows.

Planes

A plane through the origin consists of all the points $su + tv$; these are the points we can get to by going an arbitrary amount in the direction of the vector u, and then an arbitrary amount in the direction of v. If we do the same starting at position p instead of at the origin, we obtain points of the form $p + su + tv$. This then is the parametric specification of a plane that need not contain the origin. (It might contain the origin; we have not started at O, but our wanderings might bring us there.)

There are two types of question that now arise. We may want to know simply, what kind of equation will such a plane have? Or we may be dealing with particular vectors p, u, v and want to know, what exactly is the equation of this particular plane?

We begin with the first question, because the answer to it will help with the second question.

We have already seen that the plane through the origin, consisting of all points of the form $su + tv$, has some equation, $ax + by + cz = 0$. If we add p to $su + tv$ we obtain

$p + su + tv$. Thus the plane we are interested in, consisting of all the points $p + su + tv$, can be obtained by displacing a plane through the origin; the displacement is given by the vector p (Fig. 139).

Questions about displacement should always be done slowly; it is very easy to make a slip and in fact give a displacement opposite to what you intended. For instance, in 2 dimensions, the mistake is often made of supposing that $y = (x + 1)^2$ is the result of displacing the parabola $y = x^2$ one unit to the right; it is not, rather it is the result of a displacement one unit to the *left*, i.e. in the direction of the *negative* x-axis.

Suppose then that the displacement p has components f, g, h. Any point (x, y, z) in the plane we are investigating can be obtained by giving the displacement p to some point, say (x_0, y_0, z_0), in the plane through the origin. Thus $x = x_0 + f$, $y = y_0 + g$, $z = z_0 + h$. As (x_0, y_0, z_0) lies in the plane through the origin, $ax_0 + by_0 + cz_0 = 0$. Our problem is: what restrictions are placed on (x, y, z) if the equations above hold for some (x_0, y_0, z_0)? We are looking for a condition that (x, y, z) lies in a certain plane, so only x, y, z can appear in the equation we want; we have to get rid of all reference to (x_0, y_0, z_0). To do this, we can solve for x_0, y_0, z_0; we find $x_0 = x - f$, $y_0 = y - g$, $z_0 = z - h$. (This makes sense; it shows that we get from (x, y, z) back to the point (x_0, y_0, z_0) by reversing the displacement p.) Substituting in $ax_0 + by_0 + cz_0 = 0$ we obtain the required equation $a(x - f) + b(y - g) + c(z - h) = 0$. If we multiply out, and introduce the abbreviation d for $af + bg + ch$, this equation can be written $ax + by + cz = d$, a formula that covers every possible plane.

We now come to our second question. Suppose for example, we start at $(1, 2, 3)$ and allow our point to move from there for s seconds with velocity $(1, 1, 1)$ and then for t seconds with velocity $(4, 5, 7)$. However s and t are chosen, the point will end up lying in a certain plane. What is the equation of this plane? If (x, y, z) is the point reached, we have

$$x = 1 + s + 4t,$$
$$y = 2 + s + 5t,$$
$$z = 3 + s + 7t.$$

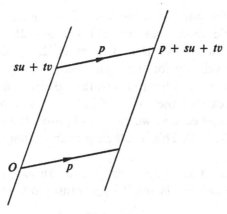

Fig. 139

These points all lie in a fixed plane, so they must satisfy some equation

$$ax + by + cz = d,$$

whatever values s and t may have. Now, from our equations for x, y, and z, we see that

$$ax + by + cz = a + 2b + 3c + (a + b + c)s + (4a + 5b + 7c)t.$$

This is to have a fixed value, d, regardless of the values of s and t. There is only one way this can happen; the coefficients of s and t must be zero. This means $a + b + c = 0$ and $4a + 5b + 7c = 0$. We can make this so by taking $a = 2, b = -3, c = 1$. (Any other solution of the two equations would, in principle, do equally well, and would lead to the same final result.) Thus we are led to $2x - 3y + z = -1$ as the equation satisfied by every point in the plane.

In dealing with lines and planes, certain questions may arise. Some problem has led us to specifications of a line and a plane. It may be important for some purpose to know how they are related. Does the line meet the plane in one point, or does it lie in the plane, or is it parallel to the plane? If we have two lines, do they meet or not? If they do not meet, is it because they are parallel lines or because they are skew? If we have three lines through a point, do they lie in a plane? If we have a point and a line, what is the equation of the plane containing them both?

There are a great variety of such questions. Often there are several ways of answering a question and it is only by practice and by experimenting with different approaches that a student, who needs such skills, can gain facility in translating geometrical problems into algebraic equations and judgement as to the best method to use in any particular case.

There are certain considerations that it is useful to bear in mind; these are discussed in the following paragraphs.

The same line can be traversed at many different speeds. Thus the points (s, s, s) and the points $(2s, 2s, 2s)$, where s takes all real values, are both parametric specifications of the line $x = y = z$. In the second specification the velocity is twice what it is in the first. Thus, for velocity (u_1, u_2, u_3) it is only the *ratios* $u_1 : u_2 : u_3$ that are important for the direction of the line.

Two lines are parallel (or possibly coincident) if they have the same direction. Thus the lines with parametric representations $(11 + t, 6 + 2t, -1 + 3t)$ and $(2 + 10t, 7 + 20t, 4 + 30t)$ are parallel. For if u denotes $(1, 2, 3)$, the velocity of the point tracing the first line, the velocity for the point tracing the second line is $10u$, so it is moving in the same direction but ten times as fast. Before we can assert that the lines are parallel we must check that they are not identical. The point $(2, 7, 4)$ lies on the second line. It is easily checked that we cannot choose t in $(11 + t, 6 + 2t, -1 + 3t)$ to make it coincide with $(2, 7, 4)$. This is sufficient to show that the lines are not identical, so they must be parallel.

We obtained the equation $ax + by + cz = d$ for an arbitrary plane by taking the plane $ax + by + cz = 0$ and displacing it by a translation. Thus the plane

$$ax + by + cz = d$$

is parallel to $ax + by + cz = 0$. Thus the constant d is important when we are talking about the *position* of a plane, but may be neglected in any question about the *orientation* of the plane.

Some questions can be translated immediately into a convenient algebraic form. For instance, is the line specified by $(1 + 4t, 2 + 5t, 3 + 6t)$ parallel or not to the plane $4x - 2y - z = 20$? This simply means – does the line meet the plane? So we try to find the point of intersection and see whether we run into any difficulty. When we substitute the co-ordinates of the moving point into the equation of the plane, we get $-3 = 20$, which can certainly not be satisfied. Thus there is no time t for which the moving point lies in the plane. This means that the line traced by the point must be parallel to the plane.

We could apply this method to the general problem, is the line specified by $(e + pt, f + qt, g + rt)$ parallel to the plane $ax + by + cz = d$? Putting the co-ordinates of the moving point into the equation of the plane we obtain

$$(ap + bq + cr)t + ae + bf + cg = d.$$

We can certainly solve this for t, except when $ap + bq + cr = 0$. So this equation indicates that something unusual is happening. If, as in our numerical example above, $ae + bf + cg \neq d$, there is no value t that satisfies the equation; the line is parallel to the plane. If $ae + bf + cg = d$, every t satisfies the equation: the moving point is always in the plane: the line lies in the plane.

We can check our conclusion by a quite different argument. It will be noticed that our condition $ap + bq + cr = 0$ does not involve d, e, f or g. It depends only on a, b, c, which specify the orientation of the plane, and p, q, r, which fix the direction of the line. In fact, our condition amounts to saying that (p, q, r), the velocity of the point, is a vector lying in $ax + by + cz = 0$, the plane through the origin parallel to the given plane. And this is reasonable on geometrical grounds. If a line is parallel to a plane, it is parallel to some line lying in that plane.

There is a certain trap to avoid when parametric specifications are being used. Suppose, for instance, we wish to find out whether the line

$$(x - 1)/4 = (y - 2)/5 = (z - 3)/6$$

meets the line $(x + 1)/2 = (y + 2)/3 = (z + 3)/4$. The first line can be specified as consisting of all the points $(1 + 4t, 2 + 5t, 3 + 6t)$, and the second one all the points $(-1 + 2t, -2 + 3t, -3 + 4t)$. If we now look for a common point by equating these, we find the x co-ordinates are equal only when $t = -1$, the y co-ordinates only for $t = -2$, and the z co-ordinates only when $t = -3$. There is no time when all three co-ordinates coincide. This seems to indicate that the lines have no common point. But this conclusion is certainly false. If we try $x = 5, y = 7, z = 9$ we find that original equations are all satisfied; the point $(5, 7, 9)$ lies on both lines.

One way to avoid this difficulty would be to avoid the parametric approach altogether. Our equations contain 4 'equals' signs; we have 4 equations for 3 unknowns. We could solve 3 of them for x, y, z and then see whether these values satisfied the remaining equation. This would be quite a good way of doing things, but the question remains – what was the fallacy in our first method?

If we look back at our parametric specifications we will see that these do make the lines have the common point $(5, 7, 9)$. The first line was specified by the moving point $(1 + 4t, 2 + 5t, 3 + 6t)$, and this point is at $(5, 7, 9)$ when $t = 1$. The second moving point was specified as $(-1 + 2t, -2 + 3t, -3 + 4t)$ and this is at $(5, 7, 9)$ when $t = 3$. Here we have the key to our difficulty; there is a place that both points pass through, but they pass *at different times*; we were asking too much when we equated the co-ordinates – that required not merely a common point on the two itineraries, but that it should be reached at the same time. *It is essential not to tie the movements of the two points together by using the same parameter t for both.* We should ask rather – is there a place that the first point reaches after s seconds and the second after t seconds? Our first line would then be specified as all the points $(1 + 4s, 2 + 5s, 3 + 6s)$ and the second as all the points $(-1 + 2t, -2 + 3t, -3 + 4t)$. Our equations would be $1 + 4s = -1 + 2t$, $2 + 5s = -2 + 3t$, $3 + 6s = -3 + 4t$. Solving the first two equations, we find $s = 1, t = 3$, and these values do satisfy the third equation. The discrepancy has been cleared up.

There is a useful device that can be used in a problem where, for instance, we are given two planes and asked to find a plane through their intersection that satisfies some further condition. This device may have been met already in 2-dimensional co-ordinate work, and we will explain it first in that context. Suppose we are dealing with the two lines, $x + 4y - 9 = 0$ and $2x + 3y - 8 = 0$. As may be checked by putting $x = 1$, $y = 2$, both these lines pass through the point $(1, 2)$. Now consider the equation $(x + 4y - 9) + k(2x + 3y - 8) = 0$. Put $x = 1, y = 2$ in this. We obtain $0 + 0k = 0$, which is true, whatever k. Thus, for any k, the equation above gives a line passing through the point $(1, 2)$.

The principle used here is perfectly general. If $f(x, y) = 0$ and $F(x, y) = 0$ are any two curves in the plane, the equation $f(x, y) + kF(x, y) = 0$ is automatically satisfied by any (x, y) for which $f(x, y) = 0$ and $F(x, y) = 0$; that is, $f(x, y) + kF(x, y) = 0$ always represents a curve that passes through all the intersections of $f(x, y) = 0$ and $F(x, y) = 0$.

Worked example

$x^2 + y^2 - 1 = 0$ is the unit circle. $xy = 0$ gives the axes. Any curve
$$x^2 + y^2 + kxy - 1 = 0$$
is bound to pass through the intersections of the unit circle with the axes. In Fig. 140, the dotted graph shows such a curve for one value of k.

The same principle applies in 3 dimensions. If we are given $13x - 11y + 73z - 2 = 0$ and $5x + 2y - 8z - 9 = 0$, the equation
$$(13x - 11y + 73z - 2) + k(5x + 2y - 8z - 9) = 0$$

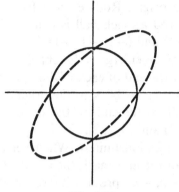

Fig. 140

is bound to represent a surface passing through the intersection of the given planes. It is clear that the equation in fact represents a plane through that intersection. The value of k might then be fixed by some further condition, for instance that the plane pass through the origin. This would lead to $k = -\frac{2}{9}$.

It is perhaps useful to repeat here what was said earlier; all the considerations so far given belong to affine geometry, to 'pure' vector theory. They do not in any way depend on the Euclidean concepts of length, angle, perpendicularity. They can therefore be used in any system of axes based on congruent parallelograms; it is not necessary for rectangles or squares to be involved in the network.

The diamond crystal, illustrating the use of oblique axes

Crystals exist in our everyday physical space, in which 'length' and 'perpendicular' have meanings, so one might expect that conventional perpendicular axes would give the best way of dealing with crystals, and that it would pay to use the full machinery of Euclidean geometry. However, this is not always so; there are some questions for which it pays to forget about right angles and scalar products, to use only the expressions of pure vector theory, $u + v$ and ku, and to work with oblique axes, that is, axes which are not perpendicular.

To illustrate this a partial investigation will be made of the structure of diamond. Naturally, one of the best ways of understanding this structure is to examine a physical model of this crystal; nevertheless, there are some problems for which we need to use co-ordinate methods. In the following work, we compare two possible approaches. We begin by using perpendicular axes; the investigation is left to the reader to complete. For comparison, we then begin again, using oblique axes.

We first describe a procedure for making a model of a diamond crystal. Let the vectors a, b, c, d be given by $a = (-1, 1, 1)$, $b = (1, -1, 1)$, $c = (1, 1, -1)$, $d = (-1, -1, -1)$ in a conventional system with perpendicular axes. The model is made from red and black balls, with rods joining them. The balls represent carbon atoms; the colours have no chemical significance, they simply facilitate the description of the model.

A red ball is placed at the origin. Rods emerge from it, representing the vectors a, b, c, d. At the end of each rod a black ball is placed. Thus we have black balls at $(-1, 1, 1)$, $(1, -1, 1)$, $(1, 1, -1)$ and $(-1, -1, -1)$.

Now from each black ball there emerge rods representing the vectors $-a, -b, -c, -d$, and a red ball is placed at the end of each of these. One of these positions requires no attention; the vector $-a$ brings us back to the origin, where there is already a red ball. We have to place red balls at the ends of the vectors $-b, -c$ and $-d$, that is at the points $(-2, 2, 0)$, $(-2, 0, 2)$ and $(0, 2, 2)$.

The construction now continues indefinitely. Whenever we have a red ball, we allow vectors a, b, c, d to sprout from it and mark their ends with black balls. From each black ball vectors $-a, -b, -c, -d$ sprout with red balls at their ends. This construction in fact leads to a reasonable framework. Models of this kind (without the distinction of red and black) can often be seen in chemical departments of schools and universities.

It is instructive to work out the positions of a few balls, as given by this construction, and to try to see some order in it. Can you find a rule for determining whether any given point, for example $(21, 13, 7)$, will have a ball at it, and if so, what the colour of the ball will be? It is desirable to try this before reading on.

*　　*　　*

We now look at this question with the help of oblique axes. The vectors a, b, c, d which represent the chemical bonds are of course not perpendicular; we cannot have 4 perpendicular vectors in 3 dimensions. We can get a co-ordinate system appropriate to the carbon atom by taking axes in the directions of three of our vectors, say a, b and c. Note that $d = -a - b - c$, so in this system d has the co-ordinates $(-1, -1, -1)$. It is purely by chance that these co-ordinates of d happen to coincide with the co-ordinates of d in the system we used first.

Let us put d on one side for the moment and consider where we can get to by starting at the origin and using only a, b, c when leaving a red ball and $-a, -b, -c$ when leaving a black ball. Positive and negative steps will be taken alternately. Thus if we used the sequence $+a, -b, +c, -b$ we would meet a black ball at position a relative to the origin, a red ball at $a - b$, a black ball at $a - b + c$, and finally a red ball at $a - 2b + c$. It is noticeable that when we leave a red ball a coefficient changes by $+1$, while when we leave a black ball a coefficient changes by -1. After two steps the sum of the coefficients is unchanged. Thus for all the red balls the coefficient sum is the same, 0, while for all the black balls the sum is 1. Thus the red balls that we can reach without using d at all will be at points $xa + yb + zc$ with $x + y + z = 0$, and the black balls that can be reached without using d will be at points with $x + y + z = 1$. (Question for investigation; does $every$ point with $x + y + z = 0$ have a red ball on it?)

Now let us gradually bring d into the picture. Suppose that at one step in a sequence such as that just considered we replace a step $-b$ by $-d$. Now $-d = a + b + c$. Thus, instead of a coefficient change of -1 we shall have an increase of $+3$. Thus replacing $-b$ by $-d$ increases the coefficient sum by 4. Accordingly we shall find some red balls at points with $x + y + z = 4$ and some black balls at points with

$$x + y + z = 5.$$

The argument can now be extended. If in a sequence of steps we replace a number of negative steps, $-a$, $-b$ or $-c$, by $-d$, each such change increases the eventual coefficient sum by 4. If, on the other hand, we replace positive steps, $+a$, $+b$ or $+c$, by $+d$, each such change reduces the coefficient sum by 4. Thus we find that red balls cannot occur anywhere except in planes of the form $x + y + z = 4n$ and black balls cannot occur except in planes $x + y + z = 4n + 1$, where n is some integer.

We will not go on to obtain a complete and detailed picture of the structure of diamond, though this can be done by vector methods. Readers interested in this question may like to work out some further details.

Exercises on lines and planes

1 An object moves with constant velocity. At time $t = 0$ it is at $(1, 1, 1)$, at time $t = 1$ at $(2, 3, 4)$. What vector specifies its velocity? Where will it be at time t? The answer to this last question provides a parametric specification of the line joining $(1, 1, 1)$ and $(2, 3, 4)$. Find a specification of this line by a pair of equations (there are many possible correct answers).

2 The equation $x - 2y + z = 0$ specifies a plane. Find at what times (if any) objects moving as specified below will lie in this plane;
Object A begins at $(0, 1, 0)$ and has velocity $(1, 1, 2)$.
Object B begins at $(0, 1, 0)$ and has velocity $(1, 1, 1)$.
Object C begins at $(1, 2, 3)$ and has velocity $(4, 5, 6)$.
In what relation to the plane do the lines described by these three moving objects stand?

3 Express the equations of the lines mentioned in questions 1 and 2 in the form

$$(x - a)/d = (y - b)/e = (z - c)/f.$$

Specify a starting position and velocity that would cause an object to move in the line

$$(x - 10)/2 = (y - 20)/3 = (z - 30)/4.$$

Does this question have one or many correct answers?

4 An object starts at the point $(5, 1, 7)$ and moves with velocity $(1, 2, 3)$. Will it pass through the point $(8, 7, 16)$ or not? What relation exists between the lines

$$(x - 5)/1 = (y - 1)/2 = (z - 7)/3$$

and

$$(x - 8)/1 = (y - 7)/2 = (z - 16)/3?$$

5 An object starts at the point $(0, 4, 6)$ and moves with velocity $(1, 0, 0)$. Another object starts at $(5, 0, 0)$ at the same time and moves with velocity $(0, 2, 3)$. Will the objects collide? Do the lines in which they move meet each other?

Lines and planes in Euclidean space of 3 dimensions

All the considerations so far given remain true in Euclid's geometry, so we start our present topic with a good deal of information and we have not much to add to it.

We will suppose, for the remainder of this section, that we are in Euclidean geometry and are using 'standard' axes, based on squares or cubes. We are thus entitled to use $u_1v_1 + u_2v_2 + u_3v_3$, the scalar product of vectors u and v, and to give it the usual geometrical or mechanical interpretation. (By the mechanical interpretation is meant 'work done'.)

Let us take some fixed vector, (p, q, r) and consider the vectors (x, y, z) perpendicular to it. The condition for perpendicularity is that the scalar product be zero;

$$px + qy + rz = 0.$$

This is the equation of a plane. It verifies, what we would expect on geometrical grounds, that all the vectors perpendicular to a fixed vector do lie in a plane. It also gives us a way of finding the normal to a plane, the equation of which we know; the normal to $px + qy + rz = 0$ is (p, q, r).

As we saw above, a plane $px + qy + rz = d$ is parallel to $px + qy + rz = 0$, so the constant d does not affect the result; when it is present, the normal is still (p, q, r).

If we need to write down the equation of the plane, through (a, b, c), perpendicular to the direction (p, q, r), we can proceed as follows. The plane must be

$$px + qy + rz = d$$

for some value of d; this ensures that (p, q, r) will be the normal. The plane will go through (a, b, c) if $pa + qb + rc = d$, so we choose this value for d. The required plane is $px + qy + rz = pa + qb + rc$.

This result could be written $p(x - a) + q(y - b) + r(z - c) = 0$, which we can interpret as a scalar product. The equation states that $(x - a, y - b, z - c)$ is perpendicular to (p, q, r). Now $(x - a, y - b, z - c)$ is the vector going from the fixed point (a, b, c) to the variable point (x, y, z), and if (x, y, z) wanders about in the plane in question, this vector ought to be perpendicular to (p, q, r). This argument provides an alternative way of arriving at the required equation.

An engineer's calculations lead to practical action, and if the calculations are incorrect, the consequences can be disastrous. It is therefore very wise not to rely on a single method blindly applied, but whenever possible to use two different methods and check whether they lead to the same conclusion. When a result has been obtained it is good to examine it, to see if it seems reasonable, and whether it can be interpreted in some way akin to what we did with the result $px + qy + rz = pa + qb + rc$. We should not be content to work blindly with equations but should try to see their meaning with the help of drawings, models and perhaps additional calculations.

For example, consider the equation $x + y + z = 1$. We know this is a plane, perpendicular to $(1, 1, 1)$. But we can also observe that the equation is satisfied by the points $(1, 0, 0)$, $(0, 1, 0)$, $(0, 0, 1)$, shown in Fig. 141 as A, B and C. So the plane contains the equilateral triangle ABC, which, as the sketch of the cube shows, is indeed perpendicular to the diagonal OG, where G is $(1, 1, 1)$. The vector $(-1, 1, 0)$ corresponds to the journey from A to B, and $(-1, 0, 1)$ to that from A to C. We could use AB and

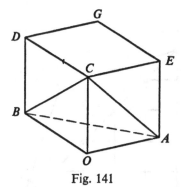

Fig. 141

AC as basis vectors for an oblique co-ordinate system in the plane *ABC*. Any point of that plane can be shown as $A + sAB + tAC$ for some s, t; that is, as

$$(1, 0, 0) + s(-1, 1, 0) + t(-1, 0, 1)$$

or $(1 - s - t, s, t)$. This checks; if $x = 1 - s - t, y = s, z = t$ we do have

$$x + y + z = 1.$$

We have in fact the parametric form, which can be reached geometrically, as above, or by a purely algebraic argument.

The lines *AB*, *BC* and *CA* are the intersections of the plane $x + y + z = 1$ with the co-ordinate planes $z = 0$, $x = 0$ and $y = 0$. Considering the intersections of a plane with the co-ordinate planes is often a useful way of seeing how the plane is situated in space.

Worked example

A drawing is to be made showing the section of some object by the plane passing through the points $(5, 0, 0)$, $(0, 5, 0)$, $(0, 0, 5)$. On this drawing it is desired to mark all the points (x, y, z) whose co-ordinates x, y, z are whole numbers. Find a parametric representation of the plane suitable for this purpose and make a diagram of the section with the points in question marked.

Solution Let A, B, C in Fig. 142 be the points $(5, 0, 0)$, $(0, 5, 0)$ and $(0, 0, 5)$. Fig. 142 shows the plane *ABC* and the co-ordinate grid in the plane $y = 0$. It is clear that if we go from *A* to *C* by the straight route, the first point with whole number co-ordinates we meet after leaving *A* will be *P*, the point $(4, 0, 1)$. If we use *u* to denote the vector *AP*, the points on the line *AC* that interest us may be written as $A, A + u, A + 2u$, $A + 3u, A + 4u$ and $A + 5u$, the last of these being *C*. Thus *u* presents itself as a vector lying in the plane *ABC* that we can profitably use when making our parametric representation. The vector $u = (-1, 0, 1)$.

Similarly, if we consider the plane *OAB*, our attention is drawn to the vector $v = (-1, 1, 0)$, which takes us from *A* in the direction *AB*.

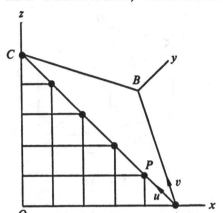

Fig. 142

If we start at A and travel for s seconds with velocity u and then for t seconds with velocity v, we shall end somewhere in the plane ABC; our position will be given by $A + su + tv$. What values of s and t will give us points with whole number co-ordinates x, y, z? To answer this we must express $A + su + tv$ in co-ordinate form as

$$(5, 0, 0) + s(-1, 0, 1) + t(-1, 1, 0);$$

this simplifies to $(5 - s - t, t, s)$. The co-ordinates here will be whole numbers if s and t are whole numbers and $s + t \leqslant 5$.

Now s and t are co-ordinates in the plane ABC in the system with origin at A and axes in the directions AC and AB. Obviously the triangle ABC is equilateral, so the axes AC and AB are at an angle of $60°$. The basis vectors u and v are of equal length. If we draw the grid for the s, t co-ordinate system and use the expression $(5 - s - t, t, s)$ to give us the x, y, z labels for the points of the grid, we obtain the diagram shown in Fig. 143. If we wish to mark any further point, the co-ordinates (x, y, z) of which are known, this is easily done, since we have $x = 5 - s - t$, $y = t$, $z = s$. The values of s and t are given directly by those of z and y. The value of x will automatically be correct, since the equations imply $x + y + z = 5$ and it is only points of this plane that appear in the section of the object that we are considering.

Exercises

1 For each of the following planes write a vector normal to it;
(a) $2x + 3y + 4z = 0$ (b) $2x + 3y + 4z = 1$ (c) $x + y + z = 17$
(d) $5x - 2y + 11z = 10$ (e) $x + y = 3$ (f) $z = 0$.

2 Find the equation of the plane through the points $(a, 0, 0)$, $(0, b, 0)$, $(0, 0, c)$. Find the normal to this plane.

3 State the vector normal to the plane $x + 2y + 3z = 56$. Give the parametric representation of the line through the origin perpendicular to this plane. Find the point where this line meets the plane.

Also give a parametric representation for the line through the point $(3, 4, 1)$ perpendicular to the plane, and find where this line meets the plane.

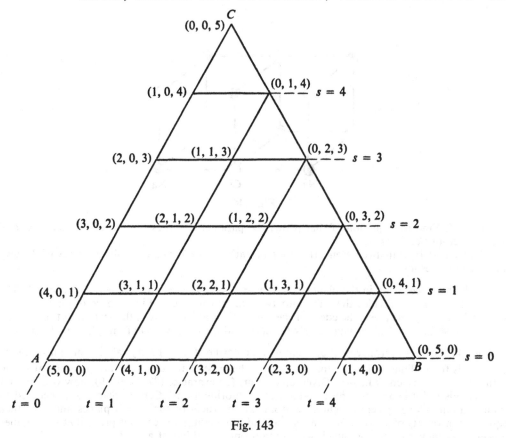

Fig. 143

4   Carry through the work of question 2 but for the plane $ax + by + cz = d$, and with (x_0, y_0, z_0) instead of $(3, 4, 1)$.

(The second part of this question leads to a formula for the perpendicular projection of any point onto any plane.)

5   By means of your result in question 4 find the point obtained by projecting (x_0, y_0, z_0) perpendicularly onto the plane $x + y + z = 0$. Hence obtain the matrix that represents the operation of perpendicular projection onto this plane, i.e. the matrix M such that if v is any point, Mv is its projection.

Check the accuracy of your work by verifying that the determinant, eigenvectors and eigenvalues of M are what they ought to be.

6   (a) Let $D = (4, 0, 0)$, $E = (0, 4, 0)$, $F = (0, 0, 4)$. Draw a diagram for the points with whole number co-ordinates in the plane DEF similar to that drawn for the plane ABC in the worked example above.

(b) Find the co-ordinates for the perpendicular projection of (x_0, y_0, z_0) onto the plane $x + y + z = 5$.

(c) Copy the diagram for triangle ABC found in the Worked example, and then mark on it the perpendicular projections onto the plane ABC of all the points in the plane DEF that were shown in your answer to (a).

7   Let $P = (x_0, y_0, z_0)$ and $K = (1, 1, 1)$. We can find the perpendicular projection of P onto the line OK by constructing the plane through P that is perpendicular to OK and taking its intersection,

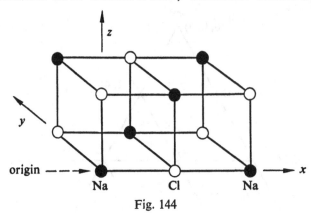

Fig. 144

N, with OK. Find the equation of that plane, the parametric representation of the line OK, and hence the co-ordinates of N.

Write the matrix M that represents the transformation $P \to N$, and check its accuracy in the way suggested in question 5.

8 A cube has faces lying in the planes $x = 1$, $y = 1$, $z = 1$, $x = -1$, $y = -1$, and $z = -1$. The plane $x + y + z = 0$ divides the cube into two congruent pieces. Find the co-ordinates of the points where this plane meets the edges of the cube. What plane figure is the section of the cube by this plane? Make a sketch showing clearly the cube and the intersection of this plane with its faces.

9 Fig. 144 shows the arrangement of sodium and chlorine atoms in a crystal of common salt, with black balls for sodium and white for chlorine. If the cubes have unit side, atoms occur at points with whole number co-ordinates. Given any point, for example, (87, 1230, 29), how can one tell whether this point is occupied by a sodium or a chlorine atom? Certain planes are occupied by sodium atoms alone, others by chlorine atoms alone. Identify some such planes and give their equations; a variety of possible orientations should be sought. How can one prove that such planes, however far they may be continued, will contain only one kind of atom?

10 If P is any point and M is its perpendicular projection onto a plane, the reflection of P in the plane is reached from P by going twice the distance $\|PM\|$ in the direction of PM. Find the co-ordinates of the reflection of (x_0, y_0, z_0) in the plane $ax + by + cz = d$.

Find the matrix that represents reflection in the plane $x + 2y + 3z = 0$. Do you expect this matrix to be (a) orthogonal, (b) symmetric?

Vector product

As was mentioned earlier, vector product is a very restricted concept; it has meaning only in Euclidean geometry of 3 dimensions. It arises very naturally in statics and dynamics and also in electromagnetic theory.

In Fig. 145 a force F acts in the plane of the paper, at a point whose position, relative to the origin, is specified by a vector r. The force exerts a twisting effect about the origin. It has a tendency to make things turn about a line through the origin perpendicular to the plane of the paper. The magnitude of this torque is pF where p is the perpendicular from O to the line of action of F. Thus the custom has arisen of representing the turning effect of F about the origin by a vector of length pF perpendicular to the plane of the paper. This vector is known as the vector product of r and F, and is written $r \wedge F$ or $r \times F$.

Fig. 145

Note that pF represents the area of the parallelogram contained between the vectors r and F, and so the magnitude of $r \times F$ can be defined as equal to the area of this parallelogram. Note incidentally that 'a length $=$ an area' does not conform to common practices in regard to 'dimensions'. We have to suppose our units fixed; the number of (say) metres in the length is to equal the number of square metres in the area of the parallelogram.

Suppose we are given two vectors u, v, and we wish to find a formula for their vector product, $w = u \times v$.

The first requirement on w is that it should be perpendicular to both u and v, just as, in our discussion above, the torque is perpendicular to the plane of r and F. Let u be a (column) vector with components a, b, c, and let v have components d, e, f. As w is perpendicular to both, we must have $u'w = 0$ and $v'w = 0$. If w has components x, y, z, this means

$$ax + by + cz = 0, \qquad dx + ey + fz = 0.$$

At the end of §27, we saw that the solution of these equations was $x = k(bf - ce)$, $y = -k(af - cd)$, $z = k(ae - bd)$, where k could be any number.

Let A stand for the area of the parallelogram between u and v, and L for the length of w. The definition of w instructs us to make $L = A$. Now the area of a parallelogram between two arbitrary vectors u, v can be calculated, but the calculation is moderately long. The area of the parallelogram involves lengths and the sine of the angle between u and v. From the dot product $u \cdot v$ we can arrive at the cosine of the angle, and from the cosine we can get the sine – a somewhat roundabout procedure.

We can bypass this by considering the volume of the cell contained by u, v, w. Since w is perpendicular to the plane of u and v, taking A as the area of the base, L will give the perpendicular height, so the volume is LA. As we want $L = A$, this means the volume must be simply L^2. We must choose k to make this so.

Now

$$L^2 = x^2 + y^2 + z^2 = k^2\{(bf - ce)^2 + (af - cd)^2 + (ae - bd)^2\}.$$

Fortunately for the simplicity of our final formula, the expression above, involving the sum of 3 squares, also appears in the expression for the volume of the cell, and cancels. For the volume of the cell is

$$\det (u, v, w) = -\det (u, w, v) = \det (w, u, v)$$

$$= \begin{vmatrix} x & a & d \\ y & b & e \\ z & c & f \end{vmatrix}$$

$$= x(bf - ce) - y(af - cd) + z(ae + bd).$$

Substituting the values found for x, y, z we obtain

$$k\{(bf - ce)^2 + (af - cd)^2 + (ae - bd)^2\}.$$

Comparing this with our expression for L^2 above, which the volume of the cell has to equal, we find $k^2 = k$. Obviously we do not want the solution $k = 0$, so we are left with the very simple conclusion, $k = 1$. Thus the components of w are simply $bf - ce$, $-(af - cd)$, $ae - bd$. There is a simple rule for obtaining these. We write first of all

$$\begin{matrix} a & b & c \\ d & e & f. \end{matrix}$$

To obtain the first component of $u \times v$, we cover the first column, a and d, and write the determinant of the matrix we then see. For the second component, we cover b and e, and write the determinant of what we then see, *but with a minus sign*. For the third component, we cover c and f, and write the determinant of what remains visible. The alternation of signs, $+, -, +$, is exactly the same as in the rules for multiplying out a determinant.

One last point remains to be settled. Suppose for example u represents a unit due east, and v a unit due north in the horizontal plane. The parallelogram is then a square of unit area, and we know $u \times v$ should be perpendicular to the horizontal plane, that is, vertical. But should it be a unit vector vertically upwards or vertically downwards? If we write $1, 0, 0$ for a, b, c and $0, 1, 0$ for d, e, f in the scheme explained above, we obtain $0, 0, 1$ for the components of $u \times v$; thus $u \times v$ is vertically upwards. But how are we to interpret our formula when u and v do not lie so conveniently, but are any two vectors? Our earlier work gives us a way to answer this question. We found k by using the equation $\det (u, v, w) = L^2$. *This means that u, v, w must always lie in such a way that $\det (u, v, w)$ is positive.*

To interpret this result geometrically, we use two very simple properties of determinants. A determinant is given by a polynomial function of the numbers $a, b, c, \ldots$ that appear in it. Accordingly, it is given by a continuous function; little changes in a, b, c and the other numbers produce only little changes in the determinant. Now the value of a continuous function can change from $+$ to $-$ only by passing through the value 0 (Fig. 146). So if we have three vectors whose determinant is positive and we change them gradually, with care never to let the volume of the cell become 0, we shall be sure of ending up with a positive determinant.

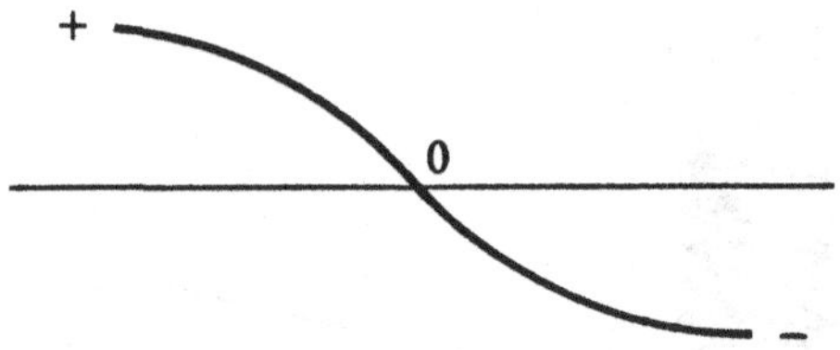

Fig. 146

Now we have seen that u to the east, v to the north and w upwards gives a positive value to the determinant. Suppose we keep the directions of u and w the same, but allow v to swing round in the horizontal plane. The determinant will become zero only if v swings so far that its direction becomes either east or west. If we forbid this, if we insist that v points somewhere between just north of east and just north of west, we have removed all possibility of a change of sign; with this condition, det (u, v, w) will certainly be positive.

Now consider a different kind of variation. Suppose we have our vectors u, v, w in any position, and that we represent them by three rods, welded together to form a rigid body. We then gradually rotate this rigid system in any manner whatever. All the vectors then change gradually, and so the value of the determinant can never jump suddenly. But, since the system is rigid, the volume of the cell is fixed in size; its magnitude cannot change. If, at the start, det (u, v, w) was K, at any later time it must be either K or $-K$. But one can only get from K to $-K$ by a sudden jump; there is no way of going gradually from K to $-K$ without passing through other numbers. But we showed earlier that gradual rotation can never make the determinant's value jump; there is only one way to meet both requirements: during rotation, the determinant's value must stay unaltered at K. Rotation does not change the magnitude, or the sign, of det(u, v, w).

Accordingly let us take our three welded rods and turn them until the rod u points due east. Then turn them, keeping u due east, until v lies in the horizontal plane through u, and in the half of that plane described above, from just north of east to just north of west. This operation can always be carried out. The third vector w equals $u \times v$, so det(u, v, w) must be positive. This means w must point vertically upwards, not vertically downwards.

The relationship of u, v, w, in the situation just described, is often embodied in the Right-Hand Rule; if we turn a screw head in such a way that a lever, originally pointing in the direction of u, comes to the direction v *by the shortest route*, then the screw will advance in the direction of w, where $w = u \times v$, as in Fig. 147.

If we now suppose our welded rods returned to their original position, this statement will remain correct; it is not affected by a rotation of the rigid object.

In all the discussion above we have assumed that u and v point in different directions. If u and v point in the same direction, the area of the 'parallelogram' between them is 0, and $u \times v = 0$. No further discussion is called for.

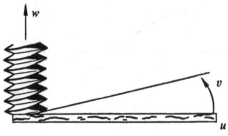

Fig. 147

Exercises

1 Show, both algebraically and from the geometrical definition, that $v \times u = -(u \times v)$.

2 If u, v, r are any three vectors, we can form the scalar product of r with the vector $u \times v$. The resulting number is written $r \cdot (u \times v)$. Show that, if r lies in the plane of u and v, then $r \cdot (u \times v) = 0$. (This comes without any calculation by a simple geometrical argument.)

3 Show, by algebra, that $r \cdot (u \times v) = \det(r, u, v)$. Is this result reasonable on geometrical grounds?

4 How do we express, by algebraic symbols, that r lies in the plane of u and v? What theorem about determinants comes by combining the statements in questions 3 and 4?

Note The arguments used in this section justify the statement made in §36, that a rotation has determinant $+1$ while a reflection has determinant -1. The essential reason for this is that any rotation can be carried out gradually. We can bring a body to any required position by a series of tiny changes. A 'tiny change' has a matrix that is very nearly the identity matrix, I. Since $\det I = +1$, and the determinant is a continuous function, the determinant of a tiny rotation must be $+1$; it cannot jump to -1. Accordingly a rotation, which can be brought about by a succession of small rotations, must also have determinant $+1$, for determinant of a product equals product of determinants.

A reflection on the other hand is a sudden change. There is, for instance, no way in which we can *gradually* change a car with right-hand drive into one with left-hand drive.

Exercises on vector products

1 Find the following vector products, and check that they do lie in the directions to be expected
  (a) $(1, 1, 0) \times (-1, 1, 0)$ (b) $(1, 1, 1) \times (1, 1, 2)$ (c) $(-1, -2, 3) \times (4, 5, -9)$
  (d) $(1, 1, 3) \times (5, 5, 16)$ (e) $(1, 2, 3) \times (5, 10, 15)$.

2 By means of vector product theory find a vector normal to the plane containing the vect(
$(2, 6, 9)$ and $(-6, -7, 6)$. What is the area of the parallelogram that has these vectors as two of sides?
  What figure in fact is that parallelogram, and how can you confirm your answer by determin its area without any use of vector products?

3 In each of the following cases find a vector normal to the plane of the two given vectors, also the area of the parallelogram with the two vectors as sides;
  (a) $(1, -1, 0)$ and $(0, 1, -1)$ (b) $(1, -1, 0)$ and $(1, 0, -1)$,   (c) $(1, 0, -1)$ and $(0, 1,$

In each case, deduce the sine of the angle between the two given vectors. Also, by means of scalar products, find the cosine of the angle between them. Check the consistency of these results.

Discuss the geometrical figure formed by the 3 vectors that are mentioned in (*a*), (*b*) and (*c*) above.

4 From any point *O* of the paper you are using draw 2 arrows to represent vectors, *u* and *v*, in the plane of the paper. Let $p = u \times v$. Indicate on your drawing where you would expect $u \times p$ and $v \times p$ to lie. Test your conclusion by the example $u = (1, 0, 0)$, $v = (1, 1, 0)$, it being understood that $z = 0$ is the plane of the paper.

5 Let $u = (1, 2, 3)$, $v = (2, 3, 4)$, $w = (3, 4, 5)$. Find $p = u \times v$, and the scalar product $w \cdot p$. Does *w* lie in the plane of *u* and *v*?

6 If a force *F* acts at the point *R* its moment about an axis through the origin is given by $U \cdot (R \times F)$, where *U* is a vector of unit length along the axis.

(*a*) Find the moment of the force $(6, 3, 7)$ acting at $(4, 2, 5)$ about the axis joining the origin to the point $(1, 2, 2)$.

(*b*) Show that if the force in (*a*) were changed to $(-4, 10, 1)$, the other data remaining unaltered, there would be no moment about the axis.

7 A set of three perpendicular vectors can be constructed in the following way. Choose any non-zero vector *u*. Choose *v* perpendicular to it. Let $w = u \times v$. Then *u, v, w* constitute a system of mutually perpendicular vectors. Construct a system of this kind corresponding to each of the vectors *u* listed here;

(*a*) $u = (1, -1, 0)$,   (*b*) $u = (1, 1, 1)$,   (*c*) $u = (1, 2, 2)$,   (*d*) $u = (1, 2, 3)$.

Note. This procedure can be used to construct examples of orthogonal matrices, if in addition we require the vectors *u* and *v* to be of unit length; we then form the matrix with columns *u, v, w*.

40 Systems of linear equations

Suppose we have the equations

$$x + 3y = 25, \tag{1}$$

$$2x + 7y = 55. \tag{2}$$

If we subtract twice equation (1) from equation (2), and leave equation (1) unaltered, we obtain

$$x + 3y = 25, \tag{3}$$

$$y = 5. \tag{4}$$

If we now subtract 3 times equation (4) from equation (3), we bring our pair of equations to the form

$$x \qquad = 10, \tag{5}$$

$$y = 5. \tag{6}$$

If we write

$$M = \begin{pmatrix} 1 & 3 \\ 2 & 7 \end{pmatrix}, \qquad v = \begin{pmatrix} x \\ y \end{pmatrix},$$

it will be seen that equations (1) and (2) have, on the left, the vector Mv, while equations (5) and (6) have simply Iv, where I is, as usual, the identity matrix.

A remarkable verbiage has grown up around this calculation, and its generalization to m equations in n unknowns. It is called the Gauss–Jordan elimination procedure, and the matrix I, appearing in equations (5) and (6) is known as the Row–Echelon normal form of the original matrix M.

Many textbooks, both for school and university courses, begin with this procedure and use it to introduce the idea of matrix. We have done the opposite; we have kept linear equations until near the end of the course, and this has been done for a very definite reason. The use of linear equations to introduce the idea of matrix has led to confusion in the minds of some students which the authors of the textbooks certainly did not foresee or intend. That approach, and the use of the term 'row–echelon normal form of the matrix M', led some students to believe that somehow the final matrix, I, was *the same as* the original matrix M. Now this is clearly false; whatever else it may be, the matrix $\begin{pmatrix} 1 & 3 \\ 2 & 7 \end{pmatrix}$ is most certainly *not* the identity matrix I.

This confusion is very likely if a student identifies a matrix with a system of equations. The equations (5) and (6) are *equivalent* to the equations (1) and (2). The pair of statements, (5) and (6), contains exactly the same information as the pair of statements (1) and (2).

Let us examine in more detail the actual relationship between the matrices. We will begin with the equations

$$x + 3y = a_1, \tag{7}$$

$$2x + 7y = a_2. \tag{8}$$

Applying the same procedure as before, we obtain first

$$x + 3y = a_1 \qquad \quad = b_1, \quad \text{say.} \tag{9}$$

$$y = -2a_1 + a_2 = b_2, \quad \text{say.} \tag{10}$$

Next we have

$$x \qquad = b_1 - 3b_2 = c_1, \quad \text{say} \tag{11}$$

$$y = \qquad b_2 = c_2, \quad \text{say.} \tag{12}$$

Equations (7) and (8) can be combined in a single vector equation, $Mv = a$, where the meaning of a should be self-explanatory. The vector b, involved in equations (9) and (10), can be written $b = Qa$, where $Q = \begin{pmatrix} 1 & 0 \\ -2 & 1 \end{pmatrix}$. Since $a = Mv$, we must have $b = QMv$, and equations (9) and (10) may be put in the form $QMv = b$. (If desired, this can be verified by actual calculation of QM.) Similarly, we have $c = Pb$, where

$P = \begin{pmatrix} 1 & -3 \\ 0 & 1 \end{pmatrix}$. Accordingly, on substituting for b, we obtain $PQMv = c$ as a concise form of equations (11), (12). Now equations (11) and (12), as we see by looking at the actual equations, are in fact $Iv = c$. Comparing the two equations we see that I can be identified, not with M, but with PQM. We have $PQM = I$. Thus $PQ = M^{-1}$.

The calculations indeed pass from $Mv = a$ to $QMv = Qa$ and finally to

$$PQM = PQa.$$

As $PQ = M^{-1}$, the overall effect is that we go from $Mv = a$ to $v = M^{-1}a$. The procedure in fact is one that can conveniently be used for calculating the inverse matrix, M^{-1}.

Many books give a procedure for calculating the inverse matrix, in which matrices only are written. The logic of this procedure may be seen by considering the process of solution described above. The only change is that we write Ia instead of a. The successive pairs of equations can then be shown as follows:

$$Mv = Ia, \tag{7}, (8)$$

$$QMv = QIa, \tag{9}, (10)$$

$$PQMv = PQIa. \tag{11}, (12)$$

That is

$$Iv = PQIa.$$

Suppose we now agree to omit the vectors v and a, since they remain unchanged throughout. We can then record the calculation as

$$M , \quad I .$$
$$QM , \quad QI .$$
$$PQM , PQI .$$
$$I , PQI (= M^{-1}).$$

In carrying out such a calculation, we do not usually think of multiplying by the matrices Q and P, but rather of operations done to the rows. For if we consider any matrix

$$K = \begin{pmatrix} d & e \\ f & g \end{pmatrix},$$

we have

$$QK = \begin{pmatrix} 1 & 0 \\ -2 & 1 \end{pmatrix} \begin{pmatrix} d & e \\ f & g \end{pmatrix} = \begin{pmatrix} d & e \\ -2d + f & -2e + g \end{pmatrix}.$$

Thus going from K to QK is achieved by *subtracting twice the first row from the second*. This in fact was how the matrix Q first came into the story. (Look back at our original solution of the equations, and see how it led us to consider Q.) The multiplication by P also indicates an operation on the rows.

The procedure for calculating M^{-1} may be summarized as follows; choose a sequence of operations on rows that will change M into I; this sequence of operations will change I into M^{-1}.

We assume here that M does have an inverse; if M^{-1} does not exist, we shall not be able to find a sequence of operations that takes us from M to I.

All the work of this section is best understood by thinking about the solution of equations. All the operations we shall consider will be calculations that any student would naturally perform if confronted with a number of linear equations.

The matrix form of this work adds no essential idea, but simply puts the material in a form suitable for an electronic computer. A computer does not record equations such as $x + 3y = 25$, $2x + 7y = 55$ in the way we do. It simply records the numbers involved at appropriate addresses, and so deals with a scheme of the type

$$\begin{matrix} 1 & 3 & 25 \\ 2 & 7 & 55. \end{matrix}$$

Such a scheme to us looks like a matrix and can indeed be handled by rules of matrix algebra. This change of notation is a very minor affair, as it was also earlier when we agreed to drop v and a from our equations and simply understand that they were supposed to be there. At any time when we are uncertain why some particular step is being taken, it will usually be sufficient to go back from the matrix symbols to the equations they represent, and the reason will become clear.

Equivalence of equations

When we operate on a system of equations, we want to end up with another system of equations that says neither more nor less than our original set: we do not want to bring in any condition that was not in the original set, for this might lead us to overlook some perfectly good solutions; on the other hand, we do not want to lose any condition, for that might lead us to accept something which in fact was not a solution of the original equations. For example, suppose we have to attack the following system of equations:

$$3x + 4y = 20, \tag{13}$$

$$4x + 5y = 27, \tag{14}$$

$$4x + 3y = 23. \tag{15}$$

Students sometimes hand in the following calculation. Subtract equation (13) from (14), and also subtract (14) from (15). This gives

$$x + y = 7, \tag{16}$$

$$x - y = 3. \tag{17}$$

These equations lead to $x = 5$, $y = 2$, and these values are handed in as a solution. But if we try these values in equations (13), (14), (15), we find that not one of the equations is satisfied. In fact, equations (13), (14), (15) are inconsistent; they have no solution. The trouble is that our argument is not reversible; equations (16) and (17) do follow from equations (13), (14), (15), but equations (13), (14), (15) cannot be deduced from equations (16), (17). So (16), (17) is *not* a system equivalent to (13), (14), (15).

We would have obtained an equivalent system if we had not merely written (16) and (17), but had copied (13) down again. Equations (13), (16), (17) are equivalent to (13), (14), (15); for we can get back to (14) by adding (13) and (16), and back to (15) by adding (13) and (17).

The situation can be shown in the language of symbolic logic as follows. In the first argument we have

$$(13) \ \& \ (14) \ \& \ (15) \Rightarrow (16) \ \& \ (17).$$

That is, (16) and (17) follow from (13), (14), (15), but we do not have any authority for going in the opposite direction. The statement shows that any solution (if it existed) of (13), (14), (15) would satisfy (16), (17). This gives us a kind of elimination procedure; it tells us that it is no use considering values x, y *unless* they satisfy (16) and (17). So $x = 5$, $y = 2$ is all we need try: either it is a solution, or there is no solution.

In the second way of doing things we have

$$(13) \ \& \ (14) \ \& \ (15) \Leftrightarrow (13) \ \& \ (16) \ \& \ (17).$$

That is, each set of statements can be deduced from the other set. The information contained in (13) & (16) & (17) is exactly the same as that in (13) & (14) & (15). Now this of course is the situation we want to achieve – to bring our equations to a simpler form, and to be sure that we have neither brought in things that appear to be solutions and in fact are not (like $x = 5$, $y = 2$ in our example) nor ruled out perfectly good solutions of the original equations; *we want our arguments to be reversible*; we want $\Leftrightarrow$, not the one-way street of $\Rightarrow$.

Given any system of equations, there are three quite simple and obvious reversible operations that we can apply to them.

(1) We may alter the order of the equations. To reverse this, we simply go back to the original order.

(2) We may replace any equation by that equation multiplied by a non-zero number. To reverse this is always possible; for instance, if the equation had been replaced by 3 times itself, we would get back by using $\frac{1}{3}$ times the new equation.

(3) We can replace any equation by that equation together with any multiple of another equation. Here a shorthand arrangement is convenient; if for example, we add 10 times equation (73) to equation (52), we may denote this by $(52) + 10(73) = (52^*)$, where (52^*) is the new equation that replaced equation (52). If we want to get back to our original equation we can do so, for $(52) = (52^*) - 10(73)$. We simply have to subtract equation (73), which is still there in the new system, from the new equation (52^*).

Our concern in the remainder of this section will be to see how we can use these simple operations to bring a system of equations to the most convenient form. The general objective can be seen from the following example. Suppose we are given some equations in 5 unknowns, and we find they are equivalent to the following 3 equations:

$$x_1 \qquad - 5x_3 \qquad + 2x_5 = 0,$$
$$x_2 + x_3 \qquad + 7x_5 = 0,$$
$$x_4 - 4x_5 = 0.$$

Here x_1, x_2 and x_4 are in a special position; x_1 occurs in the first equation only, x_2 in the second only, x_4 in the third only. Suppose we allow someone to choose any values he likes for x_3 and x_5. We can meet his requirements and still find a solution. For we can make the first equation correct by putting $x_1 = 5x_3 - 2x_5$. As x_1 occurs only in the first equation, this will have no repercussions on the other equations. Similarly, by taking $x_2 = -x_3 - 7x_5$ and $x_4 = 4x_5$ we can satisfy the other two equations. Clearly, then, there are a lot of solutions; we have 2 degrees of freedom – free choice of x_3 and x_5. If we put $x_3 = s$ and $x_5 = t$, we have the 2 parameters s and t, and the solution $x_1 = 5s - 2t$, $x_2 = -s - 7t$, $x_3 = s$, $x_4 = 4t$, $x_5 = t$. In fact these solutions fill a plane; if we put $s = 1$, $t = 0$ we get a particular solution corresponding to the point $(5, -1, 1, 0, 0)$, while for $s = 0$, $t = 1$, we get $(-2, -7, 0, 4, 1)$. If we denote these points by the (column) vectors u, v, every solution is of the form $su + tv$. The solutions fill a subspace of 2 dimensions.

Our work falls into three parts, (1) bringing the left-hand sides of the equations to the simplest form, (2) interpreting the results when the right-hand sides are all 0, (3) interpreting the results when non-zero numbers occur on the right-hand side.

Obtaining a simple form for the left-hand sides

The procedure for simplification will usually be carried out by a computer. We therefore have to be careful to make sure that the procedure can always be carried through, and that we do not overlook any possibilities that might arise.

We begin by considering the first variable, x_1, in the first equation. The situation is straightforward if we see a term ax_1, where $a \neq 0$. In this case, we divide by a, so our equation begins $x_1 + \cdots$ By subtracting suitable multiples of this equation from the other equations, we can get rid of x_1 in all equations except the first.

Worked example

Given equations $10x_1 + 4x_2 - x_3 = 0$, $3x_1 - 2x_2 + 11x_3 = 4$, $7x_1 + 22x_2 + 5x_3 = 9$. Multiply the first equation by 0.1 to obtain $x_1 + 0.4x_2 - 0.1x_3 = 0$. Subtracting 3 times this equation from the second equation, and 7 times this equation from the third equation leads to a system where x_1 has coefficient 1 in the first equation, and coefficients 0 in the other equations. (For an understanding of this point it is not necessary actually to calculate the coefficients of x_2 and x_3 in the resulting equations.)

First obstacle; suppose x_1 does not appear in the first equation. In this case, we look down the equations until we find one that does contain a term ax_1. We bring this equation up to first place, and then proceed as before.

Second obstacle; suppose we cannot do this – that is, suppose x_1 does not appear in any of the equations. In this case, we ignore x_1 and try to get x_2 in the first equation with a coefficient 1, and no x_2 in any other equation. If x_2 occurs nowhere, we go on to x_3 and so on.

Let us assume, for simplicity of description that x_1 actually does occur, so that we have our first equation $x_1 + \cdots,$ and no x_1 anywhere else. The remaining equations contain the $(n - 1)$ unknowns $x_2, x_3, \ldots, x_n$, and in effect we are faced with the same problem again, but with one variable less. So we try to arrange for a second equation beginning with $x_2 + \cdots$, and for x_2 to be absent from the remaining equations, *including the first equation of all*. If the second equation starts $x_2 + \cdots$, we certainly can get rid of x_2 everywhere else by appropriate subtractions of multiples of the second equation. However, it may happen that x_2 does not occur at all in any equation except the first; in that case there is nothing more we can do, we leave x_2 in the first equation and go on to consider x_3.

Thus we may reach a set of equations that begin with, for example, something like

$$x_1 \qquad + 2x_3 + 3x_4 = 4,$$
$$x_2 + 5x_3 + 6x_4 = 7,$$

or we may have something like

$$x_1 + 8x_2 \qquad + 2x_4 = 4,$$
$$x_3 + 3x_4 = 7.$$

In both cases, x_1 occurs only in the first equation. In the first case x_2 occurs only in the second equation; in the second case it is x_3 that occurs there only.

It could of course happen that x_2 never occurs at all. By erasing $8x_2$ in the example above, we can see how the first two equations might appear in such a case.

The process continues, always using very much the same ideas. Suppose, at a certain stage, we have dealt with a number of equations; the remaining equations do not contain $x_1, x_2, \ldots, x_{p-1}$. They do, however, contain x_p. We choose an equation that contains x_p, bring it to the top, if it is not already there, and multiply it by the number that makes the coefficient of x_p equal to 1. By using this equation, we get rid of x_p in all the other equations, whether they come before or after the equation in question. This equation has now been 'dealt with'. The equations that come after it are now free of $x_1, x_2, \ldots, x_{p-1}$ and x_p. If x_q is the first variable that occurs in them, we repeat the procedure, with x_q playing the role that x_p has just played. The process terminates when either there are no more equations, or when every variable in the remaining equations has coefficient 0. These possibilities are illustrated in the following examples.

In order not to distract attention by arithmetical complications, these equations have been 'faked' to work out with whole numbers. In real life, of course, we would expect to work to several places of decimals.

Worked example 1

Put the following system of equations into standard form:

$$x_1 + 2x_2 + 2x_3 + 5x_4 + 36x_5 = 0, \tag{18}$$

$$3x_1 + 6x_2 + 8x_3 + 19x_4 + 136x_5 = 0, \tag{19}$$

$$4x_1 + 8x_2 + 7x_3 + 19x_4 + 135x_5 = 0. \tag{20}$$

Solution The first equation already contains x_1 with coefficient 1. So we repeat equation (18) as equation (21). To get rid of x_1 in the other equations, we take $(22) = (19) - 3(18)$ and $(23) = (20) - 4(18)$. This gives

$$x_1 + 2x_2 + 2x_3 + 5x_4 + 36x_5 = 0, \tag{21}$$

$$2x_3 + 4x_4 + 28x_5 = 0, \tag{22}$$

$$- x_3 - x_4 - 9x_5 = 0. \tag{23}$$

Now x_3 is the first variable that remains in the last pair of equations. Halving equation (22) gives us equation (25) with coefficient 1 for x_3. To get rid of x_3 in the other equations we form $(24) = (21) - 2(25)$ and $(26) = (23) + (25)$. This leads to

$$x_1 + 2x_2 \qquad + x_4 + 8x_5 = 0, \tag{24}$$

$$x_3 + 2x_4 + 14x_5 = 0, \tag{25}$$

$$x_4 + 5x_5 = 0. \tag{26}$$

As equation (24) contains the next variable, x_4, already with coefficient 1, we retain this equation unaltered as equation (26a). To get rid of x_4 elsewhere, we take

$$(24a) = (24) - (26)$$

and $(25a) = (25) - 2(26)$. We thus finish with

$$x_1 + 2x_2 \qquad\qquad + 3x_5 = 0, \tag{24a}$$

$$x_3 \qquad + 4x_5 = 0, \tag{25a}$$

$$x_4 + 5x_5 = 0. \tag{26a}$$

The process is now complete, for there are no more equations to deal with.

Worked example 2

Simplify the system

$$x_1 + 2x_2 + 3x_3 = 0, \tag{27}$$

$$3x_1 + 6x_2 + 7x_3 = 0, \tag{28}$$

$$2x_1 + 4x_2 + 10x_3 = 0. \tag{29}$$

Solution In equation (27) we already have x_1 with coefficient 1, so we form simply $(28) - 3(27)$, which is $-2x_3 = 0$, and $(29) - 2(27)$, which is $4x_3 = 0$. We divide the first of these by -2 and so obtain

$$x_1 + 2x_2 + 3x_3 = 0, \tag{30}$$

$$x_3 = 0, \tag{31}$$

$$4x_3 = 0. \tag{32}$$

Forming $(30) - 3(31)$ gets rid of x_3 in the first of these equations. Also $(32) - 4(31)$ gets rid of x_3 in the third equation, but then nothing at all remains in that equation. The original 3 equations are thus equivalent to the 2 equations

$$x_1 + 2x_2 \qquad = 0, \tag{33}$$

$$x_3 = 0. \tag{34}$$

The first example we had of equations in standard form was

$$
\begin{aligned}
x_1 \quad - 5x_3 \quad + 2x_5 &= 0, \\
x_2 + x_3 \quad + 7x_5 &= 0, \\
x_4 - 4x_5 &= 0.
\end{aligned}
$$

The special variables here were x_1, x_2 and x_4. The form of the equations is easier to see if x_1, x_2, x_4 are written first, then x_3 and x_5. We can bring this about by introducing new symbols. Let $y_1 = x_1, y_2 = x_2, y_3 = x_4, y_4 = x_3, y_5 = x_5$; that is, we give priority to the special variables. The equations then appear like this;

$$
\begin{aligned}
y_1 \quad - 5y_4 + 2y_5 &= 0, \\
y_2 \quad + y_4 + 7y_5 &= 0, \\
y_3 \quad - 4y_5 &= 0.
\end{aligned}
$$

It will simplify our later discussions to suppose that, if it is needed, such a substitution has been made. We shall not bother to use the letter y; we shall simply assume that the variable at the beginning of the first equation is called x_1, that at the beginning of the second is called x_2, and so on. Then we shall always see a pattern like that of the equations last written. If there are k equations, $x_1, x_2, \ldots, x_k$ will stand alone, then $x_{k+1}, \ldots, x_n$ will occur with arbitrary coefficients. Some of these coefficients may happen to be 0, as for y_4 in the third equation above, but this is of no significance. The extreme cases, $k = n$ and $k = 0$ may occur.

The important thing to notice is that the reduction to standard form *can always be carried through*. It may be that all the coefficients in all the given equations are 0; in that case, we have no equations at all; this is a standard case, with $k = 0$. If this extreme case does not occur, then some equation contains some variable with non-zero coefficient. We write that equation first, call the variable x_1, and get the equation into the form $x_1 + \cdots$ By subtracting suitable multiples of this from the other equations we can make sure that x_1 is never mentioned again. The remaining equations now contain only the other variables, and we have the same alternative again. Are all the coefficients 0? Then our job is finished. If not, there is some variable that actually occurs in some equation; we call it x_2, get the equation into the form $x_2 + \cdots$ and so proceed.

Solutions when all right-hand sides are 0

The situation is particularly simple when only 0 occurs after the equals signs. For example, suppose we have some equations in 5 unknowns, and that this system, when reduced to standard form gives the 3 equations

$$
\begin{aligned}
x_1 \qquad\quad - ax_4 - dx_5 &= 0, \\
x_2 \qquad - bx_4 - ex_5 &= 0, \\
x_3 - cx_4 - fx_5 &= 0.
\end{aligned}
$$

As we saw earlier, we can allow someone to prescribe arbitrary values for x_4 and x_5, and still have a solution of the equations. In fact, if he chooses $x_4 = s$, $x_5 = t$ we take $x_1 = as + dt$, $x_2 = bs + et$, $x_3 = cs + ft$, as the equations force us to do. Now every solution is covered by this procedure. Suppose, for instance, we are told that g_1, g_2, g_3, g_4, g_5 is a solution of the system of equations. We will take $s = g_4, t = g_5$. This gives the correct values for x_4 and x_5. But once x_4 and x_5 have been chosen, the equations fix x_1, x_2 and x_3. So, if g_1, g_2, g_3, g_4, g_5 really is a solution, our formula is bound to give the values g_1, g_2, g_3 for x_1, x_2, x_3.

We may write our solution in the form

$$
\begin{pmatrix} x_1 \\ x_2 \\ x_3 \\ x_4 \\ x_5 \end{pmatrix} = s \begin{pmatrix} a \\ b \\ c \\ 1 \\ 0 \end{pmatrix} + t \begin{pmatrix} d \\ e \\ f \\ 0 \\ 1 \end{pmatrix}.
$$

If we call the three vectors occurring here v, u_1, u_2, the above equation may be written $v = su_1 + tu_2$. So a solution, v, results if we start at the origin, and go an arbitrary distance in the direction of the vector u_1, and then an arbitrary distance in the direction of u_2. That is to say, the solutions fill a plane, a subspace of 2 dimensions.

We can see where this number, 2, comes from. We have 5 unknowns. There are 3 equations, and in these x_1, x_2 and x_3 occur, each only once. We are left with $5 - 3 = 2$ unknowns, x_4 and x_5, that can be chosen at will.

This argument can be generalized. If we have n unknowns, and the standard form contains k equations, then $x_1, x_2, \ldots, x_k$ will occur each once only, and we shall be able to choose $x_{k+1}, \ldots, x_n$ at will; once chosen, these fix the values of $x_1, \ldots, x_k$. You should satisfy yourself that, if we choose numbers $t_1, t_2, \ldots, t_{n-k}$ for $x_{k+1}, x_{k+2}, \ldots, x_n$, then $x_1, \ldots, x_k$ are given by expressions linear in $t_1, \ldots, t_{n-k}$ and the general solution may be written in vector form as $v = t_1u_1 + t_2u_2 + \cdots + t_{n-k}u_{n-k}$, where $u_1, u_2, \ldots, u_{n-k}$ are *fixed* vectors and $t_1, t_2, \ldots, t_{n-k}$ are arbitrary parameters.

Are we now in a position to assert the theorem: *if there are k equations in the standard form for a system of equations in n variables* (with 0 only on right-hand sides), *then the solutions fill a subspace of n − k dimensions?* We are very close to having proved this, but there is one point still to settle. In the equation $v = t_1u_1 + \cdots + t_{n-k}u_{n-k}$ we have something that looks very much like the parametric specification of a subspace of $n - k$ dimensions. But what would happen if the vectors $u_1, \ldots, u_{n-k}$ were not independent? As an extreme case, if these vectors were all 0, v would be confined to a single point, the origin. If $u_1, \ldots, u_{n-k}$ were all multiples of the single vector u_1 we would have a subspace of 1 dimension only. If any one of them were a mixture of the others, the dimension would fall below $n - k$. However, none of these things can happen. If you look back at the particular example we had at the beginning of this discussion, you will see that u_1 had the components $a, b, c, 1, 0$ and u_2 had $d, e, f, 0, 1$. So we can certainly dismiss the possibility of u_1 and u_2 both being 0. Admittedly a, b, c, d, e, f might all be 0 (this is a perfectly possible case) but then we would be left with 0, 0, 0, 1, 0 and 0, 0, 0, 0, 1. There is no way of making either vector 0; we cannot get rid of these components with value 1. Nor is there any way of choosing a, b, c, d, e, f to make these vectors linearly dependent. If we try to make $u_2 = hu_1$, for some number h, the equations corresponding to the first three equations do not present any obstacle; however, when we come to the fourth and fifth components we get the equations $0 = h \cdot 1, 1 = h \cdot 0$, which it is impossible to satisfy; things turn out much the same if we try to make $u_1 = hu_2$.

In the same way, if $n - k = 3$, we find that the last three components of u_1 are $\cdots 1, 0, 0$; those of u_2 are $\cdots 0, 1, 0$; and those of u_3 are $\cdots 0, 0, 1$. By considering these components alone we can see that it is impossible for any one of u_1, u_2, u_3 to be a mixture of the other two.

The same argument applies in every case and so the theorem suggested above is in fact true.

Notice that the number k in the expression $n - k$ is the number of equations *in the standard form*, not the number of equations as originally given.

Non-zero solutions

Equations of the type we are considering always have a solution in which every variable is 0. We often want to know (for instance, if the equations arise in a search for eigenvectors) whether any non-zero vector v provides a solution. Here again, we have a very definite, simple answer: if $k < n$, *a non-zero solution certainly exists*. This looks very reasonable, for when $k < n$, the number $n - k$ must be either 1 or larger than 1; accordingly the solutions fill a line, a plane, or subspace of higher dimension, so must include vectors other than 0. We can put the matter beyond doubt by appealing to the algebra.

The general solution is $v = t_1u_1 + \cdots + t_{n-k}u_{n-k}$. If we put $t_1 = 1$ and the other parameters 0, we get $v = u_1$ and, as we saw, u_1 is never 0. Indeed we can see this even more simply by looking at the standard form of the equations. When we had in the standard form 3 equations for 5 variables, we saw that x_4 and x_5 could be chosen at will and a solution obtained by solving for x_1, x_2 and x_3. So we can assign x_4 and x_5 any non-zero values we like, say $x_4 = 2$ and $x_5 = 3$, and obtain a solution. The last two components being 2 and 3, we clearly are not getting 0 as our vector.

Now suppose we are given m equations in n variables, but not in standard form. Without making calculations we cannot be sure exactly what value k has, but there is one thing of which we can be certain; k is not going to be larger than m. Our process of reducing to standard form may decrease the number of equations, but it never increases them. So, if at the start $m < n$, we can be sure $k < n$, and so, by our theorem above, there must be a non-zero solution. Accordingly we have – *if the number of equations is less than the number of variables, a non-zero solution is bound to exist.*

It is important to remember that, at the moment, we are considering only the case in which all equations have 0 on the right-hand side. We can use matrix notation to save repeating this phrase. Our equations can be put in the concise form $Mv = 0$. If we have m equations in n variables, M will be an $m \times n$ matrix (i.e. m rows, n columns). Obviously, we may often need to refer to k, the number of equations in the standard form; k is called the *rank* of the matrix M. Our theorems may then be stated thus:

Theorem 1 The solutions of $Mv = 0$ fill a subspace of dimension $n - k$. (If $n = k$, this number is 0. By a subspace of dimension 0 we understand a single point, the origin.)

Theorem 2 If $k < n$, a non-zero solution exists. (In fact, an infinity of non-zero solutions will exist.)

Theorem 3 If $m < n$, a non-zero solution exists.

In Theorem 1 we met the number $n - k$, and this may make us wonder; what happens if $k > n$? What do we mean by a subspace of negative dimension? This question need not occupy us, for $k > n$ can never happen. Suppose we apply the procedure for getting the standard form and at some stage of it we have written as many as n equations. These equations must be simply $x_1 = 0, x_2 = 0, \ldots, x_n = 0$. This must represent the end of the process. There may still be equations to deal with, but they cannot produce anything new. For suppose such an equation is

$$a_1 x_1 + a_2 x_2 + \cdots + a_n x_n = 0.$$

Subtracting from this equation a_1 times our first equation ($x_1 = 0$), a_2 times our second equation ($x_2 = 0$), and so on, up to a_n times our last equation ($x_n = 0$), nothing at all remains.

Note. The results above allow us to prove formally some statements that were presented as 'reasonable' in §7.

The first statement is that you cannot have $n + 1$ linearly independent vectors in space of n dimensions. A set of vectors is linearly dependent if one of the vectors can be expressed as a mixture of the others. Another way of expressing this, which is often useful, is to say that vectors $u_1, u_2, \ldots, u_p$ are linearly dependent if there is an equation $c_1 u_1 + c_2 u_2 + \cdots + c_p u_p = 0$ where the numbers $c_1, c_2, \ldots, c_p$ are *not all* 0. For suppose one of these numbers, say c_q, is not 0. Then we can solve for u_q by dividing by c_q, and get u_q expressed as a mixture of the other vectors.

Now we come to our statement. Suppose we are given any $n + 1$ vectors in n dimensions, $u_1, u_2, \ldots, u_{n+1}$. Each vector is a column vector with n components. Can we find numbers $c_1, c_2, \ldots, c_{n+1}$, not all zero, for which $c_1 u_1 + c_2 u_2 + \cdots + c_{n+1} u_{n+1} = 0$? If we write this equation out in full we shall have n equations, since each vector has n components. But the unknowns, $c_1, c_2, \ldots, c_{n+1}$ are $n + 1$ in number. There are more unknowns than equations; this, as we saw earlier, means that a non-zero solution exists. So the $n + 1$ vectors are bound to be linearly dependent.

It would be very strange if this statement were not true. For $n + 1$ independent vectors form a basis for a space of $n + 1$ dimensions; if we could find such a basis in space of n dimensions it would mean that a space of $n + 1$ dimensions could be found within a space of n dimensions, which, as mentioned in §7, we certainly do not expect to be possible.

The second statement we can now prove is that any n independent vectors in space of n dimensions form a basis for that space, that is to say, every vector in the space can be expressed as a mixture of them, and this in only one way. Suppose then we have n independent vectors, $u_1, u_2, \ldots, u_n$, in n dimensions, and let v be any vector in that space. We want to prove that v must be a mixture of $u_1, u_2, \ldots, u_n$. The proof is not long. First of all, $u_1, u_2, \ldots, u_n, v$ are $n + 1$ vectors in the space of n dimensions. By our first result, these vectors must be linearly dependent, that is, we must have an equation $c_1 u_1 + c_2 u_2 + \cdots + c_n u_n + c v = 0$, where the coefficients are not all 0. If $c \neq 0$, we can solve for v, and get v as a mixture of $u_1, \ldots, u_n$, and all will be well. So our result will be proved if we show that it is impossible for c to be 0. Suppose it were; then we would have $c_1 u_1 + c_2 u_2 + \cdots + c_n u_n = 0$, and this would mean that $u_1, \ldots, u_n$ were not linearly independent, in contradiction to the information given.

Questions

1 The equation $c_1 u_1 + c_2 u_2 + \cdots + c_n u_n = 0$ would not prove $u_1, \ldots, u_n$ linearly dependent, if $c_1 = c_2 = \cdots = c_n = 0$. Why can we rule out this situation?

2 How can we complete the proof by showing that v can only be expressed in one way as a mixture of $u_1, \ldots, u_n$?

If this second statement were not true, it would mean that we could have one space of n dimensions inside another space of n dimensions, and not filling it.

Exercises

1 In each system given below reduce the equations to standard form, determine the general solution, and describe in geometrical terms the subspace filled by the solutions (e.g. a plane in 7 dimensions; a subspace of 3 dimensions in 5-dimensional space.)

(*a*) $x - y = 0$, $y - z = 0$, $z - x = 0$.

(*b*) $2x - 3y = 0$, $5x - 3z = 0$, $5y - 2z = 0$.

(*c*) $2x + 7y - 4z = 0$, $16x - 9y - 6z = 0$, $7x + 2y - 5z = 0$.

(*d*) $x + y + z = 0$, $x + 2y + 3z = 0$, $x + 4y + 5z = 0$.

(*e*) $\begin{aligned} x + y + z + u + v &= 0, \\ 2x + 3y + z + 5u &= 0, \\ 3x + 7y - z + 15u - 5v &= 0, \\ 8x + 11y + 5z + 17u + 2v &= 0. \end{aligned}$

(*f*) $\begin{aligned} x + 7y - 13z &= 0, \\ 13x + 91y - 169z &= 0, \\ 17x + 119y - 221z &= 0. \end{aligned}$

(*g*) $\begin{aligned} x_1 + 2x_2 + 2x_3 + x_4 &= 0, \\ 2x_1 + 7x_2 + 11x_3 + 3x_4 &= 0, \\ x_1 + 3x_2 + 6x_3 + 8x_4 &= 0, \\ 4x_1 + 11x_2 + 19x_3 + 21x_4 &= 0. \end{aligned}$

(*h*) $\begin{aligned} 2x_1 + 3x_2 + 4x_3 + 8x_4 + 6x_5 + 7x_6 &= 0, \\ 4x_1 + 6x_2 + 3x_3 + 11x_4 + 7x_5 + 9x_6 &= 0, \\ 8x_1 + 12x_2 + 7x_3 + 23x_4 + 15x_5 + 19x_6 &= 0, \\ 2x_1 + 3x_2 + x_3 + 5x_4 + 3x_5 + 5x_6 &= 0. \end{aligned}$

2 (*a*) What equations would we have to solve to determine whether the vectors $(1, 1, 1)$, $(2, 3, 2)$, $(4, 9, 4)$ are linearly dependent or not? Are they in fact linearly dependent?

(*b*) Are $(1, 1, 1)$, $(2, 3, 5)$, $(4, 9, 25)$ linearly dependent or not?

(*c*) What are the conditions for $(1, 1, 1)$, (a, b, c), (a^2, b^2, c^2) to be linearly dependent? What are the conditions for these vectors to form a basis in 3 dimensions?

(*d*) What are the conditions for $(1, a, a^2)$, $(1, b, b^2)$, $(1, c, c^2)$ to form a basis in 3 dimensions?

(*e*) In how many ways can a basis for space of 3 dimensions be selected from the 4 vectors $(1, 2, 4)$, $(1, 3, 9)$, $(1, 4, 16)$, $(1, 5, 25)$?

3 Let $u_1 = (1, 1, 1)$, $u_2 = (2, 3, 5)$, $u_3 = (4, 9, 25)$, $u_4 = (8, 27, 125)$. Any 4 vectors in 3 dimensions must be linearly dependent. Find an equation of the form $c_1 u_1 + c_2 u_2 + c_3 u_3 + c_4 u_4 = 0$, where c_1, c_2, c_3 and c_4 are constants, that expresses the linear dependence of u_1, u_2, u_3 and u_4.

4 In questions 2(*a*) and 2(*b*) we investigated the possible linear dependence of three vectors. It would also be possible to settle this question by finding the value of the determinant of the three vectors. Compare the work involved in answering this question (*a*) by the method of the present section, (*b*) by multiplying the determinant right out, (*c*) by using the properties of determinants. Is there any resemblance between any of these methods?

Equations with non-zero numbers on right-hand side

The equations $Mv = 0$ considered so far show two possibilities; there may be just the one solution, $v = 0$, or there may be an infinity of solutions. We are now going to consider equations of the form $Mv = h$, where h is some given vector. Here again there may be just one solution, or there may be an infinity of solutions; there is also a third possibility – there may be no solution at all.

It is obvious that this possibility exists. We have such simple examples as the pair of equations $x + y = 1$, $x + y = 2$, where the second statement contradicts the first. Clearly no solution can exist. Sometimes we need to be aware of this possibility only in order to avoid it; for instance in designing a structure we may want it to carry a large weight while itself weighing as little as possible. There is clearly a potential conflict between these aims; if we ask for too much we may be demanding the impossible – that is; producing equations with no solution. But sometimes we have a positive interest in there being no solution. In our example above, $x + y = 1$ and $x + y = 2$ are the equations of parallel lines. If in the course of some work we needed the equation of the line through $(2, 0)$ parallel to $x + y = 1$, we might check our answer, $x + y = 2$, by trying to solve the pair of equations. We now hope there will be no solution. If there were one, it would mean we had made a mistake; the lines would then have an intersection and not be parallel.

Fortunately, there is very little to add to our earlier remarks. If we apply to equations of the form $Mv = h$ the procedure already described for bringing the *left-hand side* to the standard form, this will automatically tell us whether we have 0, 1 or ∞ solutions.

The very first example in this section illustrates the case where there is just one solution. In equations (1) and (2), we had $x + 3y = 25$, $2x + 7y = 55$. Some simple operations brought the left-hand side to the standard form. By reading through the calculations there made, it will be seen that at no stage was reference made to the numbers 25 and 55 on the right of the equals signs. The successive multiplications and subtractions were determined purely by the coefficients of x and y. This is always so. The sequence of calculations made to bring $Mv = h$ to standard form is exactly what would be done to bring $Mv = 0$ to standard form. If, as in this first example, $Mv = h$ is changed by these operations to an equivalent system of equations of the form $Iv = g$, then there is clearly one solution only, namely $v = g$.

We can obtain an illustration of equations with an infinity of solutions by amending equations (27), (28), (29) in worked example 2, considered earlier in this section. We keep the coefficients of x_1, x_2 and x_3 the same, but replace the zeros on the right-hand side. Accordingly, let us consider the equations

$$x_1 + 2x_2 + 3x_3 = 13, \tag{27a}$$

$$3x_1 + 6x_2 + 7x_3 = 37, \tag{28a}$$

$$2x_1 + 4x_2 + 10x_3 = 30. \tag{29a}$$

If we now carry out the same series of operations as those used in Exercise 2, the same simplification of the left-hand sides will result, but certain numbers will automatically appear on the right. It is left to you to check that we eventually get the equations

$$x_1 + 2x_2 \qquad = 10, \tag{33a}$$

$$x_3 = 1. \tag{34a}$$

Here x_1 and x_3 are the 'special' variables, and we may put $x_2 = t$, an arbitrary parameter. We thus find $x_1 = 10 - 2t$, $x_2 = t$, $x_3 = 1$ as the general solution. These solutions fill a line; it is the line that would be described by a point starting at $(10, 0, 1)$ and travelling with velocity $(-2, 1, 0)$. In vector form it could be represented as $v = u_0 + tu_1$.

This line is not a subspace, for it does not go through the origin. It represents rather a displacement of a subspace; the points tu_1 represent a subspace, a line through the origin. If we displace every point of this subspace by the same vector u_0, we obtain the line filled by the solutions.

In other problems we will also find that the solutions fill a region obtained by displacing a subspace.

Equations with no solutions Now consider the effect on equations $(27a)$, $(28a)$, $(29a)$ of making a slight change in one of them. Suppose that in the last equation we replace 30 by 31, and so consider the equations

$$x_1 + 2x_2 + \ 3x_3 = 13, \tag{27b}$$

$$3x_1 + 6x_2 + \ 7x_3 = 37, \tag{28b}$$

$$2x_1 + 4x_2 + 10x_3 = 31. \tag{29b}$$

As before, let us carry out the sequence of operations prescribed in Worked example 2. You will find that we then reach the equations

$$x_1 + 2x_2 + 3x_3 = 13, \tag{30b}$$

$$x_3 = 1, \tag{31b}$$

$$4x_3 = 5. \tag{32b}$$

Our next instruction is to get rid of x_3 in the last equation by subtracting 4 times $(31b)$. And here something essentially new happens; we get $0 = 1$, which is impossible. There is no solution.

The standard form gives a system of equations equivalent to the original system. If the original system makes impossible demands, the standard form must do the same. The only way in which the standard form can require an impossibility is by containing one or more equations of the type $0x_1 + 0x_2 + 0x_3 = a$, where the number $a \neq 0$. Thus the fact that there are no solutions will come to light in the course of the usual procedure.

Topic for discussion

The equations $(27a)$, $(28a)$, $(29a)$ have an infinity of solutions. Each equation represents a plane. What figure do these planes form?

By changing the constant in the last equation, we obtain $(27b)$, $(28b)$, $(29b)$, a system with no solution. What figure do these planes form?

Questions

1 Reduce each of the following systems of equations to standard form. Give the general solution or state that no solution exists.

(a) $\quad 2x - y - z = 1$,
$\quad -x + 2y - z = 2$,
$\quad -x - y + 2z = 3$.

(b) $\quad 2x + 5y + 3z = 0$,
$\quad 4x + 11y + z = 6$,
$\quad 10x + 26y + 10z = 6$.

(c) $\quad x + 2y + z + 2u + v = 7$,
$\quad 3x + y + 4z + 3u + 5v = 16$,
$\quad 15x + 10y + 19z + 18u + 23v = 80$.

(d) $\quad x + 2y - 3z = 1$,
$\quad 2x + 3y + z = 8$,
$\quad 5x + 9y - 8z = 11$,
$\quad 4x + 5y + 9z = 22$.

2 Three planes are given by the equations

$$2x + 3y - 5z = 23, \quad x - 2y + z = 1, \quad 10x - 6y - 4z = 10.$$

Do these planes have any point in common? Each pair of planes intersects in a line; find a parametric representation for each such line. What can be said about the three lines in question?

3 Make up equations for three planes that are related to each other like the planes in Question 2, but involve different numbers.

4 In some experimental work it is thought probable that a parabola $y = a + bx + cx^2$ can be found which passes through the points $(1, 10)$, $(2, 23)$, $(4, 79)$, $(-1, 14)$, $(-3, 58)$. Can this in fact be done?

5 (a) What solutions (if any) have the equations $3z - 5y = 8$, $5x - 2z = 3$, $2y - 3x = -5$?
(b) What solutions (if any) have the equations $3z - 5y = 4$, $5x - 2z = 1$, $2y - 3x = 3$?
(c) What conditions must a, b, c satisfy if the equations $3z - 5y = a$, $5x - 2z = b$, $2y - 3x = c$ are to have a solution or solutions? Is it possible for these equations to have exactly one solution?
(d) Interpret (c) in terms of vector products and your answer in terms of scalar products. Show that your answer to (c) could have been predicted by purely geometrical reasoning.

6 Consider a rigid piece of material subjected to forces which lie in a plane. The material is in equilibrium (that is, it does not translate or rotate) when the resultant force acting on it is zero. The condition that the resultant force is zero can be represented by three scalar equations. For example the equation $M_A = 0$, stating that the moment of the forces about A is zero, indicates that the resultant force must pass through A. Adding the condition $R_H = 0$, (the horizontal component of the resultant is zero) would only permit a vertical force through A. Finally the equation $R_V = 0$, stating that the vertical component of the resultant is zero, would ensure no resultant at all and thus that the body was in equilibrium under the given force system. Some sets of three equations ensure equilibrium and others do not. The following example will illustrate dependent and independent sets of equations.

By isolating the frame shown in Fig. 148 as a free body the six following equations are readily obtained.

$$M_A = 0 \qquad\qquad 3X_2 + 8X_3 - 72 = 0, \qquad\qquad (1)$$
$$M_B = \qquad 4X_1 \qquad\quad + 4X_3 - 36 = 0, \qquad\qquad (2)$$
$$M_C = \qquad 8X_1 - 3X_2 \qquad\qquad = 0, \qquad\qquad (3)$$
$$M_D = 0 \quad 8X_1 \qquad\qquad\quad - 36 = 0, \qquad\qquad (4)$$
$$R_H = 0 \qquad\qquad X_2 \qquad - 12 = 0, \qquad\qquad (5)$$
$$R_V = 0 \quad X_1 \qquad - X_3 \qquad\quad = 0. \qquad\qquad (6)$$

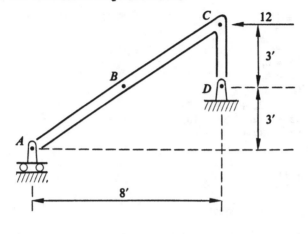

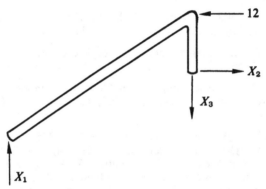

Fig. 148

Determine by the row–echelon method, and also by physical considerations, which of the following sets of equations are independent.

(*a*) equations (1), (2) and (4).   (*b*) equations (1), (2) and (3)
(*c*) equations (1), (5) and (6)   (*d*) equations (3), (4) and (5)
(*e*) equations (1), (3) and (5)   (*f*) equations (2), (4) and (5)
(*g*) equations (2), (4) and (6)

41   Null space, column space, row space of a matrix

In §40 we were concerned, almost exclusively, with certain routines for calculation. In this section we try to show pictorially certain aspects of the topics considered in §40. This also provides an opportunity for explaining the meanings of certain technical terms that may be met in the literature on matrices.

In §40, Exercise 2, with minor variations, was discussed on three separate occasions. We will therefore take the matrix occurring in this exercise as our theme. It may appear a somewhat special case, since it is a 3 × 3 matrix, and in general we are concerned with $m \times n$ matrices. A 3 × 3 matrix has the advantage that it is easily visualized in terms of physical space. Most of the features we shall observe can be generalized to the case of an $m \times n$ matrix.

Accordingly we consider the matrix M and vector v, where, as in Exercise 2,

$$M = \begin{pmatrix} 1 & 2 & 3 \\ 3 & 6 & 7 \\ 2 & 4 & 10 \end{pmatrix}, \qquad v = \begin{pmatrix} x_1 \\ x_2 \\ x_3 \end{pmatrix}.$$

We saw in Exercise 2, in its original form, that the equations $Mv = 0$ were equivalent to the pair of equations $x_1 + 2x_2 = 0$, $x_3 = 0$. Thus $x_1 = -2, x_2 = 1, x_3 = 0$ is a solution, and the subspace filled by the solutions consists of the line through the origin and the point $(-2, 1, 0)$. Any point of this line is mapped to the origin under the mapping M.

Now when a line is sent to the origin in this way, the mapping always gives a space of lower dimension. This may be proved as follows. Let u_1 be a vector for which $Mu_1 = 0$. Choose any two vectors u_2, u_3 so that u_1, u_2, u_3 is a basis for the space. Then every point can be represented as $a_1u_1 + a_2u_2 + a_3u_3$ for some numbers, a_1, a_2, a_3. M sends this point to $a_1u_1^* + a_2u_2^* + a_3u_3^*$. However, $u_1^* = 0$, so every point in the output is of the form $a_2u_2^* + a_3u_3^*$; that is, it lies in the plane containing the origin, u_2^* and u_3^*. (This proof is for 3 dimensions, but it generalizes, in an obvious way, to n dimensions.)

This consideration applies to the particular mapping M, that we are using as an example. Let $Mv = w$, where w has components y_1, y_2, y_3. The equation $Mv = w$, written in full, is the system of equations

$$x_1 + 2x_2 + 3x_3 = y_1, \tag{27c}$$

$$3x_1 + 6x_2 + 7x_3 = y_2, \tag{28c}$$

$$2x_1 + 4x_2 + 10x_3 = y_3. \tag{29c}$$

Note that these equations cover the three situations considered in §40 in relation to Worked example 2. In each situation the output w was specified, and we tried to find what inputs would give it. In the original example, we required an output with $y_1 = 0$, $y_2 = 0$, $y_3 = 0$. Then we considered the output $y_1 = 13$, $y_2 = 37$, $y_3 = 30$, and finally (the system with no solution) $y_1 = 13$, $y_2 = 37$, $y_3 = 31$.

We have seen from the argument above that the output, Mv, must lie in a certain plane, whatever the input v. This suggests the reason why we cannot find any solution in the third problem; it seems that $y_1 = 13$, $y_2 = 37$, $y_3 = 30$ specifies a point not lying in the plane containing all possible outputs; that is why we cannot find any input v to produce it.

The procedure for finding the standard form of the equations gives us a way of finding the actual equation of the plane in question. If we apply the steps given in Worked example 2 of §40 to the general equations (27c), (28c), (29c) we find,

$$x_1 + 2x_2 + 3x_3 = y_1, \tag{30c}$$

$$x_3 = -(y_2 - 3y_1)/2, \tag{31c}$$

$$4x_3 = y_3 - 2y_1. \tag{32c}$$

Now, forming (32c) − 4(31c) as instructed, we get

$$0 = (y_3 - 2y_1) + 2(y_2 - 3y_1) = y_3 + 2y_2 - 8y_1.$$

So every possible output satisfies the equation $y_3 + 2y_2 - 8y_1 = 0$. We can check this by going back to our original equations, (27c), (28c) and (29c). If we calculate $y_3 + 2y_2 - 8y_1$ we get $0x_1 + 0x_2 + 0x_3$, so whatever the input x_1, x_2, x_3 may be, the

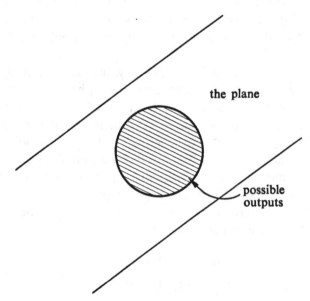

the plane

possible
outputs

Fig. 149   Conceivable situation.

output must satisfy $y_3 + 2y_2 - 8y_1 = 0$. It can be checked that $y_1 = 13$, $y_2 = 37$, $y_3 = 30$ does satisfy this equation, and so seems to be a possible output, while $y_1 = 13$, $y_2 = 37$, $y_3 = 31$ does not satisfy it and so is certainly not a possible output.

The rather cautious phrase 'seems to be a possible output' was used because a question of logic is involved. We have proved that every output lies in the plane

$$y_3 + 2y_2 - 8y_1 = 0,$$

but by itself this does not prove that every point of that plane is a possible output. For instance, it is imaginable that the possible outputs could all lie within a certain circle in that plane, as shown in Fig. 149. Then all the outputs would lie in the plane, but it would be quite false to assume that every point in the plane was automatically a possible output, for such a point might happen to lie outside the shaded circular region. *Now in fact this is not how things are*; actually, every point of the plane is an output for some input, but we must prove that this is so.

If we take equations (30c), (31c), (32c) and carry out the remaining instructions given in Exercise 2, we find the standard form to be the following:

$$x_1 + 2x_2 \qquad = -\tfrac{7}{2}y_1 + \tfrac{3}{2}y_2, \tag{33c}$$

$$x_3 = \tfrac{3}{2}y_1 - \tfrac{1}{2}y_2, \tag{34c}$$

$$0 = -8y_1 + 2y_2 + y_3. \tag{35c}$$

If the output point lies in the plane in question, equation (35c) is satisfied. In equations (33c) and (34c), whatever values y_1 and y_2 may have, we can find x_1 and x_3 to make the equations hold, and indeed we can do this for an arbitrary value of x_2. Thus if we put $x_2 = t$, we can find not merely *a* solution but an infinity of solutions depending on the parameter t. This is in agreement with our experience in §40, when we found that the output $y_1 = 13$, $y_2 = 37$, $y_3 = 30$ gave the solution $x_1 = 10 - 2t$, $x_2 = t$, $x_3 = 1$. It was pointed out in §40 that this solution was of the form $v = u_0 + tu_1$.

We can see why, if there is one solution of $Mv = w$ there must be an infinity of solutions. At the beginning of this section we noted that the input $x_1 = -2$, $x_2 = 1$, $x_3 = 0$ gave zero output. Call this input u_1. Then $Mu_1 = 0$. Now suppose $Mv = w$ has a solution u_0; this means that input u_0 gives output w; that is $Mu_0 = w$. Now consider what output comes from the input $u_0 + tu_1$. We have

$$M(u_0 + tu_1) = Mu_0 + t \cdot Mu_1.$$

This equals Mu_0 since $Mu_1 = 0$; also $Mu_0 = w$. So $M(u_0 + tu_1) = w$; the input $u_0 + tu_1$ gives the output w, whatever number may be chosen for t. The points $u_0 + tu_1$ fill a line, parallel to the vector u_1. Thus we may think of the input space as being full of lines, all pointing in the same direction; if two inputs lie on the same line, their outputs will be identical.

In the output space we have a plane; it is impossible to produce an output that does not lie in this plane, but in compensation for this, every output that does lie in the plane can be produced by infinitely many different inputs.

Thus, when we are trying to visualize the effect of this particular mapping M, two geometrical objects come naturally to our attention. In the input space there is a line through the origin, consisting of all the points tu_1. Each of these points maps to the origin in the output space; if $v = tu_1$, then $Mv = 0$. In the output space there is a plane; whatever the input may be, the output lies in this plane.

For *any* matrix M corresponding objects exist, and have special names. All the vectors v for which $Mv = 0$ constitute the *Null Space of M*. By Theorem 1 of §40, if k is the rank of the matrix M (that is, the number of equations in the standard form of the equation system $Mv = 0$) then the null space of M is a subspace of dimension $n - k$, the matrix M being of type $m \times n$. The null space of M lies in the input space, which is of n dimensions. If $k = 0$, which happens only for $M = 0$, the null space is the entire input space. If $k = n$, the null space contains only the origin, O.

In the output space, corresponding to the plane $y_3 + 2y_2 - 8y_1 = 0$ in our particular example, there is an object formed by all possible outputs, that is, all the points Mv that can be obtained from some v in the input space. This is known as the *Column Space of M*; we can at the same time show why this name is used and demonstrate that it is in fact a subspace of the output space. The essential ideas can be seen by considering our particular example, M. This matrix expresses in the form $Mv = w$, the equations (27c), (28c), (29c) considered near the beginning of this section. Now these equations can be written like this:

$$x_1 \begin{pmatrix} 1 \\ 3 \\ 2 \end{pmatrix} + x_2 \begin{pmatrix} 2 \\ 6 \\ 4 \end{pmatrix} + x_3 \begin{pmatrix} 3 \\ 7 \\ 10 \end{pmatrix} = \begin{pmatrix} y_1 \\ y_2 \\ y_3 \end{pmatrix}.$$

If we write C_1, C_2, C_3 for the 3 columns of the matrix M, this equation can be written $x_1C_1 + x_2C_2 + x_3C_3 = w$. Thus every possible output w, is a mixture of the three vectors given by the columns of M. That is why the name 'column space' is given to the geometrical object consisting of all possible outputs. We see too that this object is a subspace; taking arbitrary mixtures of any collection of vectors is bound to produce a subspace. (The formal proof would verify that, if w is in such a collection, so is tw for any number t, and that, if w_1 and w_2 are in it, so is their sum $w_1 + w_2$. Thus the collection is a vector space, hence a subspace.)

Notice, however, that C_1, C_2, C_3 is *not* a basis for the column space. If it were, the column space would be of 3 dimensions. However, C_1, C_2, C_3 are not independent; in fact $C_2 = 2C_1$, so that a mixture of C_1, C_2, C_3 is simply a mixture of C_1 and C_3; C_1 and C_3 are in fact a basis of the column space.

The column space is said to be *spanned* by C_1, C_2, C_3; this means that every vector in the column space is a mixture of C_1, C_2, C_3. When we say that a space is spanned by a certain collection of vectors we imply that the space contains every mixture of these vectors, and nothing else. As will be clear from the example of C_1, C_2 and C_3, we are *not* implying that the collection of vectors is a basis of the space.

Returning to the question of the general $m \times n$ matrix M, we know that it acts on an

n-dimensional space and maps a subspace, the null space, of $n - k$ dimensions, to zero. Seeing M wipes out $n - k$ dimensions, we rather expect its outputs to form a space of k dimensions, since $n - (n - k) = k$. We can prove that this does indeed happen. The proof depends on selecting a convenient basis for the input space. We need n independent vectors, $v_1, \ldots, v_n$, for a basis. We know the null space has $n - k$ dimensions, so we begin by selecting $n - k$ vectors that form a basis for the null space. We call these $v_{k+1}, \ldots, v_n$. We then choose k more vectors, not lying in the null space, and call these $v_1, \ldots, v_k$; these are to be chosen in such a way that the whole collection, $v_1, \ldots, v_n$, forms a basis of the whole space. (In a fully formal treatment we would have to prove that such a selection is always possible.) Any point of the input space can now be represented as $x_1 v_1 + \cdots + x_n v_n$; M will send this point to

$$x_1 v_1^* + \cdots + x_n v_n^*.$$

Now $v_{k+1}, \ldots, v_n$ all lie in the null space, so $v_{k+1}^*, \ldots, v_n^*$ are all 0. Thus the output point is simply $x_1 v_1^* + \cdots + x_k v_k^*$. We seem to have proved what we want – that the outputs fill a space of k dimensions. There is still, however, a possible objection to meet; if $v_1^*, \ldots, v_k^*$ are linearly dependent, they will span the output space, but some of them will be superfluous, the space will be of less than k dimensions. So we have to investigate whether $v_1^*, \ldots, v_k^*$ could be linearly dependent. Suppose they were. Then these would be numbers, not all 0, such that $c_1 v_1^* + \cdots + c_k v_k^* = 0$. Now the left-hand side of this equation is the output corresponding to $c_1 v_1 + \cdots + c_k v_k$, and according to the equation this output is 0. That means the input must lie in the null space, and so be a mixture of $v_{k+1}, \ldots, v_n$. But this means that $v_1, \ldots, v_n$ cannot be a basis, for the point in question could then be represented in two quite distinct ways, one as

$$c_1 v_1 + \cdots + c_k v_k$$

and one as a mixture of $v_{k+1}, \ldots, v_n$. So supposing $v_1^*, \ldots, v_k^*$ linearly dependent leads to a contradiction. Accordingly the objection has been overcome, and we know that the column space, the space formed by the outputs, is of k dimensions exactly.

Column space and row space

The number k, the rank of M, arose first of all as the number of equations in the standard form of the system $Mv = 0$. Each equation corresponds to a *row* in M; the standard form is obtained by calculating suitable mixtures of the rows.

We have just found that k is the dimension of the column space, consisting of all possible mixtures of the columns. There is evidently some connection between the rows of a matrix and the columns, since the same number k occurs in relation to both.

The column space was obtained by partitioning the matrix into column vectors, and considering the space spanned by these. In the same way, we can define the row space; we partition the matrix into row vectors, and consider the space spanned by them; this is called the row space.

Now k is in fact the dimension of the row space. An example will show why this is so. In §40, under the heading 'Solutions when all right-hand sides are 0' we considered the equations of the standard form

$$x_1 \qquad\qquad - ax_4 - dx_5 = 0,$$
$$\qquad x_2 \qquad - bx_4 - ex_5 = 0,$$
$$\qquad\qquad x_3 - cx_4 - fx_5 = 0.$$

If we form the matrix of the coefficients and partition it into rows, we get the 3 row vectors r_1, r_2, r_3 where

$$r_1 = (1, 0, 0, -a, -d),$$
$$r_2 = (0, 1, 0, -b, -e),$$
$$r_3 = (0, 0, 1, -c, -f).$$

Now the standard form is obtained from the original system of equations, $Mv = 0$, by means of mixtures of those equations, and each equation corresponds to a row in M; that is to say, r_1, r_2, r_3 are mixtures of the rows of the original matrix M. This means that r_1, r_2, r_3 lie in the row space of M. Now r_1, r_2, r_3 are clearly independent. For any mixture of r_2 and r_3 must have 0 in the first place; as r_1 begins with 1, it is impossible that r_1 should be a mixture of r_2 and r_3. Similarly, by looking at the numbers in the second place we see that r_2 cannot be obtained by mixing r_1 and r_3, while from the third place we conclude that r_3 is not a mixture of r_1 and r_2. So r_1, r_2, r_3 are independent vectors. Thus the row space of the original matrix M contains three independent vectors, so it must be of 3 dimensions or more.

But we were careful to make the process of forming the standard equations reversible. We can get back to the rows of M by taking suitable mixtures of r_1, r_2, r_3. Any mixture of the rows of M must therefore be a mixture of r_1, r_2, r_3. This means that every vector in the row space is a mixture of the 3 vectors r_1, r_2, r_3: the dimension of the row space is therefore 3 or less.

The two arguments together show that the dimension of the row space is exactly 3.

Nothing in the arguments used depends on any special property of the numbers 3 and 5. If we have any $m \times n$ matrix M, with k equations in the standard system equivalent to $Mv = 0$, we can reason in exactly the same way, and conclude that the row space will be of dimension k.

Thus we have a theorem for any matrix; the column space and the row space have the same number of dimensions.

In some books the term 'column rank' is introduced for the number of dimensions of the column space, and 'row rank' for the number of dimensions of the row space. It is then proved that these two numbers are equal, and their common value is called simply 'the rank of the matrix'.

We have already seen how to find the dimension of the row space. It is simply k, the number of equations in the standard form obtained from the equation system $Mv = 0$. When we are speaking of the 'row space' we tend to think of the rows as containing the components of vectors, rather than as giving the coefficients in equations. This therefore seems an appropriate place to mention that the row–echelon procedure, used to bring an equation system to its standard form, can equally well be used to determine the dimension of the space spanned by a set of vectors and to give a basis of that space.

Suppose, for example, we wish to study the space spanned by the three row vectors

$$r_1 = (1, 1, 1, 1, 1),$$
$$r_2 = (1, 2, 3, 4, 5),$$
$$r_3 = (3, 5, 7, 9, 11).$$

We write

$$r_1 = (1, 1, 1, 1, 1),$$
$$r_2 - r_1 = (0, 1, 2, 3, 4) = u_1, \quad \text{say,}$$
$$r_3 - 3r_1 = (0, 2, 4, 6, 8) = u_2, \quad \text{say.}$$

We have here treated the first components of the vectors in the same way that we treated the coefficients of x_1 in the equation systems. We would next replace u_2 by $u_2 - 2u_1$ in order to get a vector beginning $0, 0, \ldots$ In this example, we get a vector that is zero throughout. Thus we reach a standard form in which r_1 and u_1 alone appear. These two vectors form a basis for the space spanned by r_1, r_2 and r_3, so the space must be of 2 dimensions.

It would be possible to give as a basis the vectors $r_1 - u_1$ and u_1, that is to say

$$(1, 0, -1, -2, -3) = v_1, \quad \text{say,}$$
$$(0, 1, 2, 3, 4) = v_2, \quad \text{say.}$$

This would correspond most closely to what we did with the equations. It has the advantage that we can tell immediately what mixture of v_1 and v_2 gives any vector in the space simply by looking at the first two components. For instance r_3 begins with 3, 5; it can be no other than $3v_1 + 5v_2$. However, enough useful information can often be obtained without going to this stage.

A similar procedure is of course possible with column vectors.

Worked example

Find a basis for the column space of the matrix whose rows are r_1, r_2, r_3 as given just above.

Solution We have the vectors

$$\begin{pmatrix} 1 \\ 1 \\ 3 \end{pmatrix}, \begin{pmatrix} 1 \\ 2 \\ 5 \end{pmatrix}, \begin{pmatrix} 1 \\ 3 \\ 7 \end{pmatrix}, \begin{pmatrix} 1 \\ 4 \\ 9 \end{pmatrix}, \begin{pmatrix} 1 \\ 5 \\ 11 \end{pmatrix}.$$

Subtract the first vector from each of the others. This gives

$$\begin{pmatrix} 1 \\ 1 \\ 3 \end{pmatrix}, \begin{pmatrix} 0 \\ 1 \\ 2 \end{pmatrix}, \begin{pmatrix} 0 \\ 2 \\ 4 \end{pmatrix}, \begin{pmatrix} 0 \\ 3 \\ 6 \end{pmatrix}, \begin{pmatrix} 0 \\ 4 \\ 8 \end{pmatrix}.$$

It is now noticeable that the last 3 vectors are all multiples of the second. Thus the first 2 vectors provide a basis for the column space.

Questions

1 Let

$$M = \begin{pmatrix} 1 & 1 & 3 & 3 & 5 \\ 1 & 0 & 2 & 1 & 3 \\ 0 & 1 & 1 & 2 & 2 \end{pmatrix}.$$

Of how many dimensions is the row space of M? What number gives the rank of M? Of how many dimensions do you expect the column space of M to be? Find a basis for the column space. Find the condition or conditions for

$$\begin{pmatrix} x \\ y \\ z \end{pmatrix}$$

to be in the column space.
Let

$$u_1 = \begin{pmatrix} 1 \\ 1 \\ 1 \end{pmatrix}, \quad u_2 = \begin{pmatrix} 4 \\ 3 \\ 2 \end{pmatrix}, \quad u_3 = \begin{pmatrix} 7 \\ 5 \\ 2 \end{pmatrix}, \quad u_4 = \begin{pmatrix} 0 \\ 1 \\ -1 \end{pmatrix}.$$

For which of these does the equation $Mv = u_s$ have a solution v? ($s = 1, 2, 3$ or 4.)
   Of how many dimensions would you expect the null space of M to be? (Give your reason.) Find the equations that specify the null space, and find a basis for the null space.
   Discuss the solution set of the equation $Mv = u_s$ for each of the vectors u_s given above.

2 For

$$M = \begin{pmatrix} 1 & 1 & 2 \\ 2 & 3 & 3 \\ 1 & 2 & 1 \\ 1 & 0 & 3 \end{pmatrix}$$

find the dimension and a basis for (a) the row space, (b) the column space, (c) the null space.
   Find the condition or conditions a vector u must satisfy if $Mv = u$ is to have a solution or solutions, v.
   Show that $Mv = u$ has solutions if

$$u = \begin{pmatrix} 3 \\ 8 \\ 5 \\ 1 \end{pmatrix}.$$

Find the general solution for v in this case.
   If v_0 is a particular column vector in 3 dimensions and $u_0 = Mv_0$, what is the general solution of $Mv = u_0$?

3   Find a basis for (*a*) the row space and (*b*) the null space of the matrix

$$M = \begin{pmatrix} 1 & 1 & 1 & 0 & 0 \\ 1 & 0 & 1 & 1 & 1 \\ 2 & 1 & 2 & 1 & 1 \end{pmatrix}.$$

For each row vector, R, included in your answer to (*a*) and for each column vector, N, in your answer to (*b*) discuss the product RN. What is its value? Is it even meaningful? Is any general conclusion suggested?

42   Illustrating the Importance of orthogonal matrices

A trap to avoid

In §38, we saw that, for a symmetric matrix M, the transformation $v \rightarrow Mv$ could be pictured with the help of the curves $v'Mv =$ constant. Here the matrix M plays a double role; it specifies a transformation, and it specifies a system of curves. This double role, if not properly understood, can lead us into errors. The present section has the aim of showing how such errors arise and how to avoid them.

As an illustration we will use the matrix $M = \begin{pmatrix} 1 & -1 \\ -1 & 2 \end{pmatrix}$; the transformation associated with this has $x^* = x - y$, $y^* = -x + 2y$; the curves associated with it are $x^2 - 2xy + 2y^2 =$ constant. In our illustrations we shall use the procedure of §38; that is, the vector Mv will be shown sprouting out of the end of the vector v. In Fig. 150(*a*) we see a number of points lying on the ellipse $x^2 - 2xy + 2y^2 = 25$. Attached to each point, v, we see the vector Mv, and in each case, as we expect from §38, the vector Mv is normal to the ellipse. The table here shows (x, y), the co-ordinates of v, and (x^*, y^*), the components of Mv.

(x, y)	(5, 0)	(7, 3)	(7, 4)	(5, 5)	(1, 4)	(−1, 3)
(x^*, y^*)	(5, −5)	(4, −1)	(3, 1)	(0, 5)	(−3, 7)	(−4, 7)

Some person, other than ourselves, is also interested in this transformation, but for his purposes the important quantities are not x, y but X, Y where $X = x - y$ and $Y = y$. Accordingly he takes the table above and uses it to construct a table, showing what happens to his co-ordinates X, Y. Simply by substitution he arrives at the following specification of the transformation $(X, Y) \rightarrow (X^*, Y^*)$.

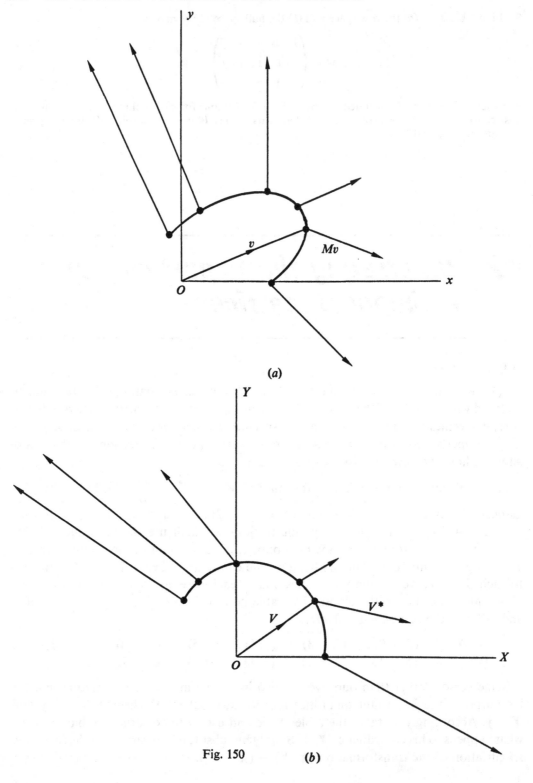

(a)

(b)

Fig. 150

(X, Y)	$(5, 0)$	$(4, 3)$	$(3, 4)$	$(0, 5)$	$(-3, 4)$	$(-4, 3)$
(X^*, Y^*)	$(10, -5)$	$(5, -1)$	$(2, 1)$	$(-5, 5)$	$(-10, 7)$	$(-11, 7)$

It may be checked that this table fits the formula $X^* = 2X - Y$, $Y^* = -X + Y$. These equations specify the transformation as it appears in his co-ordinates: these equations can of course be found directly by algebra: how to do this will be considered later in this section. For the moment it is not necessary to enquire how these equations were obtained, but only to observe that they do fit the data.

Fig. 150(b) shows the six points as they might be drawn by the person using (X, Y) co-ordinates. Attached to each vector V, with co-ordinates (X, Y), is a vector V^* with components (X^*, Y^*). It will be seen that in diagram (b) the six points lie on a circle. And this may be checked by calculation. The equations $X = x - y$, $Y = y$ imply $x = X + Y$, $y = Y$. If we substitute these values in the equation $x^2 - 2xy + 2y^2 = 25$ we obtain $(X + Y)^2 - 2(X + Y)Y + 2Y^2 = 25$ which simplifies to $X^2 + Y^2 = 25$.

Now the quadratric form $X^2 + Y^2$ corresponds to the matrix I, while the transformation, expressed in terms of X and Y, corresponds to the matrix $\begin{pmatrix} 2 & -1 \\ -1 & 1 \end{pmatrix}$.

In the original (x, y) system the transformation $v \to Mv$ and the curves $v'Mv =$ constant were specified by the same matrix M. When we go over to the (X, Y) system, we find the curves are specified by the identity matrix I while the transformation is specified by a matrix which most certainly is not the identity.

One could easily imagine someone believing that, in any system of co-ordinates, the same matrix would specify both the transformation and the curves. Such a person might calculate, quite correctly, the matrix I for the curves, and expect to find that the transformation $V \to V^*$ was also specified by I. In this particular instance, the mistake is so glaring that it would immediately be recognized, for clearly the transformation is not the identity transformation. But with some other change of co-ordinate system, the curves might be specified by some matrix that had no obvious interpretation. In such a situation the error might well escape detection, and the person might continue his calculations in the belief that the matrix he had calculated for the curves would also specify the transformation in the (X, Y) system.

The questions thus arise; if in the (x, y) system, we have the transformation $v \to Mv$ and the curves $v'Mv =$ constant, what matrices will specify the transformation and the curves in some other co-ordinate system? Why are we liable to get two matrices, one for the transformation and the other for the curves? Does this always happen, or are there special co-ordinate systems for which the two matrices remain equal?

We already have part of the answer. Exercise 7, at the end of §33, raised the question of what happens to a quadratic form when axes are changed. This question is answered as follows. We will suppose that new co-ordinates X, Y, are related to the old co-ordinates x, y, by equations of the form $x = pX + qY$, $y = rX + sY$. We can express these concisely as $v = TV$ where

$$v = \begin{pmatrix} x \\ y \end{pmatrix}, \qquad V = \begin{pmatrix} X \\ Y \end{pmatrix}, \qquad T = \begin{pmatrix} p & q \\ r & s \end{pmatrix}.$$

Note that T here establishes an alias, not an alibi. The numbers, x and y, that occur in v, are used in one system, to specify some object. The numbers, X and Y, that occur in V are used in another system to specify *the same object*. The matrix T enables us to act as interpreter between the systems; if X and Y occur in the label in the new system, by applying the matrix T we can find the label that will be understood by users of the old system.

To find how the quadratic form $v'Mv$ is represented in the new system, we need only substitute $v = TV$. Then, by the 'reversal rule for transposes', $v' = V'T'$. Accordingly, $v'Mv = V'T'MTV$. The matrix at the heart of this is $T'MT$; this is the matrix that will specify the curves when the X, Y co-ordinate system is used.

Exercise

At the beginning of this section we used the example $v'Mv = x^2 - 2xy + 2y^2$ and the equations $x = X + Y, y = Y$ for change of axes. Write the matrices M and T and verify that $T'MT$ does give the identity matrix I.

Change of axes for transformations

The principle involved, when a transformation $v \to Mv$ has to be expressed in terms of new axes, is a very general one that occurs in many branches of mathematics. It can be explained by a very simple illustration without any references to axes, co-ordinates or numerical calculations.

We will suppose we have a number of objects, labelled by Greek letters, and a mapping M involving these objects. Thus our mapping might send object α to object β, object π to object δ, object ν to object ϕ. A specification of M is available; it is a list, $\alpha \to \beta, \pi \to \delta, \nu \to \phi$ and so forth.

Now it becomes necessary for these objects to be handled by someone who uses our everyday letters as labels. He sees a package labelled N; to what object should it be mapped? He is provided with a little dictionary which tells him the Greek equivalent of every letter. He finds that ν is Greek for N. The specification of M tells him $\nu \to \phi$. But which object is ϕ? He uses his little dictionary in reverse, and finds that ϕ is the Greek for F. So, in his language, the mapping sends object N to object F; $N \to F$.

The process involved is as shown.

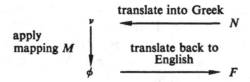

We will now express this process in symbols. Let T indicate translation from English to Greek. Thus $TA = \alpha$, $TB = \beta$, $TP = \pi$, $TD = \delta$, $TN = \nu$ and $TF = \phi$. We need, at one stage of the work to translate in the opposite direction, from Greek to English.

This is the inverse operation, T^{-1}. We have $A = T^{-1}\alpha$, $B = T^{-1}\beta$ and so on, to $F = T^{-1}\phi$.

Thus the steps in our process are as follows; $v = TN$; $Mv = \phi$; $T^{-1}\phi = F$. Notice that only in the middle stage do we actually go from one object to another. In the first step we establish that the object called N in one language is called v in the other; in the last step we are again concerned with two different names, ϕ and F, for the same object.

If we combine the results of our equations, we find

$$F = T^{-1}\phi = T^{-1}Mv = T^{-1}MT \cdot N.$$

Thus the mapping $N \to F$, which expresses in English exactly the same idea as the Greek $v \to \phi$, is given by the symbol $T^{-1}MT$. Thus $T^{-1}MT$ is the form taken in the new system by the mapping that appeared as M in the old.

Now it is clear that this account is a very general one. It does not describe at all the objects on which the mapping M acts. Nor does it depend at all on the grammar of the Greek and English languages. All that it assumes is a 1–1 correspondence between the symbols, there must be just one Greek letter corresponding to the letter N; there must be just one English letter corresponding to ϕ.

Accordingly a symbol such as $T^{-1}MT$ is likely to arise whenever we have a mapping, and two different systems in which the objects involved can be specified.

In our problem, as described earlier, we have a mapping that appears in one system as $v \to v^*$, in the other as $V \to V^*$. We have a matrix T that connects the two systems; $v = TV$, $v^* = TV^*$. And we have a mapping specified by $v^* = Mv$.

Here again we have a scheme similar to the earlier one.

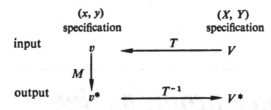

Either from this diagram, or from the equations immediately preceding it, we see that $V^* = T^{-1}MT \cdot V$. The transformation, represented by M in the (x, y) system, appears as $T^{-1}MT$ in the (X, Y) system.

Exercise

With M and T appropriate to the situation at the beginning of the section (under the heading 'A trap to avoid') check that $T^{-1}MT$ does give the matrix relating (X, Y) to (X^*, Y^*).

The role of orthogonal matrices

We now see why it is that we get led to two different matrices when we consider the effect of change of axes on the transformation $v \to Mv$ and the quadratic form $v'Mv$.

In the new axes, the transformation is specified by the matrix $T^{-1}MT$ while the quadratic form is specified by $T'MT$. As a rule, these matrices will be unequal, since usually T^{-1} is different from T'.

However, it can happen that $T' = T^{-1}$. This in fact is the condition for T to be an orthogonal matrix that we had in §36. Accordingly, if we are going to make an *arbitrary* change of axes, we must distinguish between a matrix specifying a transformation and a matrix specifying a quadratic form; however, if we restrict ourselves to using orthogonal matrices only, we can forget this complication.

It is not surprising that orthogonal matrices should appear in this connection. In §38 we used the contours $v'Mv = $ constant to give us a way of visualizing the equation $v^* = Mv$, in which M plays the role of a transformation. We showed, in effect, that if $z = \frac{1}{2}v'Mv$, then $Mv = \text{grad } z$. Now grad z was a vector giving the direction of the steepest slope. This direction is *perpendicular* to the contour through v, the point in question. On a map, one might find the direction of steepest slope by going from a point on one contour *to the nearest point* on the next contour. The words *perpendicular* and *the nearest point* involve ideas of angle and distance; that is, they belong to Euclidean geometry; they cannot be defined in terms of affine geometry, that is, of pure vector theory involving only the definitions of $u + v$ and ku. In affine geometry, we can make any change of axes – that is, we may go from any basis to any other basis. In Euclidean geometry, it is possible to use oblique axes, but this involves reconsidering all the formulas we use. If we are to go ahead without such complications, we have to stick to orthogonal matrices for change of axes; then lengths and angles are found by exactly the same formulas as in the original 'standard' system of axes.

43   Linear programming

Linear programming is a procedure for use in situations where something can be done in very many ways and we wish to find which way is the best, as judged by some agreed standard. Suppose for example a group of students have to provide food for themselves for a certain period of time. Even after ruling out articles of diet which they do not like, a very considerable choice remains. They are very scientifically minded, and want to consume an adequate amount of protein, carbohydrates, vitamins and other ingredients of a satisfactory diet. They wish to do this at the lowest possible cost. There may be certain restrictions on what they can do; some articles need to be stored in a refrigerator, and they have limited refrigerator space. They are busy, and so insist that the total time spent preparing the meals be kept below a certain amount. How do their choices appear in symbolic form? Suppose the local stores offer them 100 articles. They have to choose the amounts of each to buy, say $x_1, x_2, \ldots, x_{100}$. Each one has a certain cost, given by $c_1, c_2, \ldots, c_{100}$. Their total expenditure is

$$c_1x_1 + c_2x_2 + \cdots + c_{100}x_{100},$$

and this they want to keep as low as possible. The amount of protein contained in the various articles is specified by the numbers $p_1, p_2, \ldots, p_{100}$. They require

$$p_1x_1 + p_2x_2 + \cdots + p_{100}x_{100} \geqslant P,$$

where P is the total amount of protein needed by the student commune. Some articles might contain no protein: for them the protein coefficient would be zero. Similar inequalities would hold for the other nutritional ingredients. Each article would require a certain amount of refrigerator space, so there would be numbers $r_1, r_2, \ldots, r_{100}$ for this aspect. If some article did not require refrigeration, the number r corresponding to it would be 0. It would be necessary to have $r_1x_1 + r_2x_2 + \cdots + r_{100}x_{100} \leqslant R$, where R was the total refrigerator space. (If the refrigerator were divided into compartments with specialized purposes, this condition might have to be replaced by several different inequalities.) Each article would involve a certain time of preparation; there would be a condition $t_1x_1 + \cdots + t_{100}x_{100} \leqslant T$, where T was the total time the students felt it was reasonable to spend in the kitchen.

Such a problem then appears as that of making the linear expression

$$c_1x_1 + \cdots + c_{100}x_{100}$$

as small as possible, subject to a number of linear inequalities.

Similar problems arise in industry and in government. In a chemical plant, various raw materials may be available, containing desired substances in various proportions. The desired substances correspond to the proteins, vitamins and so forth of the dietary problem. There may be restrictions of various kinds in relation to plant available, the number of competent workers in different departments, and so forth. There may be a restriction on how much of a particular by-product the market can absorb. Similarly, in any kind of enterprise, the problem may arise of using most effectively the resources available. Very often, of course, some simplification and distortion of the actual situation may be needed to get the problem into the form of linear expressions. The engineer must use his judgement to see whether such simplification is a useful approximation to reality, or whether it will lead to entirely misleading conclusions.

Questions of linear programming may appear as minimum or maximum problems. In the diet question we wished to achieve adequate nutrition for the *least possible cost* – a minimum problem. However, the same problem could be posed in terms of a maximum – how to save as much as possible on food expenditure. Other problems may arise naturally as maximum problems – how to do something and obtain the maximum benefit.

It will be convenient to suppose that every problem we consider has been put in the form of making something a maximum. This will save us continually having to give two explanations, one for a minimum, and one for a maximum. The conversion is always possible. If some quantity f is to be made as small as possible, then $k - f$ is to be as large as possible; here k is any convenient constant.

So, from now on, we suppose our task is to make some quantity, M, as large as possible.

Fig. 151 represents a simple problem in linear programming. We have two numbers, x_1 and x_2, at our disposal, and our aim is to make $M = x_1 + 2x_2$ as large as possible, subject to the restrictions imposed by some situation in which we find ourselves. First of all, x_1 and x_2 represent quantities of some actual substances, so they cannot be negative. They are available in limited quantities; x_1 cannot be larger than 4 and x_2 cannot exceed 3. Further, there is some restriction on the total amount of these substances; we know $x_1 + x_2 \leqslant 6$. The diagram shows these restrictions graphically. The point (x_1, x_2) must lie in the first quadrant since neither x_1 nor x_2 can be negative; $x_1 \leqslant 4$ means we must keep to the left of the line ABF with equation $x_1 = 4$, or be on that line. The restriction $x_2 \leqslant 3$ similarly means that we must not go above the line DCF, while $x_1 + x_2 \leqslant 6$ means that we must keep to the left of CBE, or perhaps be on that line. Thus we are restricted to the region $OABCD$; points on the boundary of this region are permissible.

As our aim is to make $M = x_1 + 2x_2$ as large as possible, the contours $M = $ constant are shown. To avoid confusion in the figure, only a small part of each contour has been shown; the broken lines should be imagined as extended to go right across the diagram. Thus the line $M = 6$ would pass through the point E; E in fact is not a possible situation, since for it $x_1 = 6$, which is too large.

It will be clear that, for points within the permitted region, the largest value of M is taken at C. In such a simple problem, the solution can easily be obtained by this graphical method. But, as a rule, such a direct attack is not possible. Suppose, for instance, that we have to choose, not two, but three numbers. We shall have to make a

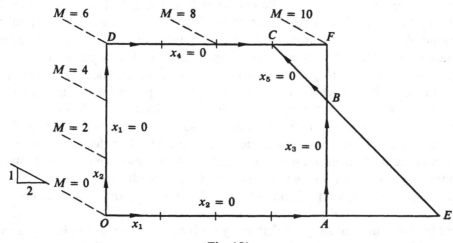

Fig. 151

model in three dimensions. The region will be bounded by plane faces; if we have many inequalities to satisfy, it will be quite hard to tell which intersections of the planes give the edges and corners of the region. If we have more than three numbers to choose, we shall not be able to make a physical model at all. And such a possibility is very real; for instance, in the manufacture of some substance, there might well be ten raw materials available, and we might wish to find the most effective and economical proportions in which to blend them.

Evidently, then, a diagram such as the one above, can serve only as a guide to suggest and illustrate methods, which in practice would involve many variables and many restrictions. This is in line with all our work in linear algebra; we can realize geometrically vector spaces of 2 or 3 dimensions; these suggest *analogies* that may be helpful when we are dealing with problems involving more than three variables. Starting from our diagram, we have to find a procedure that can be carried through purely by calculations with algebraic expressions.

In the diagram, it will be noticed that arrows have been marked on the sides of the region $OABCD$. These arrows denote the direction that leads to an increase in M. For instance, if we go from O to D, we pass in turn the contours $M = 0$, $M = 2$, $M = 4$, $M = 6$. Clearly, M is increasing, so the arrow points from O to D. If we think of the contours as being like the contours on a map, we can say that from O to D is 'uphill'.

Now, if you take a map of a piece of country, choose some point on it, and go uphill from there as far as you can, you will end at the top of some hill. However, it may be a modest mound; by descending from it, and climbing another hill, you might reach a greater height. However, in linear programming, this complication cannot arise. It can be shown that, if you start anywhere on the boundary, and proceed along the edges, always rising, you will arrive at the absolute maximum. In our diagram C is the highest point; from O you may go from O to D to C, or from O to A to B to C, but either way you arrive at C by a path that never descends. It can be proved, for problems in which linear expressions alone occur, that the maximum will always occur on the boundary of the region, and that it can be reached from any point of the boundary by following an uphill path.

In the diagram we can see that A, for example, is not the highest point, because the arrow on AB indicates that we can climb higher by going in the direction AB. On the other hand it is evident that C is the highest point, for both the arrows at C point inwards towards C. If we left C, either for D or B, we would be going against the arrows, that is to say, downhill.

Our first problem then is this; in a complicated problem, where we cannot make an effective diagram, how are we going to tell whether a point gives a maximum, like C, or is capable of improvement, like B? How are we going to translate the arrows into algebra?

This is done by a simple device known as the introduction of *slack* variables. In our example, we have the condition $x_1 \leqslant 4$. If we choose any suitable value of x_1, this falls short of 4 by $4 - x_1$; so to speak, we have $4 - x_1$ in hand, so far as this condition is concerned. We write $x_3 = 4 - x_1$, and replace the inequality $x_1 \leqslant 4$ by the equation

$x_1 + x_3 = 4$, with the understanding that no variable is ever to become negative. We deal with the other two inequalities in the same way, and so the restrictions are expressed by the equations

$$x_1 \qquad + x_3 \qquad\qquad = 4, \tag{1}$$

$$x_2 \qquad + x_4 \qquad = 3, \tag{2}$$

$$x_1 + x_2 \qquad\qquad + x_5 = 6. \tag{3}$$

In the diagram, the sides AB, CD and BC have been labelled $x_3 = 0$, $x_4 = 0$ and $x_5 = 0$ respectively. It is natural that such labels should occur on the boundary, for being on the boundary means that we are pressing hard against some restriction. On AB for instance, we have $x_1 = 4$; the condition $x_1 \leqslant 4$ does not allow us to go any further to the right; we have nothing left in reserve, and so $x_3 = 0$. Similarly on BC, we are making the fullest use of the condition $x_1 + x_2 \leqslant 6$ by taking $x_1 + x_2 = 6$, and so $x_5 = 0$ expresses that we have no slack to take in; this is verified by equation (3).

We now have five symbols x_1, x_2, x_3, x_4, x_5 and three equations, (1), (2), (3), restricting them. Accordingly, we can choose any two as variables, and the values of the others will be fixed automatically by the equations. It will appear in a moment that we can find out what is happening near any corner of the region by choosing a suitable pair of variables.

Our aim is to make $M = x_1 + 2x_2$ as large as possible. This equation tells us how the arrows run near the origin, O. For $x_1 = 0$, $x_2 = 0$ means we are at the origin. The equation $M = x_1 + 2x_2$ shows that if either x_1 or x_2 gets larger, M will increase. In other words, if we move either in the direction OA or OD, we are going uphill. So we can improve the value of M by going either from O to A or from O to D. Suppose then we go to A; how will the arrows run there? A lies on the sides of the region labelled $x_2 = 0$ and $x_3 = 0$, so we want x_2 and x_3 as variables. That means, we want to get rid of x_1 and bring in x_3, in the expression for M. Equation (1) tells us $x_1 = 4 - x_3$, so substituting this in $M = x_1 + 2x_2$ we obtain $M = 4 + 2x_2 - x_3$. Now x_3 has a negative sign; if we let x_3 grow, M will decrease, which we do not want. Our best plan is to keep $x_3 = 0$, that is, to stay on AB. But increasing x_2 will make M larger. Increasing x_2 takes us to the north, along AB. We cannot go beyond B without leaving the region, so we stop at B. Now B lies on the lines $x_3 = 0$, $x_5 = 0$, so we express M in terms of x_3 and x_5. We have already seen $x_1 = 4 - x_3$, and equation (3) gives $x_2 = 6 - x_5 - x_1$, from which we get $x_2 = 6 - x_5 - (4 - x_3) = 2 + x_3 - x_5$. Accordingly,

$$M = x_1 + 2x_2 = 8 + x_3 - 2x_5.$$

The negative coefficient of x_5 shows that it would be a mistake to let x_5 grow, but the term $+x_3$ shows that it will pay to let x_3 grow. Keeping $x_5 = 0$ and letting x_3 grow takes us along BC.

At this point it may be useful to bear in mind that (x_3, x_5) are oblique co-ordinates with origin at B, and to consider what their co-ordinate grid looks like. This is shown in Fig. 152. The lines are obtained by drawing the contours for $x_3 = 4 - x_1$ and

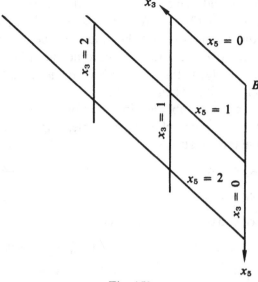

Fig. 152

$x_5 = 6 - x_1 - x_2$. It may be a useful exercise to draw the corresponding grids for (x_4, x_5) at C and (x_1, x_4) at D.

When we come to C, we find it convenient to use x_4, x_5 as variables. We find, with the help of equations (2) and (3), that $M = 9 - x_4 - x_5$. Then $x_4 = 0, x_5 = 0$ makes $M = 9$ and it would clearly be a mistake to go away from this situation; any increase in either x_4 or x_5 would drag M downwards. The negative coefficients in x_4 and x_5 thus express the same facts as did the arrows pointing in towards C.

Our origin problem was – what values of x_1 and x_2 make M a maximum? It would therefore be necessary to calculate x_1 and x_2 corresponding to $x_4 = 0, x_5 = 0$. Equations (2) and (3) easily give us $x_1 = 3, x_2 = 3$.

The equation $M = 9 - x_4 - x_5$ by itself proves that C, where $x_4 = x_5 = 0$, is the best place to be. It would be possible to search for this point by computing the expressions for M corresponding to each of the corners O, A, B, C, D and seeing in which of them only negative coefficients appeared. However, in an actual problem, the number of corners would be extremely large, and this procedure would be very inefficient. It is better to work along the lines already suggested – to choose some corner and work uphill, stage by stage, as we did from O to A to B to C.

In doing this, another problem arises. When, starting at O, we found it would pay to increase x_1, we went along the line OAE until we came to A. As the diagram showed, if we had gone any further we would have left the region of permissible choices. But in terms of algebraic calculation, the situation is not so clear; A is the point with $x_2 = 0$, $x_3 = 0$; E is the point with $x_2 = 0, x_5 = 0$; how are we to know that we should stop when we reach $x_3 = 0$ rather than when we reach $x_5 = 0$?

It is not too difficult to answer this question in terms of algebra alone, without any appeal to the diagram. Suppose we start at O and move along the line OAE at unit speed. After t seconds we shall have $x_1 = t$, $x_2 = 0$. We calculate the other variables x_3, x_4, x_5 from equations (1), (2), (3). We find $x_3 = 4 - t$, $x_4 = 3$, $x_5 = 6 - t$. The essential thing about the variables is that they must not be negative. Now x_3 becomes 0 when $t = 4$, and is negative thereafter; x_5 becomes negative when $t > 6$; x_4 never gives any trouble. Accordingly, x_3 threatens to become negative before x_5 does, and we must stop moving along the line when $x_3 = 0$.

These considerations can be embodied in a computational rule, to illustrate which we shall need a slightly more complicated example. Suppose that, after introducing slack variables, we have the equations

$$4x_1 + x_2 + x_3 \qquad\qquad\qquad = 44, \qquad (4)$$

$$3x_1 + 2x_2 \quad + x_4 \qquad\qquad = 38, \qquad (5)$$

$$2x_1 + 3x_2 \qquad\quad + x_5 \qquad = 37, \qquad (6)$$

$$x_2 \qquad\qquad\quad + x_6 \quad = 9, \qquad (7)$$

$$-x_1 + x_2 \qquad\qquad\quad + x_7 = 6. \qquad (8)$$

The point (x_1, x_2) has to lie in the region $OABCDEF$ (Fig. 153). M could still be $x_1 + 2x_2$, so that, as before, it would pay us to leave the origin and move in the direction OA. After time t, as before, we have $x_1 = t$, $x_2 = 0$. From the equations we find $x_3 = 44 - 4t$, $x_4 = 38 - 3t$, $x_5 = 37 - 2t$, $x_6 = 9$, $x_7 = 6 + t$. Thus x_3 will become negative if we continue to move after $\frac{44}{4} = 11$ seconds; x_4 if we go past $\frac{38}{3} = 12.67$ seconds; x_5 after $\frac{37}{2} = 18.5$ seconds; x_6 and x_7 never become negative. The smallest time, 11 seconds, occurs when $x_3 = 0$, so x_3 is the first variable to become negative. By pure calculation we have found a result that is evident in the figure. Thus we decide to move from O, with $x_1 = 0$, $x_2 = 0$ to A, with $x_2 = 0$, $x_3 = 0$.

The numbers we calculated were $\frac{44}{4}$, $\frac{38}{3}$ and $\frac{37}{2}$. It will be seen that the numerators

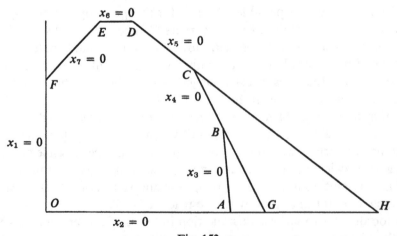

Fig. 153

44, 38 and 37 all come from the constant terms on the right; the denominators 4, 3, 2 are all coefficients of x_1. In going away from O, we decided to let x_1 grow; so the denominators are the coefficients of the variable that is allowed to grow.

In calculating these numbers, we ignore equation (7) in which the coefficient of x_1 is 0, and equation (8) in which the coefficient is negative. These correspond to $x_6 = 9 + 0t$ and $x_7 = 6 + t$, neither of which can ever become negative. This agrees with the diagram, for $x_6 = 0$ corresponds to ED which is parallel to OA; however far we go along $OAGH$, we can never cross ED to make x_6 negative, while FE is sloping away from $OAGH$, so the further we go in the direction OA, the larger and more positive x_7 gets.

We may sum up the procedure as follows: in each equation we expect to find x_1, x_2, which are zero at O, and *one other variable*. We ignore any equation in which x_1, the variable that is going to grow to make M larger, has zero or negative coefficient. In the remaining equations, we divide the constant term by the coefficient of x_1, and see for which equation the resulting quotient is least. That equation tells us which variable should become zero at the next stage. For instance, in our example, the first equation gives $\frac{44}{4}$, which is 11. If $x_1 = 11$, $x_2 = 0$, the first equation shows $x_3 = 0$. So $x_2 = 0$, $x_3 = 0$ gives us the point, A, to which we should go next.

When we get to A we find there that M can be increased if we move in the direction AB, and the same kind of question arises again, – how to tell (without using the diagram) how far we can go along AB without leaving the permissible region. We do not need any new routine to answer this question; previously we began at the origin, with $x_1 = 0$, $x_2 = 0$ and investigated how much x_1 could grow without violating the conditions; when we consider A, with $x_2 = 0$, $x_3 = 0$ the question is – how much can x_2 be increased without violating the conditions? Notice the symmetry of the diagram; there is a variable corresponding to every side of the polygon $OABCDEF$; the point O is in no way superior or significantly different from any other corner of the polygon. Admittedly, we have made the angle at O a right angle, but this is quite irrelevant; the diagram would be equally helpful if we had used oblique axes for plotting (x_1, x_2). This is a good example of the distinction we have made several times between affine and Euclidean geometry. Linear programming belongs to affine geometry; our equations can all be expressed by means of pure vector theory; scalar products, perpendiculars, distances do not enter into the question at all. Accordingly, the question of what we shall do at A, where $x_2 = 0$, $x_3 = 0$ is on exactly the same footing as the problem already considered, of what to do at O, with $x_1 = 0$, $x_2 = 0$.

This is particularly convenient for purposes of computer programming. We simply have to devise a subroutine that can be applied at any corner of the region.

As has been mentioned already, the equations (4)–(8) have a special form. In each equation we may have x_1 and x_2 (the coefficient of one of these may happen to be 0, as in equation (7)); apart from that, we have only one other variable in each equation, and that has coefficient $+1$.

Accordingly, we shall be able to apply our subroutine at corner A provided we first get the equations into the appropriate form, with x_2 and x_3 now playing the role that

x_1 and x_2 did before. That is, in each equation, x_2 and x_3 may occur, but only one other variable may appear, and that with coefficient $+1$.

If we divide equation (4) by 4, we obtain

$$x_1 + \tfrac{1}{4}x_2 + \tfrac{1}{4}x_3 = 11. \tag{4a}$$

This is a suitable form; we have x_2 and x_3, and the 'other variable' is x_1 with coefficient $+1$. In the remaining equations the 'other variables' are x_4, x_5, x_6, x_7. We must get rid of x_1 in these equations, or else we shall have two 'other variables'. This is easily done. If we subtract 3 times equation (4a) from equation (5) we eliminate x_1. Similarly we can get rid of x_1 in equations (6) and (8) by subtracting (or adding, if you prefer) suitable multiples of equation (4a). Equation (7) we leave as it is, since it is already free of x_1. Thus we find

$$\tfrac{5}{4}x_2 - \tfrac{3}{4}x_3 + x_4 \qquad\qquad = 5, \tag{5a}$$

$$\tfrac{5}{2}x_2 - \tfrac{1}{2}x_3 \qquad + x_5 \qquad\quad = 15, \tag{6a}$$

$$x_2 \qquad\qquad\qquad + x_6 \quad = 9, \tag{7a}$$

$$\tfrac{5}{4}x_2 + \tfrac{1}{4}x_3 \qquad\qquad\quad + x_7 = 17. \tag{8a}$$

We also have to express M in a form suitable for work near A. We can write our original equation for M in the form $M - x_1 - 2x_2 = 0$. We want, in this equation also, to get rid of x_1 and permit x_3 to enter in its place. If we simply add equation (4a) we obtain

$$M - \tfrac{7}{4}x_2 + \tfrac{1}{4}x_3 = 11. \tag{9a}$$

We could write this as $M = 11 + \tfrac{7}{4}x_2 - \tfrac{1}{4}x_3$, from which we see (as expected from the diagram) that M can be further increased by letting x_2 grow, but that x_3 should be held to its value 0. However, there is an established convention in linear programming to suppose the variables written on the left of the equals sign, as in equation (9a). This of course reverses the signs, and so it is important to remember that, when things are written this way, the *negative* coefficient of x_2 in equation (9a) means that it will pay to increase x_2, while the *positive* coefficient of x_3 in equation (9a) means that x_3 should be held down to 0 for the next step.

So x_2 is the variable that is to grow. Accordingly, by the rule given earlier, we divide the constant terms by the coefficients of x_2. Applying this, in equations (4a) to (8a), we get the quotients $11/\tfrac{1}{4} = 44$; $5/\tfrac{5}{4} = 4$; $15/\tfrac{5}{2} = 6$; $9/1 = 9$; $17/\tfrac{5}{4} = 13.6$. The smallest value here is 4, arising from equation (5a), which contains the 'other variable' x_4. So we should travel until x_4 becomes 0. Our next corner is that with $x_3 = 0, x_4 = 0$.

Such calculations are often presented in an abbreviated form; the main idea is simply that of omitting the variables $x_1, \ldots, x_7$, and recording only the coefficients and the constants in appropriate positions. The calculation then takes a routine shape and appears as below, except for the comments and explanations. We indicate, for instance, the variables to which the coefficients relate, and the equations in which the coefficients occurred when we first explained the procedure. Such things, of course, would not appear in routine work. In practice, of course, we might not see the steps at all, as they would be operations done inside a computer. Here, then, in more concise form, is the calculation arising from equations (4)–(8).

Equation	x_1	x_2	x_3	x_4	x_5	x_6	x_7	Constant	Quotients
(4)	4	1	1	0	0	0	0	44	← 11
(5)	3	2	0	1	0	0	0	38	12.67
(6)	2	3	0	0	1	0	0	37	18.5
(7)	0	1	0	0	0	1	0	9	–
(8)	−1	1	0	0	0	0	1	6	–
(9)	−1	−2	0	0	0	0	0	0	
	*								

Equation (9) indicates that M is given by $M - x_1 - 2x_2 = 0$. The equation always begins with M, so we do not show this term, M, in the scheme at all. Looking along row (9) we notice the negative coefficient, -1. This means that M will increase if x_1 is allowed to grow. The asterisk indicates that we are going to operate on the first column, that is, the coefficients of x_1. How much can we let x_1 grow? We saw that this can be answered by forming quotients; in each row the last entry (the constant) is to be divided by the entry in the * column. Note that these quotients can only be entered *after* we have made a suitable choice for the * column. In this case, we were not forced to choose the first column. In row (9) there is a negative number, -2, in the second column. We could, if we like, put the asterisk against the second column. Some authors suggest that we should choose the most negative number in this row. Others, probably rightly, say that the extra computer time spent in looking for the most negative number is not justified, for it does not guarantee that the calculation will be shorter. They recommend that as soon as the computer has located *any* negative coefficient in the M equation, the corresponding column should be selected and dealt with, without further delay.

No quotients are entered in the rows (7) and (8), since the numbers in the * column in these rows are zero and negative respectively.

The smallest quotient, 11, occurs in row (4). This indicates that x_1 should increase until x_3 becomes zero, for x_3 is the variable other than x_1, x_2 with a non-zero coefficient in row (4). This corresponds to the move we discussed earlier, from O with $x_1 = x_2 = 0$ to A with $x_2 = x_3 = 0$. We are going to consider whether M increases if x_2 or x_3 grows, so we want now to have M in terms of x_2 and x_3, rather than x_1 and x_2. Thus the asterisk shows the variable, x_1, that we are getting rid of in the equation for M; the arrow, opposite the smallest quotient, 11, leads us to x_3, the variable we are bringing into the the equation for M. Now x_1 has to drop back to the role of 'another variable', that is, in its column, we want to find a single 1 and the rest 0. The 1, naturally enough, must be in the row singled out by the arrow.

Our next step, then, is to multiply the 'arrow' row by $\frac{1}{4}$, so as to get 1 in the x_1 column. We then subtract suitable multiples of this row (or equation) from the other rows, so as to make all the entries in the * column 0, – apart of course from the 1 we already have. This leads us to the following table, which is simply a concise way of recording equations (4a) to (9a). This time we will omit the column headings.

									Quotients
(4a)	1	$\frac{1}{4}$	$\frac{1}{4}$	0	0	0	0	11	44
(5a)	0	$\frac{5}{4}$	$-\frac{3}{4}$	1	0	0	0	5	← 4
(6a)	0	$\frac{5}{2}$	$-\frac{1}{2}$	0	1	0	0	15	6
(7a)	0	1	0	0	0	1	0	9	9
(8a)	0	$\frac{5}{4}$	$\frac{1}{4}$	0	0	0	1	17	13.6
(9a)	0	$-\frac{7}{4}$	$\frac{1}{4}$	0	0	0	0	11	
		*							

This time we have no choice. We must put the asterisk against the second column, which is the only one with a negative number in the bottom row. The quotients can then be worked out, and the least quotient, 4, indicates the second row. We multiply this row by $\frac{4}{5}$, to get 1 in the * column, and enter this as row (5b) in our next scheme. We combine suitable multiples of the new row, (5b), with the remaining rows of our present scheme, (4a), (6a), (7a), (8a), (9a), to get 0 in the second column of the next table.

Before going to this next stage, it is worthwhile to examine the appearance of this table. The table is concerned with the action we should take after reaching position A. The point A has $x_2 = 0, x_3 = 0$. We notice the appearance of the columns corresponding to x_2 and x_3; they contain an assortment of numbers, most of which are different from 0. The third column does contain zero in the row (7a); it could happen – so to speak by chance that quite a number of zeros occurred in these columns. It does not affect the work if this should happen. Now suppose we remove these two columns, and ignore the last column. We then see a definite pattern, which is in no way due to chance; in each row except the last there is a single 1, in each column there is a single 1, everywhere else, 0. This is the hallmark for the columns not associated with the point where we are. If you are studying a record of a calculation, and enough zeros occur in the columns corresponding to our x_2, x_3 to make it hard to identify these, you should mark the columns that contain a single 1 and zeros. In this way you can discover the columns *not* corresponding to x_2, x_3 in our example.

The numbers in the last column have a meaning that should be understood. The top row, (4a), in our last table corresponds to the equation we had earlier,

$$x_1 + \tfrac{1}{4}x_2 + \tfrac{1}{4}x_3 = 11. \tag{4a}$$

At A, $x_2 = x_3 = 0$. It follows that, at A, $x_1 = 11$. In the same way, the numbers in rows (5a), (6a), (7a), (8a) in the last column tell us that, at A, we have $x_4 = 5$, $x_5 = 15$, $x_6 = 9$, $x_7 = 17$, should we need this information. The last row corresponds to our earlier equation

$$M - \tfrac{7}{4}x_2 + \tfrac{1}{4}x_3 = 11. \tag{9a}$$

Once again using the fact that $x_2 = x_3 = 0$ at A, we see that, in the situation represented by A, the quantity M that we are trying to maximize takes the value 11. So this final entry in the table tells us what progress we have made in our search for the largest possible M. Accordingly we can say, by examining this table in the way indicated, that the value $M = 11$ can be achieved by choosing $x_1 = 11$, $x_2 = 0$. The point A in fact is $(11, 0)$. It will be remembered that x_1 and x_2 are the quantities that appear in the problem as originally stated. The other variables, x_3, x_4, x_5, x_6, x_7, we have introduced to help us solve the problem.

We now return to our calculation. Carrying out the operations described, just after the last table was given, we arrive at the following table.

									Quotients
(4b)	1	0	$\frac{2}{5}$	$-\frac{1}{5}$	0	0	0	10	25
(5b)	0	1	$-\frac{3}{5}$	$\frac{4}{5}$	0	0	0	4	–
(6b)	0	0	1	-2	1	0	0	5	← 5
(7b)	0	0	$\frac{3}{5}$	$-\frac{4}{5}$	0	1	0	5	8.33
(8b)	0	0	1	-1	0	0	1	12	12
(9b)	0	0	$-\frac{4}{5}$	$\frac{7}{5}$	0	0	0	18	
			*						

This table is concerned with the situation and the action required when we reach the point B. At B, $x_3 = x_4 = 0$, and we observe the difference between the third and fourth columns and the others. We see from the final entry, 18, that by coming to B we have attained $M = 18$. From the first two numbers in the last column we see that this happens for $x_1 = 10$, $x_2 = 4$.

The only negative number in the last row is $-\frac{4}{5}$; this occurs in the third column, and indicates that x_3 should be allowed to grow away from the value 0. The arrow indicates the row (6b). Here, very conveniently, the coefficient of x_3 is already 1, so our new row, (6c), in the next table will be identical with (6b). Suitable multiples of row (6c), combined with the rows of the present table, will allow us to get zeros in the third column, except of course in row (6c). Owing to the 1 that occurs in the fifth column of row (6b), and so also in (6c), this mixing of rows will get rid of the zeros that at present occur in the fifth column. Thus in the new table, the fourth and fifth columns will stand out as differing from the others. We shall in fact have moved to C, for which $x_4 = x_5 = 0$.

In the list of quotients above, it will be noticed that no entry is made opposite row (5b). This of course comes from the rule that no entry is made when the number in the * column is negative, as $-\frac{3}{5}$ is here. It is useful to look back at the diagram and see what this means. The table relates to the point B. The asterisk indicates that x_3 is to grow; but $x_4 = 0$ is to remain. Thus we are going to leave B in the direction BC. The quotient 25 in row (4b) indicates that x_3 can grow to the value 25 without x_1 becoming negative. The quotient in row (5b), if there were one, would tell us how far we could go in the direction BC without x_2 becoming negative. But x_2 is the co-ordinate to the north, and BC is taking us somewhat west of north; the further we go along BC, the larger x_2 will be; there is no danger of x_2 ever becoming negative. In fact G, the intersection of BC with the line $x_2 = 0$ is already behind us when we set off from B with our faces towards C.

In the same way, when we get to C, and find that it pays to move on in the direction CD, there will be no quotient entries in rows $(5c)$ and $(6c)$, for the intersections of CD with $x_2 = 0$ and $x_3 = 0$ are behind us, and we are moving away from them.

The routine of linear programming can best be learned by working a fairly large number of problems, all the time having the graph before you and thinking what the meanings of the steps are. It is a defect of the method that a mistake made at any stage will spoil all the later work and make it incorrect. It is therefore wise, when learning the routine, to work with a problem where you know what should happen, and will recognize if anything goes wrong.

For instance, the problem we have been considering was composed by first of all choosing points on squared paper that would make the sides of the polygon $OABCDEF$ have convenient slopes. These points were in fact $A = (11, 0)$, $B = (10, 4)$, $C = (8, 7)$, $D = (5, 9)$, $E = (3, 9)$; $F = (0, 6)$. The equations of the sides AB, BC and so on were worked out; these gave the inequalities and hence the equations for the slack variables x_3, x_4, x_5, x_6, x_7, namely equations (4) to (8) above. From these equations, together with the equation $M = x_1 + 2x_2$, we can complete the following table, giving the values of all the variables at the points A, B, C, D, E, F.

	x_1	x_2	x_3	x_4	x_5	x_6	x_7	M
A	11	0	0	5	15	9	17	11
B	10	4	0	0	5	5	12	18
C	8	7	5	0	0	2	7	22
D	5	9	15	5	0	0	2	23
E	3	9	23	11	4	0	0	21
F	0	6	38	26	19	3	0	12

It is clear from the last column that the maximum for M will be found at D. Looking back at the table with rows $(4a)$ to $(9a)$, we see that the numbers in the final column of that table agree with the numbers in the row for A above. The column contains the numbers $11, 5, 15, 9, 17, 11$, and so gives the values of $x_1, x_4, x_5, x_6, x_7, M$ at A; it skips over $x_2 = 0$, $x_3 = 0$. In the same way, from the row for B above we can check the accuracy of the final column in the table with rows $(4b)$ to $(9b)$.

It should be understood that this way of checking is purely a learning device; it is intended as a way of avoiding errors in exercises worked to gain familiarity with the routine. It helps also to bring out the significance of the numbers in the final column of the table. In any actual problem, the labour involved in such a procedure would be much too great. In practice, the computations would be carried out by a computer programme, into which devices for recognizing the occurrence of any error would have to be incorporated.

Exercises

1 Complete the calculations for the problem discussed above; that is, obtain the tables corresponding to the points C and D. Check the accuracy of your work by means of the table above, that gives the values of the variables at the corners of the polygon.

2 Do the same problem, by going back to the beginning of the work, and placing the asterisk under the second column rather than the first. Check accuracy in the same way. By what route do you now reach the optimal point, D?

3 The first problem we considered in this section had

$$x_1 \leqslant 4, \qquad x_2 \leqslant 3, \qquad x_1 + x_2 \leqslant 6, \qquad M - x_1 - 2x_2 = 0.$$

Using the slack variables and the diagram given, solve this problem by the routine method with tables (a) using the route O, A, B, C (b) using the route O, D, C. If accuracy is checked by the method suggested, observe that the final columns give the values of x_1, x_2, x_3, x_4, x_5 but not necessarily in that order. Explain this, by writing the equations corresponding to the rows of the table.

4 On a ship a volume V is available for cargo and the weight of the cargo must not exceed W. Two commodities are available for loading. The units of money, volume and weight are so chosen that unit value of commodity A has unit volume and unit weight. In this system of measurement, unit value of commodity B has volume v and weight w. It is desired to load the ship in such a way that the cargo carried will have maximum value.

For each of the situations listed below solve this problem by means of a diagram, and also by the method of calculation explained in §43. Check that the results agree. Although no heavy calculation is needed here it is useful to draw the diagram first, as this will indicate the most economical route to the desired solution.

Situation (a): $v = 2$, $w = \frac{1}{2}$, $V = 20$, $W = 11$.
Situation (b): $v = 3$, $w = 2$, $V = 33$, $W = 26$.
Situation (c): $v = \frac{1}{2}$, $w = \frac{1}{3}$, $V = 14$, $W = 11$.
Situation (d): $v = 3$, $w = 2$, $V = 12$, $W = 20$.

5 600 tons of a certain material are available at warehouse A and 800 tons of the same material at warehouse B; 300 tons are to be delivered at a place P, 500 tons at Q and 600 tons at R. The distances in miles are as shown in the table.

	P	Q	R
A	10	20	100
B	70	90	30

It is required to arrange the delivery in such a way that the total transport required is a minimum. (A journey in which t tons are carried m miles counts as mt units of transport.)

(*Suggested method.* If x_1 tons are taken from A to P and x_2 tons from A to Q, the other quantities involved can be expressed in terms of x_1 and x_2. Certain inequalities must be satisfied. Taking $x_1 = 0$, $x_2 = 0$ provides a starting point for the calculation, but is obviously far from the best arrangement. The problem should be treated both graphically and by calculation.)

6 What difference would it make to the arrangements in question 5 if the distance from B to P were 85 miles, all the other data being as before?

7 On a ship 25 units of volume are available for cargo, the weight of which must not exceed 23 units. Unit value of commodity A has unit volume and weight 4 units; unit value of commodity B has volume 2 units and weight 2 units; unit value of commodity C has volume 3 units and weight 1 unit. Calculate the cargo that gives the greatest value.

If we have commodity A to the value x_1, commodity B to the value x_2 and commodity C to the value x_3, the restrictions on space and weight require the point (x_1, x_2, x_3) to lie in a region bounded by 5 planes. Sketch this region and find the co-ordinates of its vertices. Find the value of the cargo that corresponds to each vertex, and check that none of these values exceeds that found by linear programming.

44 Linear programming, continued

The work of §43 was based on a simple idea; start at a corner of the permitted region, and then keep moving along the edges of the region in such a way that the situation continually improves. The corner we started at was the origin. Now the origin had no claim to represent a particularly efficient arrangement. For instance, in a problem of assigning work to men in a factory, the origin would correspond to complete idleness – put no men on each machine, so nothing is produced. The only reason for choosing the origin is *to get the calculation started*; it gives us a point from which we can proceed to other, more satisfactory points. Idleness automatically satisfies the conditions; it can always be achieved; whatever restrictions may exist on manpower, supply of materials, working hours, available finance and so forth. It constitutes what is called a *feasible solution*, an arrangement that meets the restrictions under which we have to work. It is a starting point in our search for the *optimal feasible solution*, that is, the best arrangement possible in the circumstances.

However, there are problems in which the origin does not provide us with a feasible solution. Consider the problem we had earlier of feeding a student commune. The origin would correspond to leaving the supermarket empty-handed, to spending nothing on each article on the list. This would not satisfy the conditions that adequate amounts of essential nutrients were to be provided. In such a problem we do not have a feasible solution ready-made to hand. Indeed, it is possible that no feasible solution exists. Perhaps the budget committee were insufficiently aware of the rising cost of living. They may have allocated to the food buyer a sum too small to provide an adequate diet however much wisdom he may use in his spending. So the problem is – to locate a feasible solution, if such a thing exists, as a starting point for the procedure of §43, or to demonstrate that the demands made are impossible.

Fig. 154 illustrates a simple problem where impossible requirements are imposed. The restrictions are

$$x_1 + 2x_2 \leqslant 8, \qquad 3x_1 + x_2 \leqslant 9, \qquad x_1 + x_2 \geqslant 6.$$

The first two inequalities require us to be inside the quadrilateral $OABC$. The last condition requires us to be above the line DE. Evidently no point meets these requirements.

How can we reach this conclusion without the aid of a diagram? The first two conditions offer no difficulty; they are satisfied by the origin, $x_1 = 0, x_2 = 0$. The question now is – can we move around in such a way that, while still satisfying the first two conditions, we meet the third also? That is, can we keep in $OABC$, and still make $x_1 + x_2 \geqslant 6$? We decide we will do our best to meet it; 'doing our best' is itself a problem of maximizing, and we can formulate it as follows. Suppose some generous person, seeing we are in difficulties with the third condition, offers to make up whatever

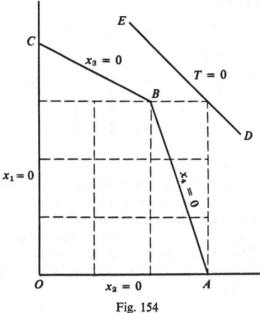

Fig. 154

amount we are short, but on the understanding that we will not ask for more assistance than is absolutely necessary. Let T stand for the deficit this person has to make good. Our problem now reads

$$x_1 + 2x_2 \leqslant 8, \qquad 3x_1 + x_2 \leqslant 9, \qquad x_1 + x_2 + T \geqslant 6,$$

where T is to be as small as possible; that is, we have to maximize $-T$.

This problem is of the kind considered in §43. We bring in slack variables x_3, x_4, x_5 and write

$$x_1 + 2x_2 + x_3 \qquad\qquad = 8, \tag{1}$$

$$3x_1 + x_2 \qquad + x_4 \qquad\quad = 9, \tag{2}$$

$$x_1 + x_2 \qquad\qquad - x_5 + T = 6. \tag{3}$$

Note here the minus sign with x_5; this is because we have $\geqslant$ in our third condition; x_5 measures the amount by which $x_1 + x_2 + T$ *exceeds* 6.

Let $N = -T$, the amount to be maximized. We have $N + T = 0$, and this may cause some dismay; in §43, the work was finished when all the signs in the bottom row (corresponding to this equation) were positive. But that was based on the fact that the variables corresponding to the corner in question were all zero, so could not be made any smaller. But in our case we are starting from the situation $x_1 = 0$, $x_2 = 0$, $T = 6$. This is feasible, since the equations show $x_3 = 8$, $x_4 = 9$, $x_5 = 0$; no variable is negative. We have not made N a maximum, as T may well be capable of decreasing from its present value of 6. We do have three zero quantities, x_1, x_2 and x_5. So the sensible thing is to express everything in terms of these. Equation (3) gives us

$$T = 6 - x_1 - x_2 + x_5,$$

accordingly $N + T = 0$ becomes

$$N - x_1 - x_2 + x_5 = -6.$$ (4)

The table now appears as shown.

x_1	x_2	x_3	x_4	x_5	T		Quotients
1	2	1	0	0	0	8	8
3	1	0	1	0	0	9	← 3
1	1	0	0	-1	1	6	6
-1	-1	0	0	1	0	-6	
*							

Observe here that the columns x_3, x_4, T each contain a single 1 and the other entries in them are all zero. Thus x_3, x_4, T are easily found in terms of the remaining variables x_1, x_2, x_5. The quantities above the line are all positive, which agrees with the observation already made that $x_1 = x_2 = x_5 = 0$ gives a feasible solution, for this makes $x_3 = 8$, $x_4 = 9$, $T = 6$. The table is now completely in the form used in §43.

If we put the asterisk under x_1, the smallest quotient occurs in the second row, and our next table is as shown below.

0	$\frac{5}{3}$	1	$-\frac{1}{3}$	0	0	5	← 3
1	$\frac{1}{3}$	0	$\frac{1}{3}$	0	0	3	9
0	$\frac{2}{3}$	0	$-\frac{1}{3}$	-1	1	3	4.5
0	$-\frac{2}{3}$	0	$\frac{1}{3}$	1	0	-3	
	*						

Continuing we obtain the next table.

0	1	$\frac{3}{5}$	$-\frac{1}{5}$	0	0	3
1	0	$-\frac{1}{5}$	$\frac{2}{5}$	0	0	2
0	0	$-\frac{2}{5}$	$-\frac{1}{5}$	-1	1	1
0	0	$\frac{2}{5}$	$\frac{1}{5}$	1	0	-1

Now the coefficients in the bottom row are all positive, and this time we really have finished. The bottom line indicates $N + \frac{2}{5}x_3 + \frac{1}{5}x_4 + x_5 = -1$. The best we can do is take $x_3 = x_4 = x_5 = 0$, but even so we are only able to get $N = -1$. We still fall short, by 1, of meeting the third condition.

The first two rows of the table indicate that, when $x_3 = x_4 = x_5 = 0$, we have $x_1 = 2$, $x_2 = 3$, that is, we are at the point B. This is the nearest we can get to the line DE. The rather small coefficients, $\frac{2}{5}$ and $\frac{1}{5}$, of x_3 and x_4, correspond to the fact that, if we leave B along BA or BC, we are indeed getting further from the line DE, but only rather slowly. At B, $x_1 + x_2 = 5$, which, as expected, falls short by 1 of the demanded value, 6.

If the third condition is replaced by $x_1 + x_2 \leqslant 4.5$, the problem becomes possible. There is then a triangle within which (x_1, x_2) must lie. In this case, by maximizing N we would arrive at a corner of this triangle. Since no 'subsidy' T would be necessary, the maximum value of N would be 0, shown by zero appearing as the last entry in the bottom row of the table.

It will be seen from the tables that T plays exactly the same role as a sixth variable, x_6. In fact this is the usual notation. A variable representing a subsidy that may be needed to rectify an unsatisfied demand is known as an *artificial variable*. Like all the other variables, it is supposed never to be negative. Our generous person may not need to subsidize us, but he does not demand that any surplus be handed to him. Accordingly, the maximum value of N can never exceed 0. When N attains the maximum value of 0, this indicates that the conditions can be fulfilled and a feasible solution has been located.

Notice that M, the quantity we eventually wish to maximize, has not been mentioned at all in this work. The procedure just explained has the purpose of locating a point within the realm of possibilities. There is no suggestion that this point will be the best point for maximizing M. It hardly could be, seeing the function defining M is nowhere mentioned in the procedure.

Exercises

1 Replacing the condition $x_1 + x_2 \geqslant 6$ by $x_1 + x_2 \geqslant 4.5$ carry through the procedure for maximizing N, and check that it does lead to a feasible solution. The work will, of course, be almost identical with that done above, so the table given there can be used as a check of numerical accuracy.

2 Find a feasible solution for $x_1 \leqslant 2, x_1 + x_2 \leqslant 4, 2x_1 + x_2 \geqslant 5$ by means of an artificial variable. After doing this, draw a diagram, and trace the path followed by the point (x_1, x_2) at the various stages of the calculation.

Beginning of solution. We write the equations

$$x_1 + x_3 = 2, \qquad x_1 + x_2 + x_4 = 4, \qquad 2x_1 + x_2 - x_5 + x_6 = 5.$$

Here x_3, x_4, x_5 are slack variables and x_6 is an artificial variable, brought in because $x_1 = 0$, $x_2 = 0$, which satisfies the first two conditions, leaves a deficit in the third condition. $N + x_6 = 0$, so $N - 2x_1 - x_2 + x_5 = -5$. Maximize N.

3 Investigate, first by calculation, then by a diagram, what happens if the third condition in question 2 is replaced by $2x_1 + x_2 \geqslant 7$.

4 What happens if, in question 2, we replace the last condition by $2x_1 + x_2 \geqslant 6$?

5 By common sense, or by a diagram, consider whether a feasible solution can be found for
(a) $x_1 \leqslant 3, x_2 \leqslant 2, x_1 + x_2 \geqslant 9$,
(b) $x_1 \leqslant 3, x_2 \leqslant 2, x_1 + x_2 \geqslant 4$.
Apply artificial variable method to these, and trace the path on the diagram followed by (x_1, x_2) at the successive improvements of N.

Problems with two artificial variables

It may happen that we have more than one inequality that is not satisfied at the origin, or at some other point where we are starting our search. Fig. 155 illustrates the problem $x_1 \leqslant 4, x_2 \leqslant 2, 2x_1 - x_2 \geqslant 1, 2x_2 - x_1 \geqslant 1$. The last two conditions are not satisfied by the origin. So we bring in slack variables x_3, x_4, x_5, x_6 and artificial variables x_7, x_8.

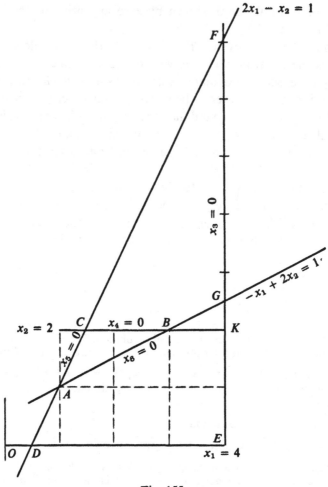

Fig. 155

We then have

$$
\begin{aligned}
x_1 \quad\quad + x_3 \quad\quad\quad\quad\quad\quad\quad\quad &= 4, \\
x_2 \quad\quad + x_4 \quad\quad\quad\quad\quad\quad &= 2, \\
2x_1 - \; x_2 \quad\quad\quad\quad - x_5 \quad\quad + x_7 \quad\quad &= 1, \\
-x_1 + 2x_2 \quad\quad\quad\quad\quad\quad - x_6 \quad\quad + x_8 &= 1.
\end{aligned}
$$

Our total deficit is $x_7 + x_8$, so we take $N = -x_7 - x_8$. Maximizing N minimizes the total deficit. Suppose we reach a situation that gives $N = 0$. As x_7 and x_8 cannot be negative, $x_7 + x_8 = 0$ implies $x_7 = 0$ and $x_8 = 0$. Accordingly we have a feasible solution, since there is no deficit in any condition.

Suppose we find the maximum value of N turns out to be negative. This would mean the conditions were impossible. For if a feasible solution exists, it makes the deficits zero, so $x_7 = 0$, $x_8 = 0$ can be achieved, and so $N = 0$ can be reached. A negative maximum for N clearly rules out this possibility.

So the maximum value for N will tell us whether the conditions are possible or not.

In this particular problem, a feasible solution does exist and is reached in two stages. The artificial variable calculation takes us from the origin to D, with $x_1 = \frac{1}{2}$, $x_2 = 0$, and then to A, with $x_1 = 1$, $x_2 = 1$, which meets all the conditions. In making the calculation, we know this because we find $N = 0$ from the bottom row.

Exercises

1 Carry out the calculation just described.

2 Find a feasible solution, or prove that none exists, for the conditions

$$x_1 \leqslant 3, \qquad x_2 \leqslant 4, \qquad 2x_1 + x_2 \geqslant 6, \qquad x_1 + x_2 \geqslant 4.$$

Is the method wasteful?

There is a point in the above procedure that may strike you as strange. The first problem in this section considered the conditions $x_1 + 2x_2 \leqslant 8$, $3x_1 + x_2 \leqslant 9$, $x_1 + x_2 \geqslant 6$. Now in fact, as may be seen from the diagram, the first two conditions made it impossible for the third one to be fulfilled. We had to bring in an artificial variable, which we then called T and would now call x_6, as a sort of subsidy to meet our deficit on this condition. Accordingly, as x_6 has to be provided to bring $x_1 + x_2$ up to 6, we might expect to write the equation $x_1 + x_2 + x_6 = 6$. But this is not the equation that our procedure tells us to write. We consider the condition $x_1 + x_2 + x_6 \geqslant 6$ and bring in the slack variable x_5 to express this inequality; so we write $x_1 + x_2 - x_5 + x_6 = 6$. In this equation, x_5 represents the *excess* of $x_1 + x_2 + x_6$ over 6. So it seems that at one and the same time we are asking for a subsidy x_6, and anticipating a surplus x_5. If x_6 were an actual subsidy, this might be regarded as a piece of adroit dishonesty. But x_6 does not represent some piece of wealth that we wish to acquire by fair means or foul. It is simply a variable in a calculation, and if we are indeed using two variables, x_5 and x_6, where one would do, we are involving ourselves in unnecessary arithmetic and extra computer time.

Now, in fact, in this particular problem it is not necessary to bring x_5 in. We could write $x_1 + x_2 + x_6 = 6$ and proceed to maximize $N = -x_6$ and be led to the point B that makes the deficit least.

Exercise

Check this statement.

There are in fact a number of situations in which we could dispense with the slack variables that represent surpluses. For instance, in our problem with two artificial variables, $x_1 \leqslant 4$, $x_2 \leqslant 2$, $2x_1 - x_2 \geqslant 1$, $2x_2 - x_1 \geqslant 1$, we could do this. We could drop x_5 and x_6, replace our earlier equations by $x_1 + x_3 = 4$, $x_2 + x_4 = 2$, $2x_1 - x_2 + x_7 = 1$, $-x_1 + 2x_2 + x_8 = 1$ and maximize $N = -x_7 - x_8$. This would lead us to the feasible solution represented by the point A in Fig. 155.

Exercise

Check this.

The reason why this simplified procedure works in this case is that the point A *just* meets the two conditions; at A we have $x_1 = 1$, $x_2 = 1$ and $2x_1 - x_2 = 1$, $2x_2 - x_1 = 1$.

In the diagram we notice that we can reach A without crossing either of the lines $2x_1 - x_2 = 1$, $2x_2 - x_1 = 1$. At A there is no surplus, and so the variables x_5 and x_6, included to meet the contingency of a surplus, can be dropped without any damage being done.

Now one might think that this would always happen. If we have difficulty in meeting a set of conditions why should we anticipate meeting them and having something to spare?

The reason can be seen from Fig. 156, which illustrates the set of conditions

$$x_1 + x_2 \leqslant 5, \qquad 3x_1 + x_2 \geqslant 3, \qquad 2x_1 + x_2 \geqslant 4.$$

Feasible solutions exist for these conditions; the point (x_1, x_2) must lie in the

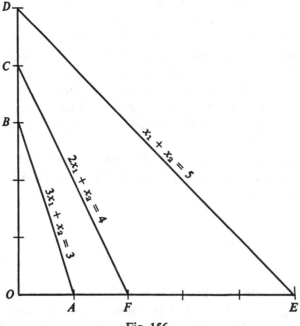

Fig. 156

quadrilateral *CDEF*, or on its boundary. The condition $3x_1 + x_2 \geqslant 3$ requires the point to lie above or to the right of the line *AB*. The condition $2x_1 + x_2 \geqslant 4$ requires it to lie above or to the right of *CF*. But any point on *CF* is already well past *AB*; this means that we can only satisfy the condition $2x_1 + x_2 \geqslant 4$ if we are prepared to satisfy $3x_1 + x_2 \geqslant 3$ with a surplus.

It is instructive to work through this problem, first by the standard method, using the equations $x_1 + x_2 + x_3 = 5$, $3x_1 + x_2 - x_4 + x_6 = 3$, $2x_1 + x_2 - x_5 + x_7 = 4$, and by the method just proposed, with x_4 and x_5 dropped. The standard method can lead you from *O* to *A* to *B* to *C* or, by another choice, from *O* to *B* to *C*. Either way, we reach the feasible solution *C*, with $N = 0$. The other method may take us from *O* to *A* to *B*, or directly from *O* to *B*, depending whether we put the asterisk under the first or second column. But, with this modified method we find we cannot get any result better than $N = -1$. The reason is that, by writing $3x_1 + x_2 + x_6 = 3$, we have committed ourselves to the region $3x_1 + x_2 \leqslant 3$; we cannot get past the line *AB*, and so we cannot get onto the line *CF*, or beyond it. Thus we have shut ourselves out from the feasible region *CDEF*.

The values of the seven variables at *C* are $x_1 = 0$, $x_2 = 4$, $x_3 = 1$, $x_4 = 1$, $x_5 = 0$, $x_6 = 0$, $x_7 = 0$. The value 1 for x_3 indicates that we are not pressing against the boundary *DE*; we have some slack in the condition $x_1 + x_2 \leqslant 5$, since, at *C*,

$$x_1 + x_2 = 4.$$

From $x_5 = x_7 = 0$ we see that we have neither a surplus nor a deficit in regard to the condition $2x_1 + x_2 \geqslant 4$; we are right on the line *CF*. From $x_6 = 0$ we see that we have no deficit in $3x_1 + x_2 \geqslant 3$, rather, as $x_4 = 1$ indicates, we have a surplus of 1, since at *C* the value of $3x_1 + x_2$ is 4, and so in excess of 3. If this excess were forbidden, it would be impossible to produce a feasible solution.

Accordingly the presence of *two* extra variables in an equation such as

$$3x_1 + x_2 - x_4 + x_6 = 3$$

is justified in certain situations. In a complicated problem it is extremely likely that we shall have to permit a surplus in some condition, so that the variables provided to meet this contingency are fulfilling a useful purpose. In any case, if it did happen in a particular problem that no surplus arose, we would not be able to recognize this simply by examining the equations. Before we could use the method without the surplus variables, we would have to apply a subroutine to determine that the situation was suitable for such a procedure, and this would almost certainly be more time-consuming than applying the standard method.

Accordingly, it does not seem that any economy of calculation could be achieved by trying to trim down the number of variables. This is satisfactory from the viewpoint of the student; we have the standard method, and do not have to bother about variations to meet special situations.

The solution as a whole

We now wish to put the pieces together. At the beginning of §43, we considered a problem where feasible solutions lay within or on a particular polygon. We showed how to work from any corner of that polygon until we arrived at the corner most suited to our purposes. In the present section we have seen how to locate a corner that can be used as a starting point, when there is no obvious corner of the permitted region. We now need to see how these two procedures fit together in detail. This is shown by means of two worked examples.

Worked example 1

It is required to maximize $M = x_1 + 4x_2$, where x_1 and x_2 are not negative and are subject to the restrictions $x_1 \leqslant 4$, $x_2 \leqslant 2$, $2x_1 - x_2 \geqslant 1$, $-x_1 + 2x_2 \geqslant 1$.

Solution   With so few and so simple conditions it is of course easy to sketch the region and spot a suitable corner. However, we are not concerned to find the solution so much as to use this simple problem to illustrate the general procedure.

We write the conditions with the help of 4 slack variables, and also 2 artificial variables needed for the last 2 inequalities. We are thus led to the table shown below.

x_1	x_2	x_3	x_4	x_5	x_6	x_7	x_8	
1	0	1	0	0	0	0	0	4
0	1	0	1	0	0	0	0	2
2	−1	0	0	−1	0	1	0	1 ←
−1	2	0	0	0	−1	0	1	1
−1	−1	0	0	1	1	0	0	−2
*								

The final row of the table is not related in any way to the quantity M that is to be maximized. We are simply trying to locate any corner of the permitted region, one that makes our deficits, x_7 and x_8, both zero. So we set out to maximize

$$N = -x_7 - x_8 = x_1 + x_2 - x_5 - x_6 - 2,$$

and the last row of the table expresses this.

If we put the asterisk under the x_1 column, the arrow will have to point to the third row. At the next stage our table becomes as shown.

0	$\frac{1}{2}$	1	0	$\frac{1}{2}$	0	$-\frac{1}{2}$	0	$3\frac{1}{2}$
0	1	0	1	0	0	0	0	2
1	$-\frac{1}{2}$	0	0	$-\frac{1}{2}$	0	$\frac{1}{2}$	0	$\frac{1}{2}$
0	$1\frac{1}{2}$	0	0	$-\frac{1}{2}$	−1	$\frac{1}{2}$	1	$1\frac{1}{2}$ ←
0	$-1\frac{1}{2}$	0	0	$\frac{1}{2}$	1	$\frac{1}{2}$	0	$-1\frac{1}{2}$
	*							

The next step gives the table below. The zero at the end of the last row makes it clear

0	0	1	0	$\frac{2}{3}$	$\frac{1}{3}$	$-\frac{2}{3}$	$-\frac{1}{3}$	3
0	0	0	1	$\frac{1}{3}$	$\frac{2}{3}$	$-\frac{1}{3}$	$-\frac{2}{3}$	1
1	0	0	0	$-\frac{2}{3}$	$-\frac{1}{3}$	$\frac{2}{3}$	$\frac{1}{3}$	1
0	1	0	0	$-\frac{1}{3}$	$-\frac{2}{3}$	$\frac{1}{3}$	$\frac{2}{3}$	1
0	0	0	0	0	0	1	1	0

that we have succeeded in making our deficit zero. The columns corresponding to x_5, x_6, x_7 and x_8 have the characteristic solid appearance, which indicates that we should take these variables as zero. We thus see that our original equations are satisfied by $x_1 = 1$, $x_2 = 1$, $x_3 = 3$, $x_4 = 1$, $x_5 = 0$, $x_6 = 0$, $x_7 = 0$, $x_8 = 0$. The fact that

$$x_7 = x_8 = 0$$

indicates that we are not working on a deficit; we have managed to meet the conditions originally specified without any need for a subsidy. We now have no further need of x_7 and x_8. Having got into the permitted region we intend to stay there; we have no intention of going into debt again. Accordingly, we now drop x_7 and x_8. In our last table, we can delete the columns corresponding to x_7 and x_8. That this step is justified can be seen by writing the equations that the table represents. Since $x_7 = 0$ and $x_8 = 0$, and since we intend to keep these values permanently, the terms containing x_7 and x_8 will simply lead to blank spaces in the equations.

Now that we are on the permitted region we can begin to think about improving our position; for the first time, we can turn our attention to M. Now $M = x_1 + 4x_2$, and we have discovered that $x_1 = 1$, $x_2 = 1$ is a feasible solution. However, x_1 and x_2 are not convenient variables in which to work from the position where we are. In §43, when we were at a corner of the permitted region, we found it was most convenient to work with the variables that were zero at that point. Now at the corner we have located we know the values of the variables. We have decided to drop x_7 and x_8; the values of the variables that still interest us are $x_1 = 1$, $x_2 = 1$, $x_3 = 3$, $x_4 = 1$, $x_5 = 0$, $x_6 = 0$, as we saw earlier. Thus the variables we want to work with are x_5 and x_6. These correspond to the 'solid looking' columns in the table – the only such columns that remain after the x_7 and x_8 columns have been deleted. This is very convenient, since our last table gives the values of x_1, x_2, x_3 and x_4 in terms of x_5 and x_6. In fact we can read off, from the third and fourth rows of that table, the equations

$$x_1 = 1 + \tfrac{2}{3}x_5 + \tfrac{1}{3}x_6$$
$$x_2 = 1 + \tfrac{1}{3}x_5 + \tfrac{2}{3}x_6,$$

from which it follows that

$$M = x_1 + 4x_2 = 5 + 2x_5 + 3x_6.$$

If we now write below the table a line corresponding to this result, we obtain the new table below.

0	0	1	0	$\frac{2}{3}$	$\frac{1}{3}$	3
0	0	0	1	$\frac{1}{3}$	$\frac{2}{3}$	1
1	0	0	0	$-\frac{2}{3}$	$-\frac{1}{3}$	1
0	1	0	0	$-\frac{1}{3}$	$-\frac{2}{3}$	1
0	0	0	0	-2	-3	5

If you wish to work entirely in terms of the tables, and not to write equations at all this last table can be reached in the following way. Write the bottom line in the form corresponding to our original equation, $M = x_1 + 4x_2$. This line will contain the numbers

$$-1 \quad -4 \quad 0 \quad 0 \quad 0 \quad 0.$$

Then operate according to the usual rules, with the asterisk first under the -1 and then under the -4. This will lead to the same result as our work above.

We now have only to apply the procedure of §43. If the asterisk is placed under the -2 in the bottom row, the arrow will have to be placed against the second row. The table is then brought to the form shown below.

0	0	1	-2	0	-1	1
0	0	0	3	1	2	3
1	0	0	2	0	1	3
0	1	0	1	0	0	2
0	0	0	6	0	1	11

All the signs in the bottom row are now positive, apart from the zeros, so we have reached the optimal solution, $x_1 = 3$, $x_2 = 2$, for which M takes its largest possible value, 11.

Worked example 2

Maximize $M = 2x_1 + x_2$ subject to the conditions $x_1 \leqslant 1$, $x_2 \leqslant 1$, $x_1 + x_2 \geqslant 1$.

Solution The permitted region is shown in Fig. 157. Here again the solution is graphically obvious, and our only purpose in using this problem is to discuss features of the procedure that arise here and might cause difficulty in a less simple situation.

In the routine treatment we begin by searching for a feasible solution. We need a slack variable for each of the conditions and an artificial variable, x_6, for the last condition. We proceed to minimize $N = -x_6$. We thus obtain the following table.

x_1	x_2	x_3	x_4	x_5	x_6		
1	0	1	0	0	0	1	←
0	1	0	1	0	0	1	
1	1	0	0	-1	1	1	
-1	-1	0	0	1	0	-1	

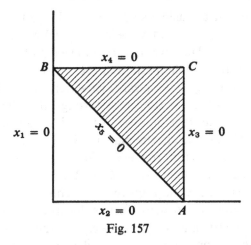

Fig. 157

The bottom line expresses the equation

$$N - x_1 - x_2 + x_5 = -1,$$

which follows from the equation corresponding to the third row of the table and the fact that $N = -x_6$.

If we put the asterisk under the x_1 column, the quotient $1/1 = 1$ occurs both in the first and the third row, so we have a choice. We have (for no special reason) chosen the first row. The next table is as shown.

1	0	1	0	0	0	1
0	1	0	1	0	0	1
0	1	-1	0	-1	1	0
0	-1	1	0	1	0	0

The last entry in the bottom row is 0, which indicates that our first objective has been attained; we have reduced our deficit, x_6, to zero. But a paradox appears to be arising; we still have -1 for the second entry in the bottom line, and this indicates that we should be able to increase N still further by putting an asterisk under this -1 and continuing by our usual procedure. But it is impossible that N should increase any further. For already $N = 0$; if N increases it will become positive. As $N = -x_6$, this means x_6 will become negative, and this is impossible, for the basic assumption is that no x ever becomes negative, and the whole routine is arranged to ensure that a negative value never arises.

Question for discussion

What does happen if we put an asterisk under the -1 and seek to continue the improvement of the value of N?

For the moment we turn our thoughts from this apparent paradox and concentrate on the fact that we have found a feasible solution, which the table shows to be $x_1 = 1$, $x_4 = 1$, the other variables x_2, x_3 and x_5 being zero. (We are no longer interested in x_6, which has done its work.) Here we have an unusual feature. Usually at the corner of the polygon which surrounds the permitted region we have *two* lines meeting and the two variables corresponding to these zero. However, in Fig. 157 it will be seen that the three lines $x_2 = 0$, $x_3 = 0$, $x_5 = 0$ all pass through A. When this happens we are said to have a *degenerate case*. It is because of this situation that we have the special features already observed, a choice in the placing of the arrow right at the outset, and a minus sign surviving in the bottom row even after $N = 0$ had been achieved.

It is now time to turn our attention to $M = 2x_1 + x_2$, the quantity we wish to maximize. We take our last table and remove the x_6 column and the bottom line, both of which were relevant only to our first objective, locating a point on the boundary of the permitted region. We write a new bottom line, corresponding to $M - 2x_1 - x_2 = 0$. (Here we are following the procedure mentioned in our first example as a possible alternative.) This gives us the table below. Now we have 3 equations (above the line)

$$
\begin{array}{rrrrrr}
1 & 0 & 1 & 0 & 0 & 1 \\
0 & 1 & 0 & 1 & 0 & 1 \\
0 & 1 & -1 & 0 & -1 & 0 \\
\hline
-2 & -1 & 0 & 0 & 0 & 0
\end{array}
$$

in 5 variables. We should be able to choose the values for 2 variables at will and solve for the other three. Examining the top three rows, we see that x_1, x_4 and x_5 each occur in one equation only, while the columns for x_2 and x_3 look reasonably 'solid'. Accordingly it is indicated that we try to get M expressed in terms of x_2 and x_3 alone. This is easily done. We need only add twice the top row to the bottom row, to bring the table to the form shown.

$$
\begin{array}{rrrrrr}
1 & 0 & 1 & 0 & 0 & 1 \\
0 & 1 & 0 & 1 & 0 & 1 \\
0 & 1 & -1 & 0 & -1 & 0 & \leftarrow \\
\hline
0 & -1 & 2 & 0 & 0 & 2 \\
 & * & & & &
\end{array}
$$

This table now looks very much like the tables we had in §43 when we were part way through the calculation. We have the solid columns for x_2 and x_3, while in the other columns we have only one non-zero entry under x_1, x_4 and x_5. The only difference is that while the first column contains the entries $1, 0, 0, 0$ and the fourth column has $0, 1, 0, 0$ (both as usual) the fifth column has $0, 0, -1, 0$. In the contexts of §43 this would have caused trouble; on putting $x_2 = x_3 = 0$ we would have obtained a negative value for x_5. The situation is saved by the zero in the sixth column. When $x_2 = x_3 = 0$, the equations represented by the first three rows reduce to $x_1 = 1$, $x_4 = 1$ and $-x_5 = 0$, so we are not confronted with any negative value.

We now proceed in the usual way. With the asterisk and arrow as indicated we next bring the table to the form shown below. This result is a little disturbing. We are still

1	0	1	0	0	1
0	0	1	1	1	1 ←
0	1	−1	0	−1	0
0	0	1	0	−1	2
				*	

at the point A with $x_1 = 1$, $x_2 = 0$. The only difference is that before we considered it as $x_2 = x_3 = 0$ while now we call it $x_3 = x_5 = 0$.

Our method will have failed if at the next step we find we are still at A, but calling it $x_2 = x_3 = 0$ once again. Fortunately this does not happen. The next step gives us the table below. We have completed our search, there are no negative entries left in the

1	0	1	0	0	1
0	0	1	1	1	1
0	1	0	1	0	1
0	0	2	1	0	3

bottom line. We have escaped from the point A and reached C with $x_1 = 1$, $x_2 = 1$, to yield the maximum value $M = 3$.

Usually things seem to work out as they have done here; in spite of degeneracy, the routine usually brings us to the solution of the problem. Theoretically the possibility does exist of the computation going round and round and never emerging from a closed cycle of situations. An example illustrating this possibility will be found on page 84 of *Mathematical Programming* by S. Vajda (Addison–Wesley, 1961). This book gives a very full bibliography of books and papers on linear programming. S. Vajda has also written a smaller book, that students will find easier to read, called *Planning by Mathematics* (Sir Isaac Pitman and Sons, 1969). This book, in about a hundred pages presents twenty-one short articles dealing with practical problems that can be handled by linear programming or related techniques. The topics range from blending aviation gasolines to betting at race-courses.

Exercises

It should be borne in mind throughout that not all linear programming objectives are attainable. An answer to a problem therefore consists either in giving a solution, or in demonstrating that the demands made are impossible.

1 Maximize $3x_1 + x_2$, subject to $x_1 + x_2 \leqslant 3$, $x_1 \geqslant 1$, $x_2 \geqslant 1$.

2 Maximize $2x_1 + x_2$, subject to $x_2 \leqslant 2$, $x_1 + x_2 \leqslant 4$, $x_1 + 2x_2 \geqslant 5$.

3 Maximize $x_1 + x_2$ subject to $3x_1 + 2x_2 \leqslant 15$, $2x_1 + 3x_2 \leqslant 15$, $3x_1 + 4x_2 \geqslant 22$.

4 Subject to $x_1 + 3x_2 \leqslant 19$, $3x_1 + x_2 \leqslant 17$, $3x_1 + 2x_2 \geqslant 15$, $x_1 + 2x_2 \geqslant 9$, minimize $x_1 + x_2$. (Maximize $-x_1 - x_2$.)

5 With the same conditions as in question 4, maximize $x_1 + x_2$.

6   Maximize $11x_1 + 17x_2$ subject to $x_1 + x_2 \leqslant 3$.

7   Minimize $11x_1 + 17x_2$ subject to $x_1 + x_2 \geqslant 3$.

8   Maximize $5x_1 + 3x_2$ subject to $x_1 \leqslant 4, -x_1 + x_2 \leqslant 1, x_1 + 2x_2 \leqslant 5, 4x_1 + 9x_2 \geqslant 23$.

9   Maximize $x_1 + 4x_2 + 3x_3$ subject to

$$x_1 + x_2 + x_3 \leqslant 11, \qquad x_1 + 2x_2 + 3x_3 \leqslant 26, \qquad x_1 - x_2 + x_3 \geqslant 5.$$

10   Maximize $x_1 + x_2 + 3x_3 + 5x_4$ subject to

$$x_1 \leqslant x_2, \qquad x_2 \leqslant x_3, \qquad x_3 \leqslant x_4, \qquad x_1 + x_2 + x_3 + x_4 \leqslant 24, \qquad x_1 + x_2 \geqslant 9.$$

11   100 tons of a certain material are available in a warehouse P and 80 tons in a warehouse Q. It is required to deliver 50 tons to a place A, 60 tons to a place B and 70 tons to a place C with the minimum use of transport. The distances in miles are given by the table below.

	A	B	C
P	100	50	20
Q	110	50	10

Find how much is delivered to each place from each warehouse in the optimal arrangement. (*Possible method.* Suppose x_1 tons go from P to A and x_2 tons from P to B. The amounts of the remaining deliveries can be calculated in terms of x_1 and x_2.)

Answers

Page 9
13 With the usual definitions of addition and multiplication of functions, (*a*), (*b*) and (*c*) but not (*d*) give linear mappings.

Page 14
1, 2, 3, 4, 7, 11, 12, 13 are true for mappings 1, 2, 3; 1 and 4 hold for mappings 4, 5. (It is possible to regard some of the others as true for 4 and 5, if limiting cases are accepted.)

Page 21
1 Three vectors OP, OQ, OR fail to provide a basis if they lie in a plane through the origin. This covers the more extreme cases where the vectors lie in a line through O, or where some or all of P, Q, R coincide with O.
2 Yes. Let v_1 = a nut, v_2 = a bolt, v_3 = a washer and v_4 = a nail. The definitions in the text then apply with v_1, v_2, v_3, v_4 as a basis. The space is of 4 dimensions.
3 (*a*) (i) fails. (*b*) is basis. (*c*) (ii) fails. (*d*) (i) and (ii) fail. (*e*) (i) fails. (*f*) (ii) fails. (*g*) is basis. (*h*) (i) and (ii) fail.

Page 24
1 (*a*) $2P + t(P + Q)$; $y = x - 2$. (*b*) $Q + t(P + Q)$; $y = x + 1$. (*c*) $3Q + t(P - Q)$; $x + y = 3$. (*d*) $P + 3Q + t(P - Q)$; $x + y = 4$. (*e*) tQ; $x = 0$. (*f*) $P + tQ$; $x = 1$. (*g*) $2Q + tP$; $y = 2$.
2 Q; P; $QFGHJ$; $(t, 1)$; $y = 1$.
3 (*a*) $SLGD$. (*b*) QLU. (*c*) QMW. (*d*) SMJ. (*e*) KGE.

Page 27
4 (*a*) $X^2 + Y^2 = 1$; note that this is *not* a circle, as the (X, Y) system uses axes that are not perpendicular. (*b*) $X^2 + 3Y^2 = 1$. (*c*) $X^2 + 2Y^2 = 1$.
6 Since every quadratic is a mixture of x^2, x and 1, the most obvious basis consists of x^2, x, 1, and the space is of 3 dimensions. $(x - 1)^2$, x^2 and $(x + 1)^2$, since every quadratic can be expressed, in one way only, in the form $p(x - 1)^2 + q(x^2) + r(x + 1)^2$, does constitute a basis.
7 $x = Y + 2Z$, $y = 3X + 4Z$, $z = 5X + 6Y$. Plane, $X = 0$.
8 $x = X + Y + Z$, $y = 10X + 20Y - 10Z$, $z = 5X - 5Y + 15Z$. (*a*) $Z = 0$. (*b*) $Y = 0$. (*c*) $X + Y = 0$.
9 $x = X + Y + Z$, $y = X + 2Y + 4Z$, $z = X + 3Y + 9Z$. (*a*) $Y = 0$. (*b*) $X = 0$. (*c*) $x - 2y + z = 0$.

Page 30
3 (*a*) $A^* = A$, $B^* = A + B$. (*b*) $A^* = B$, $B^* = A$. (*c*) $A^* = B$, $B^* = -A$. (*d*) $A^* = A + B$, $B^* = -A$.

Pages 32–3
3 $p + qx + \frac{1}{2}rx(x - 1) + \frac{1}{6}sx(x - 1)(x - 2)$.
4 Yes; $y = p(x - t)/(s - t) + q(x - s)/(t - s)$.
5 Hint: the equation corresponding to $p = 1, q = 0, r = 0$ is $y = (x - t)(x - u)/[(s - t)(s - u)]$.

Pages 36–7
1 (*a*) 2. (*b*) 6. (*c*) 2.
2 (*a*) $x \rightarrow 4x$, $x \rightarrow 8x$. (*b*) $x \rightarrow 20$, $x \rightarrow 30$. (*c*) $x \rightarrow x^4$, $x \rightarrow x^8$.
3 (*a*), (*c*), (*d*), (*f*).
4 $T^2 = I$.
5 (*a*) Yes. (*b*) Yes. (*c*) No.

7 (a) T^n: $A \to A + nB$, $B \to B$. (b) On squared paper, T represents a rotation of $45°$ combined with enlargement of scale $\sqrt{2}$ times. Thus T^n is a rotation of $n \cdot 45°$ and enlargement $(\sqrt{2})^n$. (c) $T^n = I$ when n is even, $T^n = T$ for n odd.

8 T^n: $1 \to 1 + x + x^2 + \cdots + x^n$.

9 T^n, acting on 1, gives the terms of the series for e^z as far as the term $x^n/(n!)$.

Pages 43–4

1 $(a) = (i) = (m) = (n)$. $(b) = (j) = (k) = (l)$. $(d) = (h) = (o)$. $(e) = (p) = (s)$. $(f) = (q) = (r)$. $(g) = (t)$. (c). (u). (v).

2 None for $(a), (c), (i), (m), (n), (u), (v)$. Every vector is an eigenvector with $\lambda = -1$ for $(b), (j), (k), (l)$; rotation through $180°$ has an infinity of invariant lines. $(d), (h), (o)$; $A, \lambda = 1$; $B, \lambda = -1$. $(e), (p), (s)$; $A, \lambda = -1$; $B, \lambda = 1$. $(f), (q), (r)$; $C, \lambda = 1$; $J, \lambda = -1$. $(g), (t)$; $C, \lambda = -1$; $J, \lambda = 1$.

3 All lines through the origin are invariant.

4 Horizontal and vertical axes, $A, \lambda = 1$. $B, \lambda = 0$.

5 Horizontal axis and line through origin at $45°$; $A, \lambda = 1$. $C, \lambda = 0$.

6 Horizontal axis is the only invariant line.

7 $C, \lambda = 3$. $J, \lambda = 1$. .

8 $B \to -A$. $T^3 = -I$. $T^6 = I$.

9 Reflection in OJ.

10 Reflection in OC.

11 (a) Invariant line, $y = 0$; $(t, 0)$ is eigenvector with $\lambda = 0$, for any number $t \neq 0$. (b) $x + y = 0$; $(t, -t)$. (c) $x = 0$; $(0, t)$. (d) $y = 0$; $(t, 0)$. (e) $x + y = 0$; $(t, -t)$. (f) $x = y$; (t, t).

Page 51

4 c^{-1} is at distance $1/r$ and angle $-\theta$.

5 $(r \cos \theta) + j(r \sin \theta)$.

6 No; see §14.

Page 54 (first set)

$(a) \pm(1 + j)/\sqrt{2}$. $(b) \pm j$. $(c) \pm(-1 + j)/\sqrt{2}$. $(d) -1, \frac{1}{2} \pm j(\frac{1}{2}\sqrt{3})$. $(e) 1, -\frac{1}{2} \pm j(\frac{1}{2}\sqrt{3})$. $(f) 1, j, -1, -j$. $(g) (\pm 1 + j)/\sqrt{2}$, $(\pm 1 - j)/\sqrt{2}$. (h) Combine answers to (d) and (e). $(i) \pm\{(\sqrt{3}) + j\}$. $(j) 1 + j$, $\sqrt{2} \cdot (-\cos 15° + j \sin 15°)$, $\sqrt{2} \cdot (\cos 75° - j \sin 75°)$. $(k) -2, 1 \pm j\sqrt{3}$. $(l) 3 + 3j, -3 \pm 3j$.

Page 54 (second set)

$(a) (\frac{5}{13}) - j(\frac{1}{13})$. $(b) 0.6 - j(0.8)$. $(c) -j$. $(d) \frac{1}{2} - j(\frac{1}{2})$. $(e) \cos \theta + j \sin \theta$. $(f) \cos^2 \theta - j \sin \theta \cos \theta$.

Page 62

1 (a) $\frac{1}{2}[ab + 1/(ab)]$

(b) $\frac{1}{2}[(a/b) + (b/a)]$

(c) $[ab - 1/(ab)]/(2j)$

(d) $[(a/b) - (b/a)]/(2j)$

(e) $\frac{1}{2}[a^2 + 1/(a^2)]$

(f) $[a^2 - 1/(a^2)]/(2j)$

(g) $\frac{1}{2}[a^3 + 1/(a^3)]$

(h) $[a^3 - 1/(a^3)]/(2j)$

(i) $\frac{1}{2}[a^4 + 1/(a^4)]$

(j) $[a^4 - 1/(a^4)]/(2j)$

(k) $\frac{1}{2}[ab^2 + 1/(ab^2)]$

(l) $[ab^2 - 1/(ab^2)]/(2j)$

(m) $\frac{1}{2}[(a/b^2) + (b^2/a)]$

(n) $[(a/b^2) - (b^2/a)]/(2j)$

(o) $\frac{1}{2}(ab^{-2}c^3 + a^{-1}b^2c^{-3})$

(p) $(ab^{-2}c^3 - a^{-1}b^2c^{-3})/(2j)$

(q) $\frac{1}{2}(ab^{-2}c^3d^{-4} + a^{-1}b^2c^{-3}d^4)$

(r) $(ab^{-2}c^3d^{-4} - a^{-1}b^2c^{-3}d^4)/(2j)$

2 (c) and (e) are incorrect

Page 66

1 (a) 1. (b) 1. (c) 5. (d) 5. (e) $\sqrt{2}$. (f) 3. (g) $\sqrt{(a^2 + b^2)}$. (h) 1. (i) 1.

2 (a) $\frac{1}{2}\pi$. (b) π. (c) $-\frac{1}{2}\pi$. (d) $\frac{1}{4}\pi$. (e) $\frac{3}{4}\pi$. (f) $\frac{1}{6}\pi$. (g) $-\frac{1}{3}\pi$ or $\frac{2}{3}\pi$. (h) θ. (i) θ.

3 (a) 2. (b) 3. (c) a. (d) b. (e) $\frac{1}{2}$. (f) $-\frac{1}{2}$. (g) $\cos \theta$. (h) $\sin \theta$. (i) $\cos \theta$. (j) $-\sin \theta$.

Page 68

3 Division is possible except when $a = b = 0$

Page 72

1 (a) Reflection in $y = x$; (t, t), $\lambda = 1$; $(t, -t)$, $\lambda = -1$. (b) Reflection in $y = -x$; (t, t), $\lambda = -1$;

$(t, -t))$, $\lambda = 1$. (c) Same as (a). (d) Rotation of $180°$ about O; every non-zero vector, $\lambda = -1$. (e) Same as (d). (f) Identity transformation, I; every non-zero vector, $\lambda = 1$. (g) Same as (f). (h) Same as (d). (i) Same as (f). (j) Project on y-axis, then double; $(0, t)$, $\lambda = 2$; $(t, 0)$, $\lambda = 0$. (k) Rotate $45°$ about O and enlarge $\sqrt{2}$ times. (l) Same as (f). (m) All points map to O; every non-zero vector, $\lambda = 0$. (n) Same as (m).

2 (a) False. (b) False. (c) True. (d) True. (e) True.

3 $ST = TS$ sometimes. $S + T = T + S$ always.

4 Yes.

5 (a) Perpendicular projection onto x-axis. (b) Perpendicular projection onto y-axis. (c) Rotation of $45°$ about O; compare multiplication by $(1 + j)/(\sqrt{2})$ on Argand diagram.

Pages 77–8

1 $(2x, -2y)$.

2 (x, y); I.

3 $(4x, 0)$.

4 $(4x, 0)$.

5 They are the same.

6 $(-x, -y)$.

7 $(-x, y)$.

8 Both equal $-I$.

9 $(0, 0)$; 0.

10 $(0, 2y)$.

11 $(0, 0)$; 0.

12 Both equal 0.

13 $(x - y, x - y)$.

14 $(4, 4)$; $(0, 0)$; $(0, 0)$.

15 Every point maps to $(0, 0)$.

16 (y, x); no.

17 $(x, y) \rightarrow (2y, 2x)$; no.

18 Same as (b).

19 Infinitely many solutions, including reflection in $y = mx$ for any m.

20 See 11 and 15.

Pages 84–6

1 $x^* = 2x + 5y$, $y^* = 8x + 3y$.

2 $\begin{pmatrix} x^* \\ y^* \end{pmatrix} = \begin{pmatrix} 4 & 9 \\ 16 & 25 \end{pmatrix}\begin{pmatrix} x \\ y \end{pmatrix}$.

3 (a) $\begin{pmatrix} 21 \\ 13 \end{pmatrix}$. (b) $\begin{pmatrix} 2 \\ 6 \end{pmatrix}$. (c) $\begin{pmatrix} 60 \\ 30 \end{pmatrix}$. (d) $\begin{pmatrix} 62 \\ 36 \end{pmatrix}$. (e) $\begin{pmatrix} k \\ 3k \end{pmatrix}$. (f) $\begin{pmatrix} 20m \\ 10m \end{pmatrix}$. (g) $\begin{pmatrix} k + 20m \\ 3k + 10m \end{pmatrix}$. (h) $\begin{pmatrix} -19 \\ -7 \end{pmatrix}$.

4 (a) $\begin{pmatrix} 1 & 0 \\ 0 & 1 \end{pmatrix}$. (b) $\begin{pmatrix} 0 & 0 \\ 0 & 0 \end{pmatrix}$. (c) $\begin{pmatrix} 1 & 0 \\ 0 & 1 \end{pmatrix}$. (d) $\begin{pmatrix} a + c & b + d \\ c & d \end{pmatrix}$. (e) $\begin{pmatrix} a & a + b \\ c & c + d \end{pmatrix}$.

(f) $\begin{pmatrix} ad - bc & 0 \\ 0 & ad - bc \end{pmatrix}$. (g) $\begin{pmatrix} a + 2c & b + 2d \\ 3a + 4c & 3b + 4d \end{pmatrix}$. (h) $\begin{pmatrix} a + 3b & 2a + 4b \\ c + 3d & 2c + 4d \end{pmatrix}$.

5 (a) $\begin{pmatrix} 1 & 0 \\ 0 & 1 \end{pmatrix}$. (b) $\begin{pmatrix} 4 & 0 \\ 0 & 9 \end{pmatrix}$. (c) $\begin{pmatrix} -1 & 0 \\ 0 & -1 \end{pmatrix}$. (d) $\begin{pmatrix} 1 & 0 \\ 0 & 1 \end{pmatrix}$. (e) $\begin{pmatrix} 1 & 0 \\ 0 & 1 \end{pmatrix}$.

(f) $\begin{pmatrix} 1 + bc & 0 \\ 0 & 1 + bc \end{pmatrix}$. (g) $\begin{pmatrix} 0 & 0 \\ 0 & 0 \end{pmatrix}$. (h) $\begin{pmatrix} 0 & 0 \\ 0 & 0 \end{pmatrix}$. (i) $\begin{pmatrix} 0 & 0 \\ 0 & 0 \end{pmatrix}$.

6 $\begin{pmatrix} 1 & 0 \\ 0 & 1 \end{pmatrix}$, $\begin{pmatrix} 0 & 0 \\ 0 & 0 \end{pmatrix}$.

7 $M_1 = \begin{pmatrix} 1 & 0 \\ 0 & -1 \end{pmatrix}$, $M_2 = \begin{pmatrix} -1 & 0 \\ 0 & 1 \end{pmatrix}$, $J = \begin{pmatrix} 0 & -1 \\ 1 & 0 \end{pmatrix}$.

9 (a) $\begin{pmatrix} 1 & 0 \\ 0 & -1 \end{pmatrix}$. (b) $\begin{pmatrix} -1 & 0 \\ 0 & 1 \end{pmatrix}$. (c) $\begin{pmatrix} 0 & 1 \\ 1 & 0 \end{pmatrix}$. (d) $\begin{pmatrix} -1 & 0 \\ 0 & -1 \end{pmatrix}$. (e) $\begin{pmatrix} 0 & -1 \\ 1 & 0 \end{pmatrix}$. (f) $\begin{pmatrix} 0 & 1 \\ -1 & 0 \end{pmatrix}$.

(g) $\begin{pmatrix} 1/\sqrt{2} & 1/\sqrt{2} \\ -1/\sqrt{2} & 1/\sqrt{2} \end{pmatrix}$. (h) $\begin{pmatrix} \frac{1}{2}\sqrt{3} & -\frac{1}{2} \\ \frac{1}{2} & \frac{1}{2}\sqrt{3} \end{pmatrix}$. (i) $\begin{pmatrix} \cos \alpha & -\sin \alpha \\ \sin \alpha & \cos \alpha \end{pmatrix}$. (j) $\begin{pmatrix} \cos 2\alpha & \sin 2\alpha \\ \sin 2\alpha & -\cos 2\alpha \end{pmatrix}$.

Pages 88–9

1 $\begin{pmatrix} 81 & 62 \\ 73 & 54 \end{pmatrix}$.

2 $\begin{pmatrix} 5 & 14 \\ 23 & 32 \end{pmatrix}$.

3 $\begin{pmatrix} 2 & 0 \\ 0 & 2 \end{pmatrix}$.

4 $0, 0$; yes.

5 $\begin{pmatrix} -1 & 2 \\ -2 & 1 \end{pmatrix}$, $\begin{pmatrix} 1 & 2 \\ -2 & -1 \end{pmatrix}$, $\begin{pmatrix} 0 & 2 \\ -2 & 0 \end{pmatrix}$; no.

6 0, 0; yes.   **7** No; see 4 and 6 above.   **8** Yes.

9 0.   **10** $\begin{pmatrix} 0 & 1 \\ 0 & -1 \end{pmatrix}, \begin{pmatrix} 1 & 0 \\ 0 & 1 \end{pmatrix}$.   **11** Correct.

12 $\begin{pmatrix} 4 & 2 \\ 6 & 4 \end{pmatrix}, \begin{pmatrix} 6 & 3 \\ 9 & 6 \end{pmatrix}, \begin{pmatrix} 8 & 4 \\ 12 & 8 \end{pmatrix}, \begin{pmatrix} 10 & 5 \\ 15 & 10 \end{pmatrix}$.   **13** Yes; $k = 4$.

14 0.   **15** $p = 3, q = -2$.   **16** Yes; yes.

Pages 91–2

1 $\begin{pmatrix} 13 & -21 \\ -8 & 13 \end{pmatrix}$.

2 For a symmetrical circuit, if an input v, i gives an output V, I, then an input V, $-I$ gives an output v, $-i$. The negative signs are due to the fact that input current is regarded as entering by the upper wire, while the opposite is true for output current. If $\begin{pmatrix} a & b \\ c & d \end{pmatrix}$ represents a symmetric circuit, $ad - bc = 1, a = d$.

3 $BAB = \begin{pmatrix} 2 & -1 \\ -3 & 2 \end{pmatrix}$.

4 No.

Page 101

1 (a) $\begin{pmatrix} 3t \\ t \end{pmatrix}, \begin{pmatrix} t \\ t \end{pmatrix}, \begin{pmatrix} 1 & 0 \\ 0 & -1 \end{pmatrix}$.   (b) $\begin{pmatrix} 2t \\ 5t \end{pmatrix}, \begin{pmatrix} t \\ 3t \end{pmatrix}, \begin{pmatrix} 1 & 0 \\ 0 & 2 \end{pmatrix}$.   (c) $\begin{pmatrix} t \\ t \end{pmatrix}, \begin{pmatrix} t \\ 0 \end{pmatrix}, \begin{pmatrix} 2 & 0 \\ 0 & 3 \end{pmatrix}$.

(d) $\begin{pmatrix} t \\ -t \end{pmatrix}, \begin{pmatrix} t \\ t \end{pmatrix}, \begin{pmatrix} 1 & 0 \\ 0 & 3 \end{pmatrix}$.   (e) $\begin{pmatrix} 2t \\ t \end{pmatrix}, \begin{pmatrix} t \\ -2t \end{pmatrix}, \begin{pmatrix} 3 & 0 \\ 0 & -2 \end{pmatrix}$.   (f) $\begin{pmatrix} t \\ t \end{pmatrix}, \begin{pmatrix} t \\ -t \end{pmatrix}, \begin{pmatrix} 2 & 0 \\ 0 & 0 \end{pmatrix}$.

(g) $\begin{pmatrix} t \\ t \end{pmatrix}, \begin{pmatrix} t \\ 0 \end{pmatrix}, \begin{pmatrix} 0 & 0 \\ 0 & 1 \end{pmatrix}$.

Page 104

(a) $\begin{pmatrix} t \\ -t \end{pmatrix}, \lambda = 2; \begin{pmatrix} t \\ t \end{pmatrix}, \lambda = 4$.   (b) $\begin{pmatrix} t \\ -t \end{pmatrix}, \lambda = 3; \begin{pmatrix} t \\ t \end{pmatrix}, \lambda = 5$.   (c) $\begin{pmatrix} 2t \\ t \end{pmatrix}, \lambda = 1; \begin{pmatrix} t \\ 2t \end{pmatrix}, \lambda = 4$.

(d) $\begin{pmatrix} 2t \\ t \end{pmatrix}, \lambda = -1; \begin{pmatrix} t \\ 3t \end{pmatrix}, \lambda = 4$.   (e) $\begin{pmatrix} t \\ -t \end{pmatrix}, \lambda = 0; \begin{pmatrix} t \\ 2t \end{pmatrix}, \lambda = 3$.   (f) $\begin{pmatrix} t \\ 0 \end{pmatrix}, \lambda = 2; \begin{pmatrix} t \\ t \end{pmatrix}, \lambda = 3$.

(g) $\begin{pmatrix} t \\ -t \end{pmatrix}, \lambda = a - b; \begin{pmatrix} t \\ t \end{pmatrix}, \lambda = a + b$.   (h) Real eigenvectors only when $b = 0$.

Pages 113–4

4 (a), (c), (f) and (g) have no inverses.

7 (a) The plane maps to $y = 0$, which is an invariant line with $\lambda = 1$. The line $x = 0$ maps to the origin; it is an invariant line with $\lambda = 0$. (b) $y = x$, $\lambda = 1$; $x = 0$, $\lambda = 0$. (c) $y = x$, $\lambda = 1$; $x + y = 0$, $\lambda = 0$. (d) $y = x$, $\lambda = 2$; $x + y = 0$, (e) $y = 2x$, $\lambda = 1$; $x + 2y = 0$, $\lambda = 0$.

Pages 125–7

1 (a) -1. (b) 1. (c) 0. (d) 0. (e) 0.

4 (a) 0. (b) 0.

5 (a) 240. (b) -10. (c) 10. (d) 0. (e) 10.

6 (a) 0. (b) -2.

7 (a) -30. (b) 0. (c) 0.

8 (a) $\begin{pmatrix} 1 \\ 1 \\ 1 \end{pmatrix}$. (b) $\begin{pmatrix} 1 \\ 2 \\ 3 \end{pmatrix}$. (c) $\begin{pmatrix} 1 \\ -2 \\ 1 \end{pmatrix}$. (d) $\begin{pmatrix} 1 \\ -2 \\ 1 \end{pmatrix}$. (e) $\begin{pmatrix} 1 \\ -2 \\ 1 \end{pmatrix}$.

9 Many answers.

10 $\begin{pmatrix} 1 \\ -2 \\ -2 \end{pmatrix}, \begin{pmatrix} -2 \\ 1 \\ -2 \end{pmatrix}, \begin{pmatrix} -2 \\ -2 \\ 1 \end{pmatrix}.$

11 $\lambda = 2 \quad \lambda = 0 \quad \lambda = -2$
$\begin{pmatrix} 1 \\ \sqrt{2} \\ 1 \end{pmatrix} \quad \begin{pmatrix} 1 \\ 0 \\ -1 \end{pmatrix} \quad \begin{pmatrix} 1 \\ -\sqrt{2} \\ 1 \end{pmatrix}.$

12 $\lambda = 3 \quad \lambda = 0 \quad \lambda = -3$
$\begin{pmatrix} 1 \\ -2 \\ -2 \end{pmatrix} \quad \begin{pmatrix} -2 \\ 1 \\ -2 \end{pmatrix} \quad \begin{pmatrix} -2 \\ -2 \\ 1 \end{pmatrix}.$

13 $\lambda = -3 \quad \lambda = 4 \quad \lambda = 6$
$\begin{pmatrix} 1 \\ -1 \\ 1 \end{pmatrix} \quad \begin{pmatrix} 1 \\ 1 \\ 0 \end{pmatrix} \quad \begin{pmatrix} 1 \\ -1 \\ -2 \end{pmatrix}.$

14 $\lambda = 3 \quad \lambda = -3 \quad \lambda = 9$
$\begin{pmatrix} 1 \\ -2 \\ -2 \end{pmatrix} \quad \begin{pmatrix} -2 \\ 1 \\ -2 \end{pmatrix} \quad \begin{pmatrix} -2 \\ -2 \\ 1 \end{pmatrix}.$

15 $\lambda = 1$
$\begin{pmatrix} 1 \\ 1 \\ 1 \end{pmatrix}$; for $\lambda = 0$, every non-zero vector in the plane $x + y + z = 0.$

16 $\lambda = 1$
$\begin{pmatrix} 1 \\ 1 \\ 2 \end{pmatrix}$; for $\lambda = 0$, every non-zero vector in the plane $x + y + 2z = 0.$

Pages 134–6

1 (a) $\begin{pmatrix} 6 & 15 & 15 & 6 \\ 3 & 9 & 11 & 5 \end{pmatrix}.$ (b) $\begin{pmatrix} 6 & 15 \\ 3 & 9 \end{pmatrix}.$ (c) $\begin{pmatrix} 3 & 7 & 11 \\ 3 & 7 & 11 \\ 1 & 3 & 5 \end{pmatrix}$ (d) $\begin{pmatrix} 1 & 3 & 5 \\ 2 & 6 & 10 \\ 3 & 9 & 15 \end{pmatrix}.$ (e) $(22).$

(f) $\begin{pmatrix} a^2 + b^2 + c^2 & ax + by + cz \\ ax + by + cz & x^2 + y^2 + z^2 \end{pmatrix}$ (g) $\begin{pmatrix} a^2 + x^2 & ab + xy & ac + xz \\ ab + xy & b^2 + y^2 & bc + yz \\ ac + xz & bc + yz & c^2 + z^2 \end{pmatrix}.$

3 $\begin{pmatrix} 3 & S_1 \\ S_1 & S_2 \end{pmatrix}, \begin{pmatrix} 4 & S_1 & S_2 \\ S_1 & S_2 & S_3 \\ S_2 & S_3 & S_4 \end{pmatrix}.$

4 Both products are $\begin{pmatrix} 1 & 0 \\ 0 & 1 \end{pmatrix}.$

7 $\begin{pmatrix} 1 & 0 & -2 & -3 \\ 0 & 1 & -4 & -5 \\ 0 & 0 & 1 & 0 \\ 0 & 0 & 0 & 1 \end{pmatrix}.$

8 $\begin{pmatrix} a_1u_1 + a_2u_2 + a_3u_3 & a_1v_1 + a_2v_2 + a_3v_3 \\ b_1u_1 + b_2u_2 + b_3u_3 & b_1v_1 + b_2v_2 + b_3v_3 \end{pmatrix}, \begin{pmatrix} a'u & a'v \\ b'u & b'v \end{pmatrix}.$

9 $M = (u + 2v, 3u + 4v, 5u + 6v)$. Det $M = 0$, since each column represents a vector lying in the plane of u and v. Same argument for $\det(PQ).$

Page 148

2 $u_1v_1 + u_2v_2 + \cdots + u_nv_n = 0.$

Page 155

1 (a) $1, \sqrt{2}, \sqrt{2}, \frac{1}{2}.$ (b) $0, \sqrt{6}, \sqrt{50}, 0.$ (c) $1, \sqrt{3}, 1, 1/\sqrt{3}.$ (d) $35, \sqrt{50}, \sqrt{50}, 0.7.$ (e) $-9, 3, 3\sqrt{2},$ $-1/\sqrt{2}.$ (f) $49, 7, 7\sqrt{2}, 1/\sqrt{2}.$

5 $(-\frac{1}{3}, \frac{5}{3}, \frac{1}{3}).$

6 (a) $(2,2,1).$ (b) $(1,1,2).$ (c) $(3,1,8).$ (d) $(-8, -2, -3).$ (e) $(-3, 2, 7).$

Page 158

2 $ax^2 + 2hxy + by^2.$

3 $0.$

4 $x^2 + y^2 + z^2 - 2xy - 2xz - 2yz.$

5 $M = \begin{pmatrix} 1 & 3 \\ 3 & 1 \end{pmatrix}$.

Page 161

7 $V'T'STV$; symmetric (see question 6). **8** $U'T'TV$; $T'T = I$.

Pages 183–5

1 1, 1, 90°; yes; its transpose.

2 The first three; the second; the first and third.

5 The first two must be; the third is not usually, but might be.

6 $L = \begin{pmatrix} (1 - k^2)/(1 + k^2) & 2k/(1 + k^2) \\ -2k/(1 + k^2) & (1 - k^2)/(1 + k^2) \end{pmatrix}$.

8 $(0, 0, 0)$, $(1, 0, 1)$, $(-\frac{1}{3}, \frac{1}{3}, \frac{4}{3})$, $(0, -1, 1)$.

9 The matrix corresponds to a rotation combined with an enlargement of scale by a factor 9; $(0, 0, 0)$, $(0, 9, 9)$, $(-11, 4, 5)$, $(-3, -3, 12)$.

10 Both are rotations; axes $(1, 1, 2)$, $(1, 1, 3)$.

11 Reflection in the plane $x + y + z = 0$, so determinant negative.

12 Yes; yes.

13 Axis $(0, 2, 3)$; $P^* = (1, -3, 2)$; $\cos POP^* = -\frac{6}{7}$; yes.

Pages 210–11

1 Unit eigenvectors are $(1/\sqrt{2}, 1/\sqrt{2})$, $\lambda = 9$; $(-1/\sqrt{2}, 1/\sqrt{2})$, $\lambda = 1$; $X^2 + (Y^2/9) = 1$; $(1/\sqrt{2}, 1/\sqrt{2})$, $(-3/\sqrt{2}, 3/\sqrt{2})$.

2 $(1/\sqrt{2}, 1/\sqrt{2})$, $\lambda = 1$; $(-1/\sqrt{2}, 1/\sqrt{2})$, $\lambda = 9$; $(X^2/9) + Y^2 = 1$. $(3/\sqrt{2}, 3/\sqrt{2})$, $(-1/\sqrt{2}, 1/\sqrt{2})$; by a rotation of 90°.

3 $(1/\sqrt{2}, 1/\sqrt{2})$, $\lambda = 9$; $(-1/\sqrt{2}, 1/\sqrt{2})$, $\lambda = -1$; $X^2 - (Y^2/9) = 1$; $(1/\sqrt{2}, 1/\sqrt{2})$; hyperbola.

4 Unit vectors along new axes are $(-2/\sqrt{5}, 1/\sqrt{5})$, $(1/\sqrt{5}, 2/\sqrt{5})$; $4X^2 - Y^2 = 4$; hyperbola.

5 $(X^2/4) + (Y^2/9) = 1$; axes as in 4.

6 $(10x_0 - 6y_0, -6x_0 + 5y_0)$; $(2, 3)$.

7 $(22x_0 - 3y_0, -3x_0 + 14y_0)$; $(13, 39)$.

8 $(56, 70)$, $(56, 70)$.

9 (a) $X^2 + 4Y^2 + 7Z^2 = 28$; unit vectors along axes $(-\frac{1}{3}, \frac{2}{3}, \frac{2}{3})$, $(\frac{2}{3}, -\frac{1}{3}, \frac{2}{3})$, $(\frac{2}{3}, \frac{2}{3}, -\frac{1}{3})$; ellipsoid. (b) $3X^2 - 3Y^2 + 9Z^2 = 1$; axes as in (a); hyperboloid. (c) $Y^2 + Z^2 = \frac{1}{3}$; circular cylinder; axis, $(1, 1, 1)$. (d) $2X^2 - Y^2 - Z^2 = 1$; hyperboloid of revolution; axis $(1, 1, 1)$. (e) $-2X^2 + Y^2 + 4Z^2 = 4$; hyperboloid; axes as in (a). (f) $10X^2 + Y^2 + Z^2 = 1$; flying saucer; axis $(2, 2, -1)$. (g) $X^2 + 4Y^2 + 7Z^2 = 28$; axes $(-\frac{2}{3}, \frac{1}{3}, \frac{2}{3})$, $(\frac{1}{3}, -\frac{2}{3}, \frac{2}{3})$, $(\frac{2}{3}, \frac{2}{3}, \frac{1}{3})$; ellipsoid. (h) $14Z^2 = 25$; pair of planes, perpendicular to $(1, 2, 3)$. (i) $-X^2 + Z^2 = 1$; hyperbolic cylinder; axes as in (g). (j) $-X^2 + Z^2 = 1$; hyperbolic cylinder; axes as in (a). (k) $X^2 + 2Y^2 + 3Z^2 = 1$; ellipsoid; axes as in (a).

10 (a), (f), (g), (k).

Page 225

1 $(1, 2, 3)$; $(1 + t, 1 + 2t, 1 + 3t)$; $y = 2x - 1$, $z = 3x - 2$.

2 $t = 2$; never; always; intersects; parallel; in plane.

3 $x - 1 = (y - 1)/2 = (z - 1)/3$; $x = y - 1 = z/2$; $x = y - 1 = z$; $(x - 1)/4 = (y - 2)/5 = (z - 3)/6$; $(10, 20, 30)$, $(2, 3, 4)$; many.

4 Yes; they are the same line. **5** No; yes.

Page 228

1 (a) $(2, 3, 4)$. (b) $(2, 3, 4)$. (c) $(1, 1, 1)$. (d) $(5, -2, 11)$. (e) $(1, 1, 0)$. (f) $(0, 0, 1)$.

2 $(x/a) + (y/b) + (z/c) = 1$; $(1/a, 1/b, 1/c)$.

3 $(1, 2, 3)$; $(t, 2t, 3t)$; $(4, 8, 12)$; $(3 + t, 4 + 2t, 1 + 3t)$; $(6, 10, 10)$.

Page 234

4 $r = au + bv$ for some a, b; $\det(au + bv, u, v) = 0$.

Pages 234–5

1 (a) (0, 0, 2). (b) (1, −1, 0). (c) (3, 3, 3). (d) (1, −1, 0). (e) (0, 0, 0).
2 (99, −66, 22); 121.
3 (a) (1, 1, 1), $\sqrt{3}$; $\frac{1}{3}\sqrt{3}$, $-\frac{1}{3}$. (b) (1, 1, 1), $\sqrt{3}$; $\frac{1}{3}\sqrt{3}$, $\frac{1}{3}$. (c) (1, 1, 1), $\sqrt{3}$; $\frac{1}{3}\sqrt{3}$, $\frac{1}{3}$.
5 (−1, 2, −1), 0, yes.
6 (a) 1.

Page 248

1 (a) (t, t, t); line in 3 dimensions. (b) $(3t, 2t, 5t)$; line in 3 dimensions. (c) Same as (b). (d) (0, 0, 0);
point. (e) $x + 2z - 2u + 3v = 0, y - z + 3u - 2v = 0$; subspace of 3 dimensions in 5 dimensions.
(f) $x + 7y - 13z = 0$; plane in 3 dimensions. (g) $(-11t, 9t, -4t, t)$; line in 4 dimensions. (h)
$x_1 + \frac{3}{2}x_2 + 2x_4 + x_5 = 0, x_3 + x_4 + x_5 = 0, x_6 = 0$; subspace of 3 dimensions in 6 dimensions.
2 (a) Yes. (b) Independent. (c) $a = b$ or $a = c$ or $b = c$; a, b, c all distinct. (d) a, b, c all distinct.
(e) 4.
3 $30u_1 - 31u_2 + 10u_3 - u_4 = 0$.
4 (a) resembles (c).

Pages 250–2

1 (a) No solution. (b) $(-14t - 15, 5t + 6, t)$. (c) No solution. (d) $(13 - 11t, 7t - 6, t)$.
2 No; $(7 + t, 3 + t, t), (1 + t, t, t), (4 + t, 5 + t, t)$; parallel lines.
4 Yes; $y = 7 - 2x + 5x^2$.
5 (a) $(1 + 2t, -1 + 3t, 1 + 5t)$. (b) None. (c) $2a + 3b + 5c = 0$; no.
6 (a), (c), (e), (f).

Pages 260–1

1 $2, 2, 2$; $\begin{pmatrix} 1 \\ 1 \\ 0 \end{pmatrix}$, $\begin{pmatrix} 0 \\ -1 \\ 1 \end{pmatrix}$; $x = y + z$; u_3, u_4; $5 - 2 = 3$;

$x_1 + 2x_3 + x_4 + 3x_5 = 0, x_2 + x_3 + 2x_4 + 2x_5 = 0$; $\begin{pmatrix} -2 \\ -1 \\ 1 \\ 0 \\ 0 \end{pmatrix}$, $\begin{pmatrix} -1 \\ -2 \\ 0 \\ 1 \\ 0 \end{pmatrix}$, $\begin{pmatrix} -3 \\ -2 \\ 0 \\ 0 \\ 1 \end{pmatrix}$;

for $Mv = u_3$, $x_1 = 5 - 2x_3 - x_4 - 3x_5$, $x_2 = 2 - x_3 - 2x_4 - 2x_5$, with x_3, x_4, x_5 arbitrary; for
$Mv = u_4$, $x_1 = 1 - 2x_3 - x_4 - 3x_5$, $x_2 = -1 - x_3 - 2x_4 - 2x_5$.
2 (a) 2; (1, 0, 3), (0, 1, −1). (b) 2; $\begin{pmatrix} 1 \\ 0 \\ -1 \\ 3 \end{pmatrix}$, $\begin{pmatrix} 0 \\ 1 \\ 1 \\ -1 \end{pmatrix}$. (c) 1; $\begin{pmatrix} -3 \\ 1 \\ 1 \end{pmatrix}$; $u_3 = u_2 - u_1$,

$u_4 = 3u_1 - u_2$; $\begin{pmatrix} 1 - 3t \\ 2 + t \\ t \end{pmatrix}$; $v_0 + t\begin{pmatrix} -3 \\ 1 \\ 1 \end{pmatrix}$.

3 (a) (1, 1, 1, 0, 0), (1, 0, 1, 1, 1). (b) $\begin{pmatrix} -1 \\ 0 \\ 1 \\ 0 \\ 0 \end{pmatrix}$, $\begin{pmatrix} -1 \\ 1 \\ 0 \\ 1 \\ 0 \end{pmatrix}$, $\begin{pmatrix} -1 \\ 1 \\ 0 \\ 0 \\ 1 \end{pmatrix}$.

Page 279

4 Take A of value x_1, B of value x_2. Then (a) $x_1 = 8, x_2 = 6, M = 14$. (b) $x_1 = 26, x_2 = 0$,
$M = 26$. (c) $x_1 = 0, x_2 = 28, M = 28$. (d) $x_1 = 12, x_2 = 0, M = 12$.

5 $x_1 = 100$, $x_2 = 500$.
6 $x_1 = 300$, $x_2 = 300$.
7 For $(0, 11, 1)$ $M = 12$. Other vertices are $(0, 0, 0)$, $(5\frac{3}{4}, 0, 0)$, $(0, 11\frac{1}{2}, 0)$. $(0, 0, 8\frac{1}{3})$, $(4, 0, 7)$.

Page 283
1 $x_1 = 2.25$, $x_2 = 2.25$.
2 $x_1 = 2$, $x_2 = 1$.
3 No solution; $x_1 = 2$, $x_2 = 2$ minimizes deficit.
4 $x_1 = 2$, $x_2 = 2$.
5 (*a*) No solution. (*b*) $x_1 = 3$; $x_2 = 1$.

Page 285
2 $x_1 = 2$, $x_2 = 2$.

Pages 293–4
1 $x_1 = 2$, $x_2 = 1$.
2 $x_1 = 3$, $x_2 = 1$.
3 No solution.
4 $x_1 = 3$, $x_2 = 3$.
5 $x_1 = 4$, $x_2 = 5$.
6 $x_1 = 0$, $x_2 = 3$.
7 $x_1 = 3$, $x_2 = 0$.
8 No solution.
9 $x_1 = 2$, $x_2 = 3$, $x_3 = 6$.
10 $x_1 = 4\frac{1}{2}$, $x_2 = 4\frac{1}{2}$, $x_3 = 4\frac{1}{2}$, $x_4 = 10\frac{1}{2}$.
11 $x_1 = 50$, $x_2 = 50$.

Index